AMERICAN MATHEMATICAL SOCIETY
COLLOQUIUM PUBLICATIONS
VOLUME XXXV

COEFFICIENT REGIONS FOR SCHLICHT FUNCTIONS

BY
A. C. SCHAEFFER
AND
D. C. SPENCER

WITH A CHAPTER ON

THE REGION OF VALUES OF THE DERIVATIVE OF A SCHLICHT FUNCTION

BY
ARTHUR GRAD

PUBLISHED BY THE
AMERICAN MATHEMATICAL SOCIETY
531 WEST 116TH STREET, NEW YORK CITY
1950

PREFACE

Instead of investigating various isolated extremal problems in the theory of schlicht functions, the authors have concentrated their efforts during the last three years on the investigation of the family of extremal schlicht functions in the large and this monograph is a presentation of the results of this research. For the sake of completeness and readability it has been found desirable to include in some places work that has been published elsewhere by ourselves or others. As most of the material is new, we have tried to point out carefully the material which already exists in published form.

In the calculus of variations there are two classical approaches: (a) study of specific problems using local variations; (b) study of a whole class of extremal problems and the investigation of the structure of the class as a whole. Variational methods in conformal mapping have been developed systematically in the last few years, beginning with a paper by M. Schiffer in 1938. The various publications on this subject during the last ten years have been mainly concerned with results of type (a) whereas we have tried to develop in this monograph a systematic approach to results of type (b), but we do not believe that our approach is the only one.

Since the investigation of extremal problems in conformal mapping embraces a rather wide field of research, we have confined ourselves to extremal problems relating to a finite number of the coefficients in the Taylor expansion of a function which is regular and schlicht inside the unit circle. Results of type (b) then concern the study of the region of values of the first n coefficients considered as a point in multi-dimensional euclidean space. This problem is only one of a host of problems that can be formulated in the theory of schlicht functions, and indeed a much more general problem is mentioned in Chapter I. The authors have chosen to investigate the coefficient problem not only because of its classical interest but also because it seems likely that the methods developed in this special case can be extended to many other problems. Dr. A. Grad has added a chapter in which he investigates the region of possible values of the derivative of a schlicht function at a fixed point inside the unit circle and his solution provides another example of these methods. A somewhat different version of his work has already appeared in hektographed form.

We have tried to make this monograph self-contained to as large an extent as practicable, and for this reason we have tried to keep the proofs and phraseology on as elementary a level as possible. This has lengthened the proofs in only a few cases and altogether it has increased the total length only slightly. We feel that sufficient background for reading this monograph is provided by a knowledge which is comparable to that contained in standard books on the theory of functions.

October, 1948.

A. C. Schaeffer and D. C. Spencer
Purdue University and Stanford University

ACKNOWLEDGMENTS

We are greatly indebted to the Mathematical Sciences Division of the Office of Naval Research for financial assistance in writing this monograph, and especially for support in carrying out computations. In particular, we are indebted to Dr. Mina Rees, Director of the Mathematical Sciences Division, for the interest she has shown in this work.

We wish to express our gratitude to Princeton University and to the University of Wisconsin for providing funds to print the colored frontispiece plates. Also we wish to thank Stanford University for funds made available in 1946 to make preliminary computations of the tables in the Appendix.

Finally, we are indebted to Mrs. Dolly Crane for her expert editorial assistance, and to Mr. Roy Nakata for the preparation of the diagrams and for his help in setting formulas into the final draft of the manuscript.

September, 1950

A. C. SCHAEFFER AND D. C. SPENCER
University of Wisconsin and Princeton University

TABLE OF CONTENTS

LIST OF LEMMAS AND THEOREMS

LIST OF PLATES AND DIAGRAMS

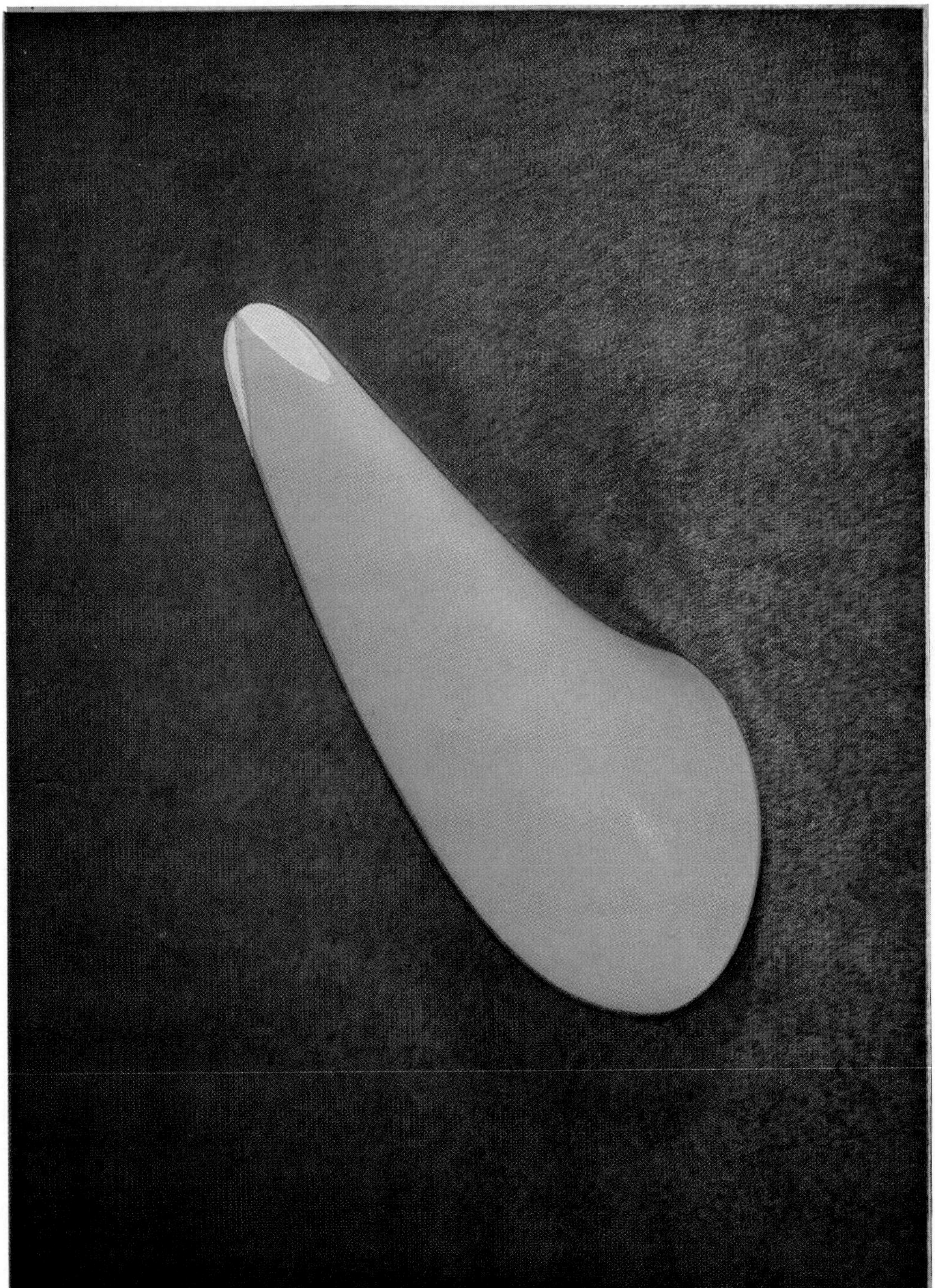

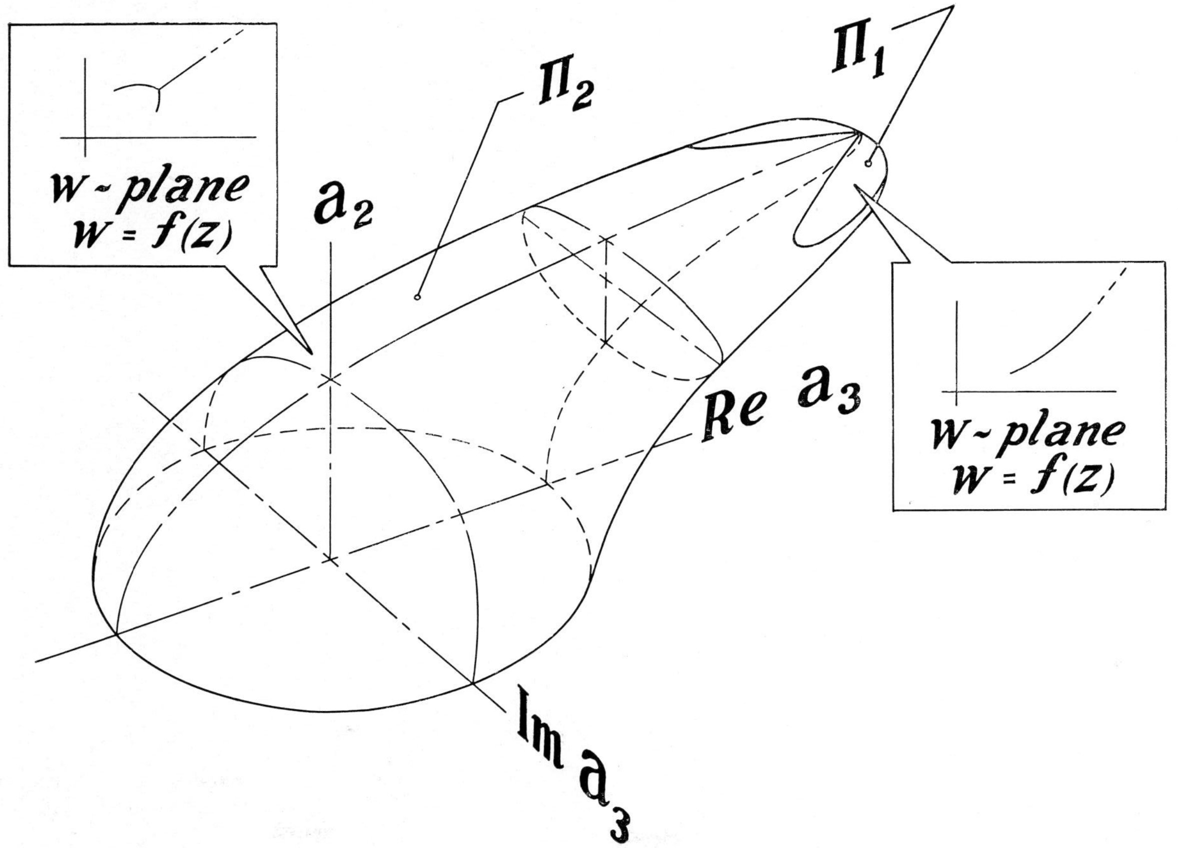

Plate I

The region of (a_2, a_3) when a_2 is real. (One half of solid shown.)

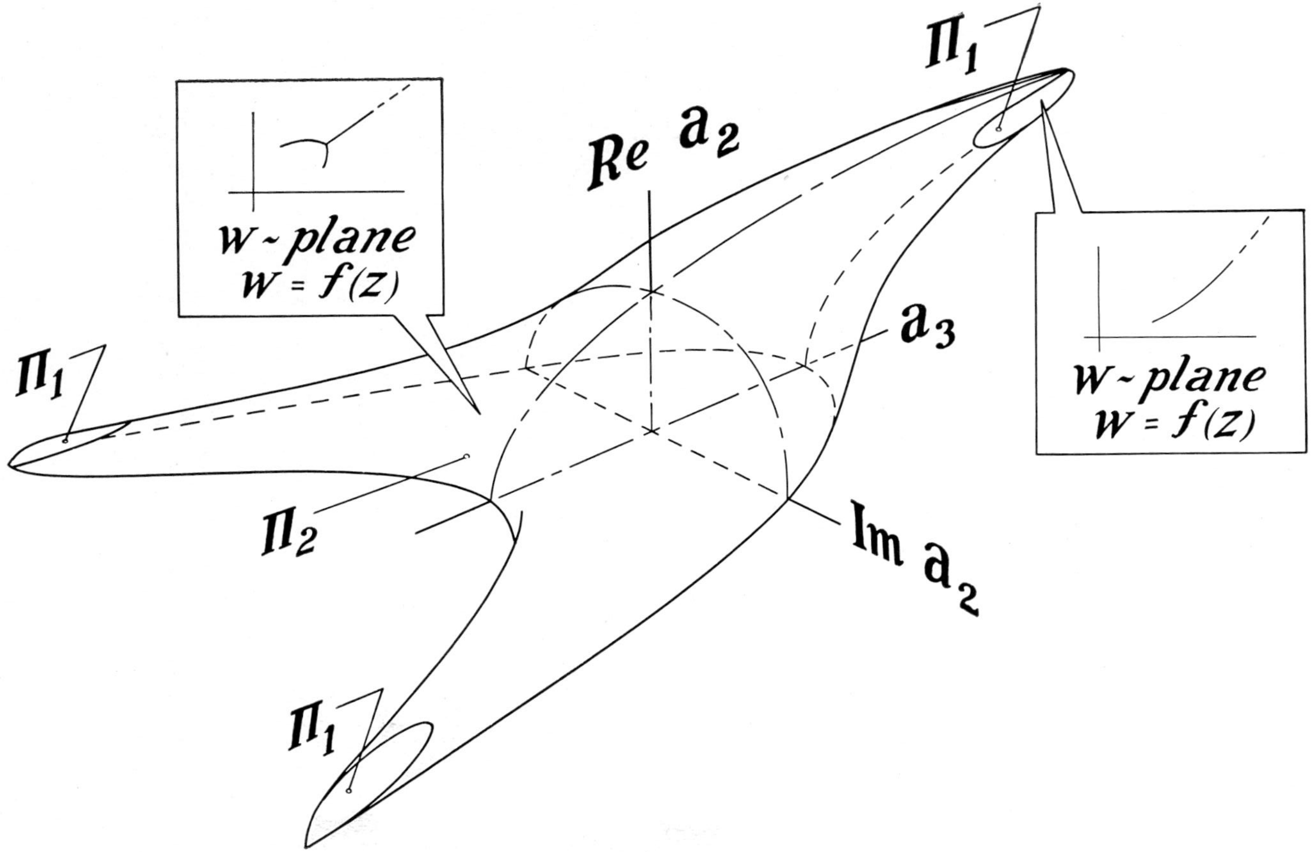

Plate II

The region of (a_2, a_3) when a_3 is real. (One half of solid shown.)

Chapter I

HISTORY OF SCHLICHT FUNCTIONS AND ELEMENTARY PROPERTIES OF THE nTH REGION

1.1. We begin with a brief history of the theory of schlicht functions, and we mention only those results which bear directly on the coefficient problem or on the problem of the region of values of $f'(z)$.

A function $f(z)$ is said to be schlicht in a domain if for any two points z_1 and z_2 of it we have $f(z_1) = f(z_2)$ only if $z_1 = z_2$. We shall be concerned with functions which are regular and schlicht in the unit circle $|z| < 1$ and which are normalized by the condition that the function vanishes at the origin and has a first derivative there equal to 1. The class of functions

$$f(z) = z + a_2 z^2 + a_3 z^3 + \cdots$$

which are regular and schlicht in $|z| < 1$ will be denoted by $\mathfrak{S}$.

The starting point in the investigation of schlicht functions was a paper by P. Koebe in 1907 on the uniformization of algebraic curves (see [10])[1] in which he proved, in particular, that there is a constant k (Koebe's constant) such that the boundary of the map of $|z| < 1$ by any function $w = f(z)$ of class $\mathfrak{S}$ is always at a distance not less than k from $w = 0$. A related result is that there exist bounds for the modulus of the derivative of $f(z)$ at any point in $|z| < 1$, these bounds depending only on $|z|$. These properties may be derived from the fact that the family $\mathfrak{S}$ is compact or, in other words, that $\mathfrak{S}$ is a normal family in the sense of Montel [15]. Actually, with the introduction of some such metric as

$$d(f_1, f_2) = \sup_{|z| = 1/2} |f_1 - f_2|$$

$\mathfrak{S}$ becomes a compact metric space.

Koebe's result soon attracted the attention of others (Plemelj [18]; Gronwall [7b, c]; Pick [17]; Faber [5]; Bieberbach [1]). Gronwall [7a] first gave the so-called "area-principle" which asserts that if the function

$$g(z) = 1/z + \sum_{\nu=1}^{\infty} b_\nu z^\nu$$

is schlicht in $|z| < 1$ and regular except at $z = 0$ where there is a simple pole, then

$$\sum_{\nu=1}^{\infty} \nu |b_\nu|^2 \leqq 1.$$

The image of $|z| < r$, $r < 1$, by $w = g(z)$ leaves uncovered a certain domain

[1] Square brackets refer to the bibliography at the end of the book.

of the w-plane, and the area-principle is an expression of the fact that the area of this domain is positive. Gronwall's paper seems to have attracted little or no attention, but in 1916 the area-principle was rediscovered and used to obtain the precise values of the constants in Koebe's results ([1], [5]). It was found that $k = 1/4$ and that

$$\frac{1 - |z|}{(1 + |z|)^3} \leqq |f'(z)| \leqq \frac{1 + |z|}{(1 - |z|)^3}. \tag{1.1.1}$$

Here the validity of the upper bound for all $|z| < 1$ gives the precise inequality[2]

$$|a_2| \leqq 2, \tag{1.1.2}$$

a result which was proved at the same time. Equality occurs in (1.1.1) (either side) or in (1.1.2) only for the function (Koebe function)

$$f(z) = \frac{z}{(1 + e^{i\varphi}z)^2}, \qquad \varphi \text{ real.} \tag{1.1.3}$$

Because of the extremal character of the function (1.1.3) so far as the inequalities (1.1.1), (1.1.2), and some others are concerned and because the nth coefficient of this function is equal to n in modulus, it was conjectured in 1916 or shortly thereafter that

$$|a_n| \leqq n, \qquad n = 2, 3, \cdots.$$

It was approximately at this time that Bieberbach proposed the so-called coefficient problem for schlicht functions. This is the problem of finding for each n, $n \geqq 2$, the precise region V_n in euclidean space of $2n - 2$ real dimensions occupied by points $(a_2, a_3, \cdots, a_n)$ corresponding to functions of class $\mathfrak{S}$. Several years earlier Carathéodory [2] and others had focused interest on this type of question by solving the coefficient problem for functions

$$p(z) = 1 + \sum_{\nu=1}^{\infty} c_\nu z^\nu \tag{1.1.4}$$

which are regular in $|z| < 1$ and have positive real parts there.

The paper [13] of Löwner forms a landmark in the historic development of this subject. Löwner gave a representation of the coefficients a_ν of a class of schlicht functions which lie everywhere dense in $\mathfrak{S}$, the representation being in terms of integrals of a function $\kappa(t)$, $|\kappa(t)| = 1$. For example, the formulas for a_2, a_3 are:

$$a_2 = -2 \int_0^\infty e^{-\tau}\kappa(\tau)\, d\tau, \tag{1.1.5}$$

$$a_3 = -2 \int_0^\infty e^{-2\tau}\kappa(\tau)^2\, d\tau + 4\left(\int_0^\infty e^{-\tau}\kappa(\tau)\, d\tau\right)^2. \tag{1.1.6}$$

---

[2] On the other hand, (1.1.2) implies (1.1.1) (both the upper and lower bounds). See [12b].

It follows at once from (1.1.5) that $|a_2| \leqq 2$, and it is readily shown from (1.1.6) (as Löwner pointed out) that

$$|a_3| \leqq 3. \tag{1.1.7}$$

The inequality (1.1.7) seems to be beyond the power of the methods used by Löwner's predecessors.

In 1925 Littlewood [12a] proved that

$$|a_n| < e \cdot n, \qquad n = 2, 3, \cdots, \tag{1.1.8}$$

and two years later Prawitz [19] gave a generalization of the area-principle. The method used by Littlewood may also be regarded as a somewhat different extension of the area-principle.

So far as the coefficient problem is concerned, mention must be made of the paper by Rogosinski [21] in which he introduced the class of functions

$$f(z) = z + a_2 z^2 + a_3 z^3 + \cdots$$

which are regular in $|z| < 1$ and have the property that they assume real values if and only if z is real. Rogosinski called such functions typically-real. All coefficients of a typically-real function are real and Im (f) and Im (z) have the same sign in $|z| < 1$. Schlicht functions of $\mathfrak{S}$ with all coefficients real form probably the most important subclass of these functions. We say that a class of functions is convex if, given any two functions f_1, f_2 of the class, the weighted mean

$$\frac{\lambda_1 f_1 + \lambda_2 f_2}{\lambda_1 + \lambda_2}$$

also belongs to the class no matter how the positive weights λ_1, λ_2 are chosen. Typically-real functions clearly form a convex class, and so do the functions (1.1.4) having positive real parts. The relation between these two classes is extremely simple. In fact, if f is typically-real, then

$$p(z) = \frac{1 - z^2}{z} f(z) \tag{1.1.9}$$

is a function of positive real part with real coefficients and conversely. Hence the coefficients c_k of p are connected with the coefficients a_k of f by the formulas

$$c_k = a_{k+1} - a_{k-1}, \qquad a_k = \begin{cases} c_1 + c_3 + \cdots + c_{k-1}, & k \text{ even}, \\ 1 + c_2 + c_4 + \cdots + c_{k-1}, & k \text{ odd}. \end{cases} \tag{1.1.10}$$

Thus the nth coefficient region of typically-real functions (that is to say, the region of points $(a_2, a_3, \cdots, a_n)$ belonging to these functions) is a simple linear map of the region of points $(c_1, c_2, \cdots, c_{n-1})$. Since the latter region is known (see [2]), the coefficient problem for typically-real functions is solved. We remark that the nth coefficient region for typically-real functions is the smallest convex region containing the nth coefficient region of schlicht functions with real coefficients. For typically-real functions we have (as Rogosinski [21] pointed out)

$$|a_n| \leqq n, \qquad n = 2, 3, \cdots, \tag{1.1.11}$$

and this estimate is a fortiori true of schlicht functions with real coefficients.[3] Equality is attained for the function (1.1.3) where $\varphi = 0$ or π.

Star-like schlicht functions also form an important subclass of $\mathfrak{S}$. A star-like schlicht function $w = f(z)$ maps $|z| < 1$ onto a domain in the w-plane having the property that any point of it can be connected to the origin $w = 0$ by a straight line lying entirely in the interior. A necessary and sufficient condition for $f(z)$ of class $\mathfrak{S}$ to be star-like is that

$$p(z) = z\frac{f'(z)}{f(z)}$$

be a function with real part positive in $|z| < 1$. Thus the coefficient regions V_n^* of star-like schlicht functions are also connected in a simple way with the coefficient regions of functions (1.1.4) having positive real part, and so the coefficient problem for these functions is solved. For star-like functions the estimate (1.1.11) is true, equality being attained for the star-like function (1.1.3) (see [6b]).

The coefficient problem for schlicht functions was first seriously considered by Peschl [16] in 1937. Peschl considered curves of Löwner type issuing from boundary points of the nth region V_n and extending to points of V_n^*, the coefficient region of star-like functions. He obtained qualitative results concerning V_n and also found the region of (a_2, a_3) when both a_2, a_3 are real. In the following years variational methods were introduced into the theory and new tools were thus provided ([6c,d,e], [22], [23], [24], and [25b]). Elementary variations were applied by Marty [14] to obtain one or two necessary conditions for functions $w = f(z)$ of class $\mathfrak{S}$ maximizing $|a_n|$. The systematic investigation of schlicht functions by the variational method began, however, with the paper by Schiffer [24a].

For many years Löwner's method provided one of the most powerful attacks in investigating schlicht functions. In recent years the variational approach has been developed to a point where it is comparable to Löwner's method in effectiveness and in some cases it seems to have led further. For example, not only has the variational method yielded many of the important results previously obtained only from Löwner's method (see [22a,b]), but it has also led to new results such as the regions of variability discussed below (see also [22c,f]). However, the variational approach has by no means displaced Löwner's method; rather it has complemented it. Löwner's integral representation for the coefficients,

[3] The estimate (1.1.11) for functions of $\mathfrak{S}$ with real coefficients was given independently and almost simultaneously by Dieudonné [4], Rogosinski [21], and Szász [26].

as well as the condition of Dieudonné [4] and the interesting set of conditions given by Grunsky [8] (see also [24f]), may be interpreted as giving necessary and sufficient conditions on a set of numbers $(a_2, a_3, \cdots, a_n)$ in order that they should be the coefficients of a function of class $\mathfrak{S}$, the conditions being expressed in terms of infinitely many parameters. The method developed here expresses conditions in terms of finitely many parameters, as we shall show.

The variational method gives necessary conditions in order that a function $w = f(z)$ of class $\mathfrak{S}$ should extremalize an arbitrary function of its first n coefficients. The method of Teichmüller [27] complements this result by proving that the necessary conditions are sufficient.

The variational method is not only applicable to problems involving the coefficients but is also applicable to a wide class of problems in conformal mapping. These more general problems are briefly described in **1.3** below and are discussed in greater detail in [22d]. The authors have confined themselves mainly to the coefficient problem although similar methods are applicable, at least in principle, to the wider class of problems discussed in **1.3.**

One of the oldest problems concerning functions of class $\mathfrak{S}$ is that of determining the possible values of $f'(z_1)$ at a fixed point z_1, $|z_1| < 1$. As mentioned above (formula (1.1.1)), precise bounds for $|f'(z_1)|$ were found in 1916. In 1936 Golusin [6a], using Löwner's method, found the precise bounds for $|\arg f'(z)|$ when f belongs to $\mathfrak{S}$, namely

$$(1.1.1)' \qquad |\arg f'(z)| \leqq \begin{cases} 4 \arcsin |z|, & |z| \leqq \dfrac{1}{2^{1/2}}, \\ \pi + \log \dfrac{|z|^2}{1 - |z|^2}, & \dfrac{1}{2^{1/2}} < |z| < 1. \end{cases}$$

Here the multi-valued functions may be assumed to have their principal values. The inequality (1.1.1)′ complements (1.1.1) and constitutes the so-called "rot-a tion theorem" for schlicht functions. The method employed by Golusin to prove (1.1.1)′ was later systematized by Robinson [20], who used it in finding the region of points $(|f(z)|, |f'(z)|)$, z fixed.

It is convenient to consider the region of values of $f'(z_1)$ in the plane of log $f'(z_1)$. The inequalities (1.1.1) and (1.1.1)′ (translated to the logarithmic plane) place this region in a rectangle. In [22d] the authors derived a differential equation for functions $f(z)$ whose derivative at the point z_1 lies on the boundary of the region of values. This differential equation involves essentially one real parameter, and its solutions may be implicitly expressed in terms of elementary functions. Dr. Grad has added a chapter (Chapter XV) in which he determines the region of values of $f'(z_1)$, z_1 any fixed interior point of the unit circle. Dr. Grad's solution determines the exact region inside the above rectangle which is occupied by values $\log f'(z_1)$, z_1 fixed.

1.2. Before proceeding with the discussion of the coefficient regions V_n, a more precise definition should be given. The point $(a_2, a_3, \cdots, a_n)$ is said to belong to the region V_n in $(2n - 2)$-dimensional real euclidean space with coordinates Re (a_2), Im (a_2), $\cdots$, Re (a_n), Im (a_n) if there is a function

$$f(z) = \sum_{\nu=1}^{\infty} b_\nu z^\nu$$

of class $\mathfrak{S}$ such that

$$b_\nu = a_\nu, \qquad \nu = 2, 3, \cdots, n.$$

We say that the point belongs to the function and that the function belongs to the point. Only the region V_2 is given from earlier results; it is simply the circle

$$|a_2| \leqq 2.$$

In the following pages it will be shown that for any n, $n \geqq 2$, the boundary of V_n can be expressed in terms of finitely many parameters. In fact, the boundary will be dissected into finitely many portions $\mathbf{\Pi}_1, \mathbf{\Pi}_2, \cdots, \mathbf{\Pi}_N$ and the coordinates a_k of the point $(a_2, a_3, \cdots, a_n)$ on any one of these portions $\mathbf{\Pi}_k$ will be functions of a finite number of parameters. The number of parameters defining $\mathbf{\Pi}_k$ will not exceed $2n - 3$. If the number of parameters is equal to $2n - 3$, then $\mathbf{\Pi}_k$ is a hypersurface of dimension $2n - 3$. In some cases, however, $\mathbf{\Pi}_k$ will depend on fewer than $2n - 3$ parameters; and if this is the case, $\mathbf{\Pi}_k$ will be a manifold of lower dimension which represents, for example, the intersection of two or more manifolds of higher dimension. In addition, we shall investigate certain geometric properties of the regions V_n.

In the case of the region V_3, there are essentially two hypersurfaces $\mathbf{\Pi}_1$, $\mathbf{\Pi}_2$ of dimension 3 which together with their 2-dimensional intersection make up the boundary. The parametric formulas for $\mathbf{\Pi}_1$ and $\mathbf{\Pi}_2$ are in terms of elementary functions. Tables have been computed for the boundary of the region V_3 and are included in the Appendix. This region is 4-dimensional, but it has a rotational property that makes it possible to define its entire structure from certain 3-dimensional cross-sections. If

$$f(z) = z + a_2 z^2 + a_3 z^3 + \cdots$$

is a function of class $\mathfrak{S}$, then the functions

$$\text{(1.2.1)} \qquad e^{-i\theta} f(e^{i\theta} z) = z + a_2 e^{i\theta} z^2 + a_3 e^{2i\theta} z^3 + \cdots, \qquad \theta \text{ real},$$

and

$$\text{(1.2.2)} \qquad \bar{f}(\bar{z}) = z + \bar{a}_2 z^2 + \bar{a}_3 z^3 + \cdots,$$

$$\bar{a}_k = \text{complex conjugate of } a_k,$$

also belong to class $\mathfrak{S}$. Thus if any one of the points

$$(a_2, a_3), \ (a_2 e^{i\theta}, a_3 e^{2i\theta}), \ (\bar{a}_2, \bar{a}_3)$$

belongs to V_3, so do the others; and, in particular, the entire domain can be constructed from any cross-section $\arg(a_2) = \text{constant}$ or $\arg(a_3) = \text{constant}$. If $(\mathrm{Re}(a_2), \mathrm{Re}(a_3), \mathrm{Im}(a_3))$ is a point of the cross-section $\mathrm{Im}(a_2) = 0$, then by a rotation (1.2.1) with $\theta = \pi$ and by a reflection (1.2.2) it is seen that the points $(-\mathrm{Re}(a_2), \mathrm{Re}(a_3), \mathrm{Im}(a_3))$ and $(\mathrm{Re}(a_2), \mathrm{Re}(a_3), -\mathrm{Im}(a_3))$ also belong to this cross-section. Thus the cross-section $\mathrm{Im}(a_2) = 0$ is symmetric about the planes $\mathrm{Re}(a_2) = 0$ and $\mathrm{Im}(a_3) = 0$.

Table I gives the part of the boundary of the cross-section $\mathrm{Im}(a_2) = 0$ which lies in

$$\mathrm{Re}(a_2) \geqq 0, \ \mathrm{Im}(a_3) \geqq 0.$$

Plate I on page xii shows the corresponding solid. It is one-half of the entire cross-section $\mathrm{Im}(a_2) = 0$. The yellow and blue surfaces correspond to $\mathbf{\Pi}_1$ and $\mathbf{\Pi}_2$. The function $w = f(z)$ belonging to a point of the yellow surface $\mathbf{\Pi}_1$ maps $|z| < 1$ onto the w-plane minus a single curved analytic slit, whereas the function $w = f(z)$ belonging to a point of the blue surface $\mathbf{\Pi}_2$ maps $|z| < 1$ onto the w-plane minus a ray $\arg(w) = \text{constant}$ extending from $w = \infty$ to some finite point where there is a fork composed in general of two prongs which form angles $2\pi/3$ with the ray. We remark that to any boundary point of the region V_3 (more generally of the region V_n) there corresponds a unique boundary function $w = f(z)$; that is, boundary points and boundary functions correspond in a one–one way.

Table II gives the part of the boundary of the cross-section $\mathrm{Im}(a_3) = 0$ which lies in

$$\mathrm{Re}(a_2) \geqq 0, \ \mathrm{Im}(a_2) \geqq 0,$$

and this is one-fourth of the boundary since the cross-section $\mathrm{Im}(a_3) = 0$ is symmetric about the planes $\mathrm{Re}(a_2) = 0$ and $\mathrm{Im}\,(a_2) = 0$. Plate II on page xiv shows one-half of the entire cross-section $\mathrm{Im}(a_3) = 0$, namely the half $\mathrm{Re}(a_2) \geqq 0$.

A detailed discussion of the domain V_3 is given in Chapter XIII.

We remark that if we know the regions V_n, then the coefficient problem is solved for any simply-connected domain D containing $z = 0$. The coefficient problem for a domain D containing $z = 0$ is that of finding the regions of values of the coefficients in the development about $z = 0$ of functions

$$f(z) = z + a_2 z^2 + a_3 z^3 + \cdots \tag{1.2.3}$$

which are regular and schlicht in D. If D is the full plane or the plane punctured at one point, there is at most one function $f(z)$ which is regular and schlicht in D and of the form (1.2.3) near $z = 0$. For example, if D is the plane minus the point $z = 1$, the only schlicht function of form (1.2.3) near $z = 0$ is $f(z) = z/(1 - z)$. If D is not the full plane or the punctured plane, there is a function $z = \varphi(\zeta)$ which maps $|\zeta| < 1$ onto D with $\zeta = 0$ going into $z = 0$. Near $\zeta = 0$ let

$$z = \varphi(\zeta) = \beta_1 \zeta + \beta_2 \zeta^2 + \beta_3 \zeta^3 + \cdots, \qquad \beta_1 > 0. \tag{1.2.4}$$

Writing

$$\{\varphi(\zeta)\}^{\kappa} = \sum_{\nu=\kappa}^{\infty} \beta_\nu^{(\kappa)} \zeta^\nu,$$

we have

$$g(\zeta) = f(\varphi(\zeta)) = \sum_{\nu=1}^{\infty} b_\nu \zeta^\nu$$

where

$$b_\nu = \beta_\nu^{(1)} + \beta_\nu^{(2)} a_2 + \cdots + \beta_\nu^{(\nu)} a_\nu, \qquad \nu = 1, 2, \cdots \tag{1.2.5}$$

Since the coefficients β_ν (and so the $\beta_\nu^{(\kappa)}$) are given and since we know all possible values for the numbers $b_\nu/b_1 = b_\nu/\beta_1$, by hypothesis, the coefficient regions of the a_ν can be determined from (1.2.5).

1.3. The coefficient problem—that is, the problem of finding the domains V_n—is only one of a wide class of problems concerning the family $\mathfrak{S}$ of schlicht functions. In fact, let R be a closed set lying in $|z| < 1$ and let $\psi_\nu(\tau)$ be a measure function defined in the space R (see [22d]). Given an integer n, there is a number $M = M(n)$ such that

$$|f^{(\nu)}(z)| \leqq M, \qquad \nu = 0, 1, 2, \cdots, n, \tag{1.3.1}$$

for all z of R. Here $f^{(\nu)}(z)$ denotes the νth derivative of a function f of class $\mathfrak{S}$. The inequality (1.3.1) is a consequence of the fact that the class $\mathfrak{S}$ is compact. Let $F_\nu(\zeta_0, \bar{\zeta}_0, \cdots, \zeta_n, \bar{\zeta}_n)$ denote a complex-valued function which is continuous together with its first order partial derivatives in an open set containing the closed set $|\zeta_\nu| \leqq M$ ($\nu = 0, 1, 2, \cdots, n$). Given the functions $F_1, F_2, \cdots, F_m$ and the measure functions $\psi_1, \psi_2, \cdots, \psi_m$, let

$$P_\nu = \int_R F_\nu(f(z), \overline{f(z)}, \cdots, f^{(n)}(z), \overline{f^{(n)}(z)})\, d\psi_\nu, \tag{1.3.2}$$

for $\nu = 1, 2, \cdots, m$. If $f(z)$ belongs to $\mathfrak{S}$, the point

$$P_f = (P_1, P_2, \cdots, P_m)$$

is a point in a euclidean space of $2m$ real dimensions, and the point P_f is said to belong to $f(z)$. As f ranges over $\mathfrak{S}$, the point P_f belonging to f ranges over a set which we call $\boldsymbol{D}_n$. Since $\mathfrak{S}$ is compact, the set $\boldsymbol{D}_n$ is closed and bounded. The problem is to find the region $\boldsymbol{D}_n$.

Many of the problems on schlicht functions which concern the values taken in the interior of the unit circle, as opposed to boundary-value problems, are contained in this general formulation. The nth coefficient region V_n is a particular region $\boldsymbol{D}_n$, and the region of values of the derivative $f'(z_1)$ at a fixed point z_1 in $|z| < 1$ is a particular region $\boldsymbol{D}_1$. Although we shall discuss here only the special problem of the regions V_n and Dr. Grad in Chapter XV will discuss the region of values of $f'(z_1)$, we remark that the methods used are applicable at least to some extent to any problem contained in the above general formulation. In the cases where P_ν as defined by (1.3.2) reduces to the form

$$P_\nu = \sum_{j=1}^{k} F_\nu(f(z_j), \overline{f(z_j)}, \cdots, f^{(n)}(z_j), \overline{f^{(n)}(z_j)}),$$

the methods seem to be closely related to those used in the coefficient problem.

1.4. Before taking up the investigation of the domains V_n, we note certain of their elementary properties.

LEMMA I. *The following statements are equivalent:*
(i) $(a_2, a_3, \cdots, a_n)$ *is an interior point of* V_n;
(ii) *there is a bounded function of class* $\mathfrak{S}$ *belonging to the point* $(a_2, a_3, \cdots, a_n)$;
(iii) *there is a function* $w = f(z)$ *of class* $\mathfrak{S}$ *belonging to the point* $(a_2, a_3, \cdots, a_n)$, *the closure of whose values* w *in* $|z| < 1$ *does not fill the whole closed* w*-plane.*

PROOF: If (i) is true, there is an $\epsilon > 0$ such that all points $(c_2, c_3, \cdots, c_n)$ satisfying the inequality

$$\sum_{\nu=2}^{n} |c_\nu - a_\nu|^2 \leqq \epsilon^2$$

belong to V_n. In particular, the point

$$(\rho a_2, \rho^2 a_3, \cdots, \rho^{n-1} a_n)$$

belongs to V_n for some $\rho > 1$, and so there is a function

$$g(z) = \sum_{\nu=1}^{\infty} b_\nu z^\nu$$

of class $\mathfrak{S}$ with

$$b_\nu = a_\nu \rho^{\nu-1}, \qquad \nu = 2, 3, \cdots, n.$$

Clearly the function

$$\rho g\left(\frac{z}{\rho}\right) = \sum_{\nu=1}^{\infty} b_\nu \rho^{-\nu+1} z^\nu$$

belongs to class $\mathfrak{S}$ and it is bounded for $|z| \leqq 1$. Moreover, its coefficients are

$$b_\nu \rho^{-\nu+1} = a_\nu, \qquad \nu = 2, 3, \cdots, n,$$

and so it belongs to the point $(a_2, a_3, \cdots, a_n)$. It has thus been shown that (i) implies (ii). It is plain that (ii) implies (iii).

Assume next that (iii) is true. Let w_0 be an exterior point of the map of $|z| < 1$ by $w = f(z)$. Then there is a $\delta > 0$ such that the circle $|w - w_0| \leqq \delta$ lies outside the map of $|z| < 1$ by $w = f(z)$. There is an $\epsilon > 0$ such that the function

$$R(w) = \frac{\epsilon_2 w^2 + \epsilon_3 w^3 + \cdots + \epsilon_n w^n}{(1 - w/w_0)^n}$$

satisfies the differential inequality

$$|R'(w)| \leqq 1/2\pi$$

in the region $|w - w_0| \geqq \delta$ whenever

$$\sum_{\nu=2}^{n} |\epsilon_\nu| \leqq \epsilon. \tag{1.4.1}$$

Now any two points w_1 and w_2 in the region $|w - w_0| \geqq \delta$ can be joined by an arc of length equal to or less than $\pi |w_1 - w_2|$; so we have the inequality

$$|R(w_1) - R(w_2)| \leqq \int |R'(w)\, dw| \leqq |w_1 - w_2|/2.$$

The function

$$G(w) = w + R(w)$$

is schlicht in $|w - w_0| \geqq \delta$ provided that (1.4.1) is satisfied, for if w_1 and w_2 are any two distinct finite points of this region, we have

$$|G(w_1) - G(w_2)| \geqq |w_1 - w_2| - |w_1 - w_2|/2 > 0.$$

The function $G(w)$ can be expanded in the series

$$\begin{aligned} G(w) &= w + \epsilon_2\{w^2 + b_{23}w^3 + b_{24}w^4 + \cdots\} \\ &\qquad + \epsilon_3\{w^3 + b_{34}w^4 + b_{35}w^5 + \cdots\} \\ &\qquad + \cdots \\ &= w + \epsilon_2 w^2 + (\epsilon_2 b_{23} + \epsilon_3)w^3 + \cdots \end{aligned}$$

where the coefficients b_{jk} depend on w_0 as well as on the integers j, k, n. The equations

$$\begin{aligned} \alpha_2 &= \epsilon_2, \\ \alpha_3 &= \epsilon_2 b_{23} + \epsilon_3, \\ &\cdots\cdots\cdots\cdots, \\ \alpha_n &= \epsilon_2 b_{2n} + \epsilon_3 b_{3n} + \cdots + \epsilon_{n-1} b_{n-1,n} + \epsilon_n \end{aligned} \tag{1.4.2}$$

have a solution $(\epsilon_2, \epsilon_3, \cdots, \epsilon_n)$ for every set of numbers $\alpha_2, \alpha_3, \cdots, \alpha_n$, and there is a positive number A depending on w_0 and n such that

$$\sum_{\nu=2}^{n} |\epsilon_\nu| \leqq A \sum_{\nu=2}^{n} |\alpha_\nu|.$$

Thus, given $\alpha_2, \alpha_3, \cdots, \alpha_n$, with

$$\sum_{\nu=2}^{n} |\alpha_\nu| \leqq \frac{\epsilon}{A}, \tag{1.4.3}$$

there is a function

$$G(w) = w + \alpha_2 w^2 + \alpha_3 w^3 + \cdots + \alpha_n w^n + \cdots$$

which is schlicht and regular in the region $|w - w_0| \geqq \delta$ except for a simple pole at $w = \infty$.

Here and henceforth let $a_\nu^{(k)}$ denote the coefficient of z^ν in the Taylor expansion about $z = 0$ of the kth power of $f(z)$. That is,

$$f(z)^k = \sum_{\nu=k}^{\infty} a_\nu^{(k)} z^\nu, \qquad a_k^{(k)} = 1,$$

where

$$f(z) = \sum_{\nu=1}^{\infty} a_\nu z^\nu, \qquad a_1 = 1$$

Now let

$$h(z) = G(f(z)) = f(z) + \alpha_2 f(z)^2 + \alpha_3 f(z)^3 + \cdots$$
$$= \sum_{\nu=1}^{\infty} a_\nu z^\nu + \alpha_2 \sum_{\nu=2}^{\infty} a_\nu^{(2)} z^\nu + \alpha_3 \sum_{\nu=3}^{\infty} a_\nu^{(3)} z^\nu + \cdots .$$

The function

$$h(z) = \sum_{\nu=1}^{\infty} a'_\nu z^\nu$$

is schlicht and regular in $|z| < 1$ and $a'_1 = 1$. Moreover

$$\begin{aligned} a'_2 &= a_2 + \alpha_2, \\ a'_3 &= a_3 + \alpha_2 a_3^{(2)} + \alpha_3, \\ &\cdots\cdots\cdots\cdots\cdots\cdots, \\ a'_n &= a_n + \alpha_2 a_n^{(2)} + \alpha_3 a_n^{(3)} + \cdots + \alpha_{n-1} a_n^{(n-1)} + \alpha_n . \end{aligned} \tag{1.4.4}$$

If the points $(a'_2, a'_3, \cdots, a'_n)$ and $(a_2, a_3, \cdots, a_n)$ are given, there will be solutions $\alpha_2, \alpha_3, \cdots, \alpha_n$ of these equations, and there is a number B, depending only on $a_2, a_3, \cdots, a_n$, such that

$$\sum_{\nu=2}^{n} |\alpha_\nu| \leqq B \sum_{\nu=2}^{n} |a'_\nu - a_\nu|.$$

Given the function $w = f(z)$ of class $\mathfrak{S}$ belonging to the point $(a_2, a_3, \cdots, a_n)$ and mapping $|z| < 1$ onto a domain lying in $|w - w_0| \geqq \delta$, let $(a'_2, a'_3, \cdots, a'_n)$

be any point satisfying

$$\sum_{\nu=2}^{n} | a'_\nu - a_\nu | \leqq \epsilon A^{-1} B^{-1}.$$

Define $\alpha_2, \alpha_3, \cdots, \alpha_n$ by equations (1.4.4). The α_ν then satisfy the inequality (1.4.3); so there is a function

$$G(w) = w + \alpha_2 w^2 + \alpha_3 w^3 + \cdots + \alpha_n w^n + \cdots$$

which is schlicht and regular in $| w - w_0 | \geqq \delta$ except for a simple pole at $w = \infty$. The function

$$h(z) = G(f(z)) = z + a'_2 z^2 + a''_3 z^3 + \cdots + a'_n z^n + \cdots$$

belongs to class $\mathfrak{S}$, and this shows that (iii) implies (i).

We define a domain to be an open, connected set and we define a closed domain to be the closure of a domain.

LEMMA II. *The set V_n is a bounded closed domain which contains the origin $a_2 = 0, \cdots, a_n = 0$ in its interior. The closed domain V_n is topologically equivalent to the closed $(2n - 2)$-dimensional full sphere.*

PROOF: The inequality (1.1.8) shows that the set V_n is bounded. It is closed because the class $\mathfrak{S}$ of functions f is compact. The function $f(z) = z$ belongs to $\mathfrak{S}$ and it is bounded; so Lemma I shows that the origin is an interior point of V_n. If

$$f(z) = \sum_{\nu=1}^{\infty} a_\nu z^\nu \tag{1.4.5}$$

belongs to the family $\mathfrak{S}$, so does the function

$$\frac{1}{\rho} f(\rho z) = \sum_{\nu=1}^{\infty} a_\nu \rho^{\nu-1} z^\nu$$

for each ρ in the range $0 \leqq \rho \leqq 1$. But $f(\rho z)/\rho$ is bounded in $| z | \leqq 1$ if $0 \leqq \rho < 1$, and hence by Lemma I the point

$$(a_2 \rho, a_3 \rho^2, \cdots, a_n \rho^{n-1}) \tag{1.4.6}$$

is an interior point of V_n for $0 \leqq \rho < 1$. If $(a_2, a_3, \cdots, a_n)$ is an interior point of V_n, the function (1.4.5) belonging to it can be supposed to be bounded, and so the path (1.4.6) connecting this point to the origin will lie entirely in the interior of V_n. Hence the interior is connected and is therefore a domain. If $(a_2, a_3, \cdots, a_n)$ is a boundary point of V_n, the path (1.4.6) will be interior to V_n for $0 \leqq \rho < 1$ and so every boundary point is a limit of interior points; this means that V_n is a closed domain.

To prove that V_n is topologically equivalent to the closed $(2n - 2)$-dimensional full sphere, suppose that $a_\nu = r_\nu e^{i\theta_\nu}$ and let

$$c_\nu = r_\nu^{1/(\nu-1)} e^{i\theta_\nu}, \qquad \nu = 2, 3, \cdots, n.$$

Then the point $(a_2, a_3, \cdots, a_n)$ of V_n determines a point $(c_2, c_3, \cdots, c_n)$. As $(a_2, a_3, \cdots, a_n)$ ranges throughout V_n, the corresponding point $(c_2, c_3, \cdots, c_n)$ sweeps out a region C_n, and the mapping defined is clearly one–one and bicontinuous, that is, topological. The point $(a_2 \rho, a_3 \rho^2, \cdots, a_n \rho^{n-1})$ defined by (1.4.6) goes into the point $(c_2 \rho, c_3 \rho, \cdots, c_n \rho)$. If $(c_2, c_3, \cdots, c_n)$ belongs to C_n, then the point $(c_2 \rho, c_3 \rho, \cdots, c_n \rho)$ is an interior point of C_n for $0 \leqq \rho < 1$. Thus C_n is a star-like region and each ray cuts the boundary of C_n in exactly one point. Any sufficiently small $(2n - 2)$-dimensional full sphere with center at the origin $c_2 = 0, c_3 = 0, \cdots, c_n = 0$ is contained entirely in the interior of C_n. The segment of each ray extending from the origin to a boundary point of C_n can be mapped linearly on the radius of the unit $(2n - 2)$-dimensional full sphere having the same direction in space. The mapping of C_n onto the unit full sphere is clearly one–one and bicontinuous; this proves that C_n, and so V_n, is topologically equivalent to the closed $(2n - 2)$-dimensional full sphere.

If the point $(a_2, a_3, \cdots, a_n)$ is an interior point of V_n, there will be many functions $f(z)$ of class $\mathfrak{S}$ belonging to the point. For Lemma I asserts that there is one such function $f(z)$ which is bounded by some constant, say M, in which case $w = f(z)$ maps $|z| < 1$ onto a domain lying in $|w| < M$. If γ is any number satisfying $|\gamma| < 1/(n+1)$, the function

$$G(w) = w + \gamma M^{-n} w^{n+1}$$

is schlicht in $|w| < M$. For if w_1 and w_2 are any two points in $|w| < M$, we have

$$G(w_1) - G(w_2) = (w_1 - w_2)\left\{1 + \gamma M^{-n} \cdot \sum_{\nu=0}^{n} w_1^\nu w_2^{n-\nu}\right\}$$

and the expression on the right vanishes only if $w_1 = w_2$. Thus the function

$$g(z) = G(f(z)) = f(z) + \gamma M^{-n} f(z)^{n+1}$$

is of class $\mathfrak{S}$ and belongs to the point $(a_2, a_3, \cdots, a_n)$. It will be shown in the sequel (Theorem IV) that to each point on the boundary of V_n there corresponds only one function $f(z)$ of class $\mathfrak{S}$. It will then follow that the statement "there is more than one function belonging to the point $(a_2, a_3, \cdots, a_n)$" is equivalent to any of the statements of Lemma I.

CHAPTER II

VARIATIONS OF SCHLICHT FUNCTIONS

2.1. Given any function f of class $\mathfrak{S}$, neighboring functions of f in $\mathfrak{S}$ can be constructed. One method of constructing such functions has been given in [22e] and we shall describe a modification of the method here.

We begin with a lemma which will also be required in subsequent chapters. Let there be $2m$, $m \geqq 1$, open arcs

$$I_1', I_2', \cdots, I_m', \qquad I_1'', I_2'', \cdots, I_m''$$

of the unit circle $|z| = 1$ which are disjoint and are identified in pairs in the following manner:

(i) There is a function $g_\nu(z)$ which is regular and schlicht in a neighborhood of each point of I_ν' and as z' describes I_ν' clockwise, the point $z'' = g_\nu(z')$ describes I_ν'' counterclockwise. The arcs I_ν' and I_ν'' are said to be identified.

(ii) No pair of identified arcs separates another identified pair. This condition means that if $j \neq \nu$, there is an arc of $|z| = 1$ which includes I_j' and I_j'' but does not contain I_ν' or I_ν''.

The part of $|z| = 1$ complementary to the arcs $I_1', \cdots, I_m''$ is the sum of their end points plus a certain number $k \geqq 0$ of unidentified arcs I_j.

If z_0' is an interior point of an arc I_ν' on $|z| = 1$, let z_0'' be the point of I_ν'' which is identified with z_0', that is, $z_0'' = g(z_0')$. Then a complete neighborhood of z_0' is mapped into a complete neighborhood of z_0'' by the function g. By (i) we see that the half-neighborhood at z_0' which is interior to $|z| < 1$ is mapped into the exterior half-neighborhood at z_0'' while the half-neighborhood at z_0' exterior to the unit circle is mapped into the interior half-neighborhood at z_0''. We denote the half-neighborhood at z_0' which is interior to the unit circle by N' and the half-neighborhood at z_0'' which is interior to the unit circle by N'', and we define a local uniformizer τ at z_0' or z_0'' by the formula

$$\tau = \begin{cases} g(z') - g(z_0'), & z' \subset N', \\ z'' - z_0'', & z'' \subset N''. \end{cases}$$

In the plane of the local uniformizer τ the two half-neighborhoods N' and N'' together with an arc of $|z| = 1$ form a single complete neighborhood of the point $\tau = 0$.

The unit circle $|z| < 1$ together with the identified arcs I_ν', I_ν'', $\nu = 1, 2, \cdots, m$, becomes a Riemann surface in which a pair of identified points z' and z'', $z' \subset I'$, $z'' \subset I''$, is considered to be a single point of the Riemann surface. The Riemann surface is then an open set whose boundary consists of the unidentified arcs on $|z| = 1$ (if any) plus end points of the identified arcs. In later applications it will sometimes be possible to introduce local uniformizers at points corresponding to the end points of identified arcs, in which case these points will be considered interior points of the Riemann surface. A surface which has a local uniformizer at each of its points is a Riemann surface. The uniformizers at points corresponding to identified points on $|z| = 1$ have already been defined. At points in $|z| < 1$ the variable itself may be taken as the local uniformizer. The Riemann surface which we have defined is of schlicht type (genus zero), but this fact will not be used explicitly. By schlicht type we mean that any closed Jordan curve in the Riemann surface divides the surface into two parts. By way of explanation of a Jordan curve on the Riemann surface in the z-plane, we remark that if it meets an identified arc I'_ν at a point z', then, since z' and z'' correspond to the same point of the Riemann surface, it may reappear at z'' and extend into the interior of the unit circle from z''.

A function F defined in a neighborhood of a point of the surface will be said to be regular-analytic there if it has a development in terms of the local uniformizer which is of the form

$$F = b_0 + b_1\tau + b_2\tau^2 + \cdots .$$

LEMMA III. *There exists on the Riemann surface a function $w = F$ which is regular at every point of it except for a simple pole of residue 1 at $z = 0$ and which maps the Riemann surface onto the w-plane minus finitely many slits parallel to the real axis, some of which may be single points. The function F is unique up to an additive constant.*

Lemma III is a special case of the uniformization principle, but we give a direct proof of the lemma here in order to avoid appealing to the general theory of Riemann surfaces. The proof is an extension of one given in [22e] and has some points in common with Weyl's proof of the Dirichlet principle for a Riemann surface. We remark that the existence of functions which, like F, take the same value at identified points of a pair of arcs has been used by R. Courant in his work on the Plateau problem (see [3]).

PROOF: Let $\mathcal{F}$ denote the family of functions F which are regular in $|z| < 1$ except for a simple pole of residue 1 at $z = 0$ and which satisfy:

(a) $$F(z) = 1/z + \sum_{\nu=0}^{\infty} a_\nu z^\nu, \qquad |z| < 1,$$

(b) $$|a_1 - 1|^2 + \sum_{\nu=2}^{\infty} \nu |a_\nu|^2 < \infty,$$

(c) $$\operatorname{Re}\{F(z')\} = \operatorname{Re}\{F(z'')\},$$

for almost all z' on the identified arcs of $|z| = 1$, $z'' = z''(z')$ being the point identified with z'. The example $F = 1/z - z$ shows that there is at least one function in $\mathcal{F}$. By (b) the radial limit of $F(z)$ exists almost everywhere on $|z| = 1$, and in this sense we speak of the values of F on $|z| = 1$.

Let

$$M = \inf \left\{ | a_1 - 1 |^2 + \sum_{\nu=2}^{\infty} \nu \, | a_\nu |^2 \right\}$$

for all functions of $\mathcal{F}$. There is then a sequence of functions

$$F_m(z) = \frac{1}{z} + \sum_{\nu=0}^{\infty} a_{\nu,m} z^\nu, \qquad | z | < 1,$$

of $\mathcal{F}$ such that

$$| a_{1,m} - 1 |^2 + \sum_{\nu=2}^{\infty} \nu \, | a_{\nu,m} |^2 < M + \frac{1}{m}$$

for $m = 1, 2, \cdots$. Without loss of generality let $a_{0,m} = 0$. Since the coefficients $a_{\nu,m}$ are uniformly bounded in m and ν, there is a subsequence of functions, which by renumbering we suppose is the whole sequence, such that

$$\lim_{m \to \infty} a_{\nu,m} = a_\nu \qquad (\nu = 1, 2, \cdots).$$

Using these limits a_ν, define

$$F(z) = \frac{1}{z} + \sum_{\nu=1}^{\infty} a_\nu z^\nu.$$

Clearly

$$| a_1 - 1 |^2 + \sum_{\nu=2}^{\infty} \nu \, | a_\nu |^2 \leqq M; \tag{2.1.1}$$

so (a) and (b) are fulfilled. The example $F = z - 1/z$ shows that $M \leqq 4$. To show that F satisfies (c), let ϵ be a fixed positive number and let k be a fixed positive integer. If m is large enough, we have

$$\sum_{\nu=1}^{\infty} | a_{\nu,m} - a_\nu |^2 = \sum_{\nu=1}^{k-1} + \sum_{\nu=k}^{\infty} < \epsilon + \sum_{\nu=k}^{\infty} | a_{\nu,m} - a_\nu |^2.$$

The last sum is less than or equal to

$$2 \sum_{\nu=k}^{\infty} \{ |a_{\nu,m}|^2 + |a_\nu|^2 \} \leqq \frac{2}{k} \sum_{\nu=k}^{\infty} \nu \{ | a_{\nu,m} |^2 + | a_\nu |^2 \} \leqq 2 \left\{ \frac{2M + 1/m}{k} \right\}.$$

It readily follows, by first choosing k large and then m large, that

$$\lim_{m \to \infty} \sum_{\nu=1}^{\infty} | a_{\nu,m} - a_\nu |^2 = 0.$$

The Riesz-Fischer theorem asserts that this sequence of functions $F_m(e^{i\theta})$ converges in mean to $F(e^{i\theta})$,

$$\lim_{m \to \infty} \int_{-\pi}^{\pi} | F_m(e^{i\theta}) - F(e^{i\theta}) |^2 \, d\theta = 0.$$

Hence some subsequence of the functions $F_m(e^{i\theta})$ must converge to $F(e^{i\theta})$ for almost all real θ. Since $F_m(e^{i\theta})$ satisfies (c) for almost all z' on $|z| = 1$, so does $F(z)$. Thus F belongs to $\mathfrak{F}$, and this implies in particular that inequality (2.1.1) is actually an equality.

Next, let $\chi(e^{i\theta})$ be a real-valued function which is continuous together with its first and second order derivatives, and which takes the same values at identified points on $|z| = 1$. Let

$$\Phi(z) = \sum_{\nu=0}^{\infty} b_\nu z^\nu$$

be a function whose real part on $|z| = 1$ is equal to $\chi(e^{i\theta})$,

$$\operatorname{Re}\{\Phi(e^{i\theta})\} = \chi(e^{i\theta}).$$

Then

$$b_\nu = o(\nu^{-2})$$

as $\nu \to \infty$; so the function

$$F(z) + \epsilon\Phi(z) = \frac{1}{z} + \sum_{\nu=0}^{\infty} (a_\nu + \epsilon b_\nu) z^\nu$$

belongs to $\mathfrak{F}$ provided ϵ is a real number. Hence

$$M \leqq |a_1 + \epsilon b_1 - 1|^2 + \sum_{\nu=2}^{\infty} \nu\,|a_\nu + \epsilon b_\nu|^2$$

or

$$M \leqq M + 2\epsilon \operatorname{Re}\left\{b_1(\bar{a}_1 - 1) + \sum_{\nu=2}^{\infty} \nu \bar{a}_\nu b_\nu\right\} + \epsilon^2 \sum_{\nu=1}^{\infty} \nu\,|b_\nu|^2.$$

Since ϵ can be either positive or negative, we obtain, letting $\epsilon \to 0$,

(2.1.2) $$\operatorname{Re}\left\{b_1 - \sum_{\nu=1}^{\infty} \nu \bar{a}_\nu b_\nu\right\} = 0.$$

Write

$$F(e^{i\theta}) = u(e^{i\theta}) + iv(e^{i\theta})$$

where u and v are real. Then v belongs to L^2; so the Fourier series of $v(e^{i\theta})$ and $e^{i\theta}\Phi'(e^{i\theta})$ can be integrated formally. Now

$$v(e^{i\theta}) \sim \frac{i}{2}\left\{e^{i\theta} - e^{-i\theta} + \sum_{\nu=1}^{\infty} (\bar{a}_\nu e^{-i\nu\theta} - a_\nu e^{i\nu\theta})\right\}$$

and

$$e^{i\theta}\Phi'(e^{i\theta}) \sim \sum_{\nu=1}^{\infty} \nu b_\nu e^{i\nu\theta};$$

so

$$\frac{i}{\pi}\int_{-\pi}^{\pi} v(e^{i\theta})e^{i\theta}\Phi'(e^{i\theta})\,d\theta = b_1 - \sum_{\nu=1}^{\infty} \nu\bar{a}_\nu b_\nu. \tag{2.1.3}$$

The real part of Φ on $|z| = 1$ is χ; so the real part of $ie^{i\theta}\Phi'(e^{i\theta})$ is $d\chi/d\theta$. Taking real parts in (2.1.3) and using (2.1.2), we obtain

$$\int_{-\pi}^{\pi} v(e^{i\theta})\,d\chi(e^{i\theta}) = 0.$$

Let I'_ν and I''_ν be a pair of identified arcs on $|z| = 1$, and suppose that the points $e^{i\theta}$ of I'_ν satisfy $\theta'_1 < \theta < \theta'_2$. Let $\theta'_1 < \alpha' < \theta'_2$, and let $e^{i\alpha''}$ be the interior point of I''_ν which is identified with the interior point $e^{i\alpha'}$ of I'_ν. Let $(\alpha' - \delta', \alpha' + \epsilon')$ be a small interval at α' and let $(\alpha'' - \epsilon'', \alpha'' + \delta'')$ be the corresponding interval on I''_ν. Here $\alpha'' - \epsilon''$ corresponds to $\alpha' + \epsilon'$ while $\alpha'' + \delta''$ corresponds to $\alpha' - \delta'$. Define

$$\chi = \begin{cases} 0, \alpha'' + \delta'' < \theta < \alpha' - \delta', \\ 1, \alpha' + \epsilon' < \theta < \alpha'' - \epsilon''. \end{cases}$$

In the interval $(\alpha' - \delta', \alpha' + \epsilon')$ let χ increase monotonically from 0 to 1 in such a manner that it is continuous together with its first and second order derivatives, and define χ correspondingly in $(\alpha'' - \epsilon'', \alpha'' + \delta'')$. The function $\chi(e^{i\theta})$ then takes equal values at identified points according to condition (ii). We have

$$\int_{-\pi}^{\pi} v(e^{i\theta})\,d\chi(e^{i\theta}) = \int_{I'_\nu} \{v(e^{i\theta'}) - v(e^{i\theta''})\}\,d\chi(e^{i\theta}) = 0$$

where $e^{i\theta''}$ denotes the point on I''_ν which is identified with the point $e^{i\theta'}$ on I'_ν. Letting χ tend to a step-function in a suitable way, we see that

$$v(e^{i\theta'}) = v(e^{i\theta''})$$

at all points of the Lebesque set. Thus

$$F(z') = F(z''), \qquad z'' = g(z'),$$

for almost all z' on I'_ν.

Let I_ν be an unidentified arc of $|z| = 1$, and let the points of I_ν be $e^{i\theta}$, $\theta_1 \leqq \theta \leqq \theta_2$. Let $\theta_1 < \varphi_1 < \varphi_2 < \theta_2$. We choose χ to be monotonically increasing from 0 to 1 as θ varies from $\varphi_1 - \epsilon$ to $\varphi_1 + \epsilon$, equal to 1 from $\varphi_1 + \epsilon$ to $\varphi_2 - \epsilon$, and monotonically decreasing from 1 to 0 as θ varies from $\varphi_2 - \epsilon$ to $\varphi_2 + \epsilon$. Elsewhere we define χ to be zero. We suppose that χ has continuous first and second order derivatives. Letting χ tend to a step-function in a suitable manner, we see that almost everywhere

$$v(e^{i\varphi_1}) = v(e^{i\varphi_2}).$$

Thus $v = \text{Im}\,(F)$ is equal to a constant almost everywhere over any unidentified arc of $|z| = 1$.

Finally, let z_0 be an interior point of an identified arc on $|z| = 1$ and let τ be a local uniformizer at z_0. The half-neighborhood at z_0 interior to $|z| < 1$ and the corresponding half-neighborhood at the point on $|z| = 1$ identified with z_0 appear in the τ-plane as half-neighborhoods joined together across a circular arc whose two edges are the images of the identified arcs on $|z| = 1$. When expressed as a function of τ, F is regular in the interior of each half-neighborhood in the τ-plane. Moreover, as τ approaches a point of the circular arc from either side, F tends to the same limit for almost all points of the arc. Since the values of F on the circular arc belong to L^2, we see that F in one of the half-neighborhoods in the τ-plane is the analytic continuation of F from the other half-neighborhood. Therefore F is regular and single-valued throughout the neighborhood of $\tau = 0$ and we have

$$F = \sum_{\nu=0}^{\infty} b_\nu \tau^\nu.$$

Thus F is regular at interior points of the identified arcs I'_ν, I''_ν and

$$F(z') = F(z''), \qquad z'' = g(z'). \tag{2.1.4}$$

In a similar fashion, F is regular at interior points of unidentified arcs and has on each such arc a constant imaginary part.

If I'_ν is one of a pair of identified arcs, let K'_ν be an open arc whose closure is interior to I'_ν and define K''_ν to be the set of points of I''_ν identified with K'_ν. The arcs of $|z| = 1$ complementary to K'_1, K''_1, $\cdots$, K'_m, K''_m are regarded as being unidentified. This new identification also defines a Riemann surface. Hence by the preceding argument there is a function

$$F^* = \frac{1}{z} + \sum_{\nu=1}^{\infty} a^*_\nu z^\nu$$

which is regular in $|z| \leqq 1$ except at the end points of the arcs K'_ν, K''_ν and at $z = 0$ where it has a simple pole of residue 1. Moreover, the relation (2.1.4) is satisfied by F^* when z' is an interior point of any arc K'_ν. On each unidentified arc the function F^* has a constant imaginary part. The function F^* is introduced as a device in the proof. The lemma will first be proved for the function F^*, and then in the general case by a limiting process.

We now prove that the function F^* is schlicht on the Riemann surface obtained by identification of the arcs K'_ν, K''_ν. Let z'_0 be an end point of an arc K' and let z''_0 be the end point of the arc K'' which is identified with z'_0. A half-neighborhood N' of z'_0 interior to $|z| = 1$ is mapped by the function $g(z)$ onto a half-neighborhood at z''_0 which is exterior to $|z| = 1$. The image of N' plus a half-neighborhood at z''_0 interior to $|z| = 1$ together with an arc of $|z| = 1$ near z''_0 makes a full neighborhood at z''_0. The pieces of the unidentified arcs near z'_0 and z''_0 fit together at z''_0 to form a circular arc of radius 1 with end points at z''_0. This circular arc is to be regarded as a cut in the complete neighborhood at z''_0. A bilinear transformation which carries $|z| = 1$ into a straight line followed by a square-root transformation carries the cut neighborhood at z_0'' into a half-neighborhood in the plane of a variable t bounded by a portion of the line Re $(t) = 0$ corresponding to the cut and lying in Re $(t) > 0$. We suppose z_0'' maps into $t = 0$. The half-neighborhoods N' and N'' map into quadrants joined along the real axis in the t-plane. On each of the two segments into which $t = 0$ divides the line Re $(t) = 0$ the function has a constant imaginary part. Then for a suitable value of the real constant c the function $F^* - c \log t$ has the same constant imaginary part on each of the two segments. Hence, by the reflection principle, $F^* - c \log t$ is single-valued in a complete neighborhood of $t = 0$. It follows that

$$F^* = c \log t + \sum_{-\infty}^{\infty} b_\nu t^\nu$$

near $t = 0$. Since the map by F^* of the half-neighborhood at $t = 0$ lying in Re $(t) > 0$ has a finite area by (2.1.1), we conclude that $c = 0$ and that all coefficients of negative powers of t are zero. This shows that F^* is continuous in the closed unit circle $|z| \leqq 1$ except at $z = 0$ where it has a simple pole.

The unidentified arcs map by $w = F^*$ into finitely many straight-line segments parallel to the real axis in the w-plane. The complement of these segments is a domain which we call D. Let w_0 be a point in D, and suppose that there is a positive integer k such that every neighborhood of w_0 contains a point which is taken at least k times by $w = F^*$ in the Riemann surface corresponding to $|z| < 1$ and the identified open arcs K_ν', K_ν''. Then a simple limiting process together with Rouché's theorem shows that w_0 is taken at least k times in the Riemann surface. We here use the fact that F^* is continuous in the neighborhood of the unidentified arcs. On the other hand, if the value w_0 is taken by F^* precisely k times, then each point in a complete neighborhood of w_0 is taken at least k times by F^* in the Riemann surface. Since D is connected and contains the point at infinity and since every sufficiently large value of w is taken precisely once in the Riemann surface, it readily follows that $w = F^*$ maps the Riemann surface in a one-one way onto D.

The function F^* is schlicht in its Riemann surface $\mathfrak{R}^*$ corresponding to the arcs K_ν', K_ν'', and we now show that F is schlicht in the Riemann surface $\mathfrak{R}$ corresponding to I_ν', I_ν''. Let all m pairs of identified arcs K_ν', K_ν'' increase monotonically and tend to the corresponding arcs I_ν', I_ν''; then by inequality (2.1.1) (since $M \leqq 4$) we see that there is a corresponding subsequence of functions F^* which converges to a schlicht function in $|z| < 1$ and that the convergence will be uniform in every fixed circle $|z| < r < 1$. Let z_0' be an interior point of I_ν' and let z_0'' be the point of I_ν'' identified with z_0'. By introducing the local uniformizer defined above, half-neighborhoods at z_0' and z_0'' appear in the τ-plane as half-neighborhoods at $\tau = 0$ joined across a circular arc of radius 1, and F^* is schlicht in a full neighborhood of $\tau = 0$. If the number r is sufficiently large, the functions F^* will converge uniformly in a subdomain of the neighborhood at $\tau = 0$ to a schlicht function. By using well known theorems concerning schlicht functions, it again follows that the functions F^* converge throughout the neighborhood of $\tau = 0$ to a function F_1 which is schlicht in that neighborhood. Thus there is a function

$$F_1 = \frac{1}{z} + \sum_{\nu=1}^{\infty} b_\nu z^\nu$$

which is schlicht in the Riemann surface arising from identification of the arcs I'_ν, I''_ν.

The function

$$F = \frac{1}{z} + \sum_{\nu=1}^{\infty} a_\nu z^\nu$$

which was constructed above identifies the arcs I'_ν, I''_ν and so it must identify the arcs K'_ν, K''_ν. Owing to the minimizing property of the function

$$F^* = \frac{1}{z} + \sum_{\nu=1}^{\infty} a^*_\nu z^\nu,$$

we have

$$M^* = |a_1^* - 1|^2 + \sum_{\nu=2}^{\infty} \nu \, |a_\nu^*|^2 \leqq |a_1 - 1|^2 + \sum_{\nu=2}^{\infty} \nu \, |a_\nu|^2 = M.$$

Since the coefficients a_ν^* tend to the coefficients b_ν, we obtain

$$|b_1 - 1|^2 + \sum_{\nu=2}^{\infty} \nu \, |b_\nu|^2 \leqq |a_1 - 1|^2 + \sum_{\nu=2}^{\infty} \nu \, |a_\nu|^2 = M.$$

Since the function F_1 identifies the arcs I'_ν, I''_ν, the reverse inequality is also true and so

$$|b_1 - 1|^2 + \sum_{\nu=2}^{\infty} \nu \, |b_\nu|^2 = |a_1 - 1|^2 + \sum_{\nu=2}^{\infty} \nu \, |a_\nu|^2 = M.$$

Since the function $F + \epsilon(F - F_1)$ satisfies the conditions (a), (b), and (c) with respect to the arcs I'_ν, I''_ν for any complex number ϵ, we have

$$M \leqq |a_1 + \epsilon c_1 - 1|^2 + \sum_{\nu=2}^{\infty} \nu \, |a_\nu + \epsilon c_\nu|^2$$

where $c_\nu = a_\nu - b_\nu$. It follows that

$$\sum_{\nu=1}^{\infty} \nu \, c_\nu \bar{a}_\nu - c_1 = 0$$

or

$$\sum_{\nu=1}^{\infty} \nu \bar{a}_\nu (a_\nu - b_\nu) - (a_1 - b_1) = 0. \tag{2.1.5}$$

Since the coefficients of F_1 also give the minimum, it follows by symmetry that

$$\sum_{\nu=1}^{\infty} \nu\bar{b}_\nu(b_\nu - a_\nu) - (b_1 - a_1) = 0. \tag{2.1.6}$$

Adding (2.1.5) and (2.1.6), we have

$$\sum_{\nu=1}^{\infty} \nu(a_\nu - b_\nu)(\bar{a}_\nu - \bar{b}_\nu) = \sum_{\nu=1}^{\infty} \nu \,|\, a_\nu - b_\nu \,|^2 = 0.$$

Hence $a_\nu = b_\nu$, $\nu = 1, 2, \cdots$.

The function F^* maps the Riemann surface $\mathfrak{R}^*$ onto the plane minus at most N slits parallel to the real axis where N is bounded above by a number depending only upon m. As the functions F^* tend to $F_1 = F$, a subsequence of the domains complementary to the parallel slits tends to a limiting domain D_1. It is clear that D_1 is then the complement of at most N slits parallel to the real axis, some of which may be points, and $F_1 = F$ maps the Riemann surface $\mathfrak{R}$ onto D_1.

If there are two functions F satisfying the conditions of Lemma III, let them be F_1 and F_2. Then $w_1 = F_1$ and $w_2 = F_2$ map the Riemann surface onto parallel-slit domains lying in the planes of w_1 and w_2 respectively. Thus there is a mapping $w_2 = g(w_1)$ from the parallel-slit domain E_1 in the w_1-plane onto the parallel-slit domain E_2 in the w_2-plane. Let $h(w_1) = g(w_1) - w_1$. This function is regular and single-valued in E_1. Let w_0 be any point in E_1 which is not a zero of $h'(w_1)$. Writing $h(w_1) = u + iv$, prolong the locus $v =$ constant in one direction from the point w_0, say in the direction u increasing. If the locus meets a zero of $h'(w_1)$, then there is at least one locus $v =$ constant emerging from this zero on which u is increasing. We then proceed along this locus. This locus cannot form a closed Jordan curve in E_1 because u is monotonic on the locus and so could not be single-valued on the Jordan curve. If the locus does not strike a boundary component (slit or isolated point) or does not tend to a single boundary component, then it may be continued as u increases toward the upper bound u_0 of Re $h(w_1)$ in E_1. For otherwise there would be a value u_1 of u, $u_1 < u_0$, and a sequence of u could be found tending to u_1 for which the corresponding points on the locus tend to a unique limit point p in E_1. If p is not a zero of $h'(w_1)$, then a complete neighborhood of p is mapped in a one-one manner onto a square in the plane of $\zeta = h(w_1)$ with sides parallel to the coordinate axes and center at the image of p. Any piece of the locus lying in the neighborhood of p maps into a straight line parallel to the real axis in the ζ-plane crossing the square from left to right. It is clear in this case that the locus can be continued with u increasing beyond u_1. If p is a zero of $h'(w_1)$, then the loci $v =$ constant emerging from p can be used to dissect the neighborhood of p into angular regions where the mapping by $h(w_1)$ is schlicht. In this case a slight modification of the above argument shows that the locus can be continued with u increasing beyond u_1. Thus if the locus does not strike a boundary component or does not tend to a single boundary component, then u can be increased up to the value u_0. By the maximum modulus principle we see in any case that the locus either meets a boundary component or tends to a single boundary component.

Let Jordan curves be drawn around each boundary component of E_1, and let these curves be mutually exterior. They map into closed Jordan curves in the w_2-plane which are mutually exterior because each may be joined to infinity by a Jordan arc lying in the exterior of all the others. Each Jordan curve in the w_2-plane contains a boundary component in its interior and the number of components in the w_2-plane is equal to the number of components in the w_1-plane. Thus each Jordan curve in the w_2-plane contains precisely one boundary component. Each curve and its map can be chosen within a given distance ϵ greater than zero from the boundary component in its interior. Thus $\text{Im}\, h(w_1)$ approaches a constant value in the complete neighborhood of each boundary component. It follows that $v = \text{Im}\, h(w_1)$ can take only a finite number of different values in E_1 and is therefore identically equal to a constant. Hence $h(w_1)$ is identically equal to a constant.

2.2. Let Γ be an analytic Jordan arc which is regular even at its end points α and β, and let D be a simply-connected domain containing Γ in its interior. Let $p_\epsilon(z)$ be regular in D for $|\epsilon| \leqq \epsilon_0$ and there satisfy the conditions

$$(2.2.1) \qquad \begin{gathered} |p_\epsilon(z)| \leqq M, \\ |p_{\epsilon'}(z) - p_{\epsilon''}(z)| \leqq M\,|\epsilon' - \epsilon''|, \\ p_\epsilon(\alpha) = p_\epsilon(\beta) = 0. \end{gathered}$$

If ϵ is small, $|\epsilon| \leqq \epsilon_1$, $0 < \epsilon_1 < \epsilon_0$, then as z moves over Γ from α to β, the point $z + \epsilon p_\epsilon(z)$ describes an analytic Jordan arc Γ_ϵ joining α and β. The two arcs Γ and Γ_ϵ form the complete boundary of an infinite simply-connected domain G_ϵ which may be twice covered in a neighborhood of Γ such that in moving from α to β along Γ and from β to α along Γ_ϵ the area to the left is constantly in the domain G_ϵ.

LEMMA IV. *Let $p_\epsilon(z)$ satisfy conditions* (2.2.1) *and let G_ϵ be as defined above. For each sufficiently small ϵ there is a function $f_\epsilon(z)$ which is schlicht and regular in G_ϵ except at $z = \infty$ where it has the development $f_\epsilon(z) = z + a_1(\epsilon)/z + \cdots$ and $w = f_\epsilon(z)$ maps G_ϵ onto the exterior of a Jordan arc in the w-plane in such a way that corresponding points of Γ and Γ_ϵ are brought together, that is,*

$$f_\epsilon(z + \epsilon p_\epsilon(z)) = f_\epsilon(z)$$

where z lies on Γ on the boundary of G_ϵ. Moreover,

$$f_\epsilon(z) = z + \frac{\epsilon}{2\pi i}\int_\alpha^\beta \frac{p_\epsilon(u) f'_\epsilon(u)}{f_\epsilon(u) - f_\epsilon(z)}\, du,$$

and, as ϵ approaches zero,

$$f_\epsilon(z) = z + \frac{\epsilon}{2\pi i}\int_\alpha^\beta \frac{p(u)}{u - z}\,du + o(\epsilon)$$

where $p(u) = p_0(u)$ and the $o(\epsilon)$ holds uniformly in G_ϵ. The path of integration lies in D, but may depend on z.

PROOF: By elementary transformations the domain G_ϵ can be mapped onto a single-sheeted domain bounded by two analytic Jordan arcs, and this in turn can be mapped onto the interior of the unit circle $|\zeta| < 1$ with $z = \infty$ going into $\zeta = 0$. The arcs Γ and Γ_ϵ map onto two complementary arcs I' and I'' of the unit circle, and the correspondence between Γ and Γ_ϵ defined by $z + \epsilon p_\epsilon(z)$ becomes a correspondence $g_\epsilon(\zeta)$ between points of I' and I''. The function $F(\zeta) = F_\epsilon(\zeta)$ whose existence is asserted in Lemma III is transformed into a function which after a magnification and translation can be written

$$f_\epsilon(z) = z + \frac{b_1}{z} + \frac{b_2}{z^2} + \cdots. \tag{2.2.2}$$

Moreover, $f_\epsilon(z)$ satisfies the functional relation

$$f_\epsilon(z + \epsilon p_\epsilon(z)) = f_\epsilon(z) \tag{2.2.3}$$

when z lies on Γ on the boundary of G_ϵ.

The function $f_\epsilon(z)$ is schlicht in G_ϵ and maps Γ and Γ_ϵ onto the opposite edges of an arc T_ϵ which is analytic at points corresponding to interior points of Γ and Γ_ϵ. By Lemma III the points α and β map into slits parallel to the real axis which may be single points. We show that for small ϵ the points α and β actually map into single points and the function $f_\epsilon(z)$ is continuous at α and β.

If ρ is the distance of the point z on Γ from α and r is the distance of the point $z + \epsilon p_\epsilon(z)$ on Γ_ϵ from α, then

$$r = \rho\,|1 + \epsilon p_\epsilon(z)/(z - \alpha)|.$$

If ϵ is sufficiently small, then for all small ρ we have

$$\frac{1}{2} \leqq \frac{dr}{d\rho} \leqq \frac{3}{2}; \tag{2.2.4}$$

in particular, $\rho/2 \leqq r \leqq 3\rho/2$. Completely analogous statements are true with respect to distances from the point β. We take $|\epsilon|$ so small that the interior angles of G_ϵ at α and β lie between $3\pi/2$ and $5\pi/2$.

For simplicity suppose that α lies at the origin and the positive real axis bisects the interior angle of G_ϵ at α. The portions of Γ and Γ_ϵ in a small neighborhood of α then lie in the left half-plane if ϵ is small. If z_1 is a point on Γ and $|z_1| = \rho$, draw the circular arc of radius ρ from z_1 with increasing amplitude until it meets the negative imaginary axis at $-i\rho$. The point $z_2 = z_1 + \epsilon p_\epsilon(z_1)$ lies on Γ_ϵ, and if $|z_2| = r$, draw the circular arc of radius r beginning at z_2 and continuing with decreasing amplitude until the positive imaginary axis is met at ir. Let a circular arc of radius $(r + \rho)/2$ with center at $i(r - \rho)/2$ be drawn in the right half-plane connecting the points ir and $-i\rho$. The three circular arcs together form a curve C_ρ which connects z_1 and z_2 and lies entirely in G_ϵ if ρ is sufficiently small.

To show that there is a sequence of ρ tending to zero such that the length of the map of C_ρ tends to zero, we suppose the contrary and arrive at a contradiction. Suppose that there is a $\lambda > 0$ such that

$$\int |f_\epsilon'(z)|\, ds \geqq \lambda \qquad (0 < \rho \leqq \rho_1)$$

where the integration is along C_ρ and ds is the element of arc length. Then

$$\lambda^2 \leqq \left\{\int |f_\epsilon'(z)|\, ds\right\}^2 \leqq \left\{\int |f_\epsilon'(z)|^2\, ds\right\}\cdot\left\{\int ds\right\},$$

each integration being over C_ρ. The last integral on the right is the length of C_ρ, and since the central angle of C_ρ is less than $5\pi/2$ and $r \leqq 3\rho/2$, it follows that this length is less than $15\pi\rho/4 \leqq 4\pi\rho$. Dividing each side by $4\pi\rho$ and integrating with respect to ρ from 0 to ρ_1, we have

$$\text{(2.2.5)} \qquad \int_0^{\rho_1} \frac{\lambda^2}{4\pi\rho}\, d\rho \leqq \int_0^{\rho_1} \int |f_\epsilon'(z)|^2\, ds\, d\rho,$$

where the inner integration on the right side is over C_ρ. On the area covered by arcs of radius ρ the differential $dsd\rho$ is the element of area. On the area covered by arcs of radius r the element of area is $dsdr$, but $dsd\rho \leqq 2dsdr$ according to (2.2.4). If a point (x, y) lies on one of the semi-circles of radius $(r + \rho)/2$ with centers at $i(r - \rho)/2$, one can write

$$x = \frac{r + \rho}{2}\cos\theta, \qquad y = \frac{r - \rho}{2} + \frac{r + \rho}{2}\sin\theta,$$

where θ is the argument of the vector from the center of the circle to the point $x + iy$. The relation between (ρ, θ) and (x, y) is one-one since the points where each circle cuts the y-axis, ir and $-i\rho$, tend monotonically to zero as ρ tends to zero. Now $ds = 2^{-1}(r + \rho)d\theta$; so $dxdy = Jdsd\rho$ where the Jacobian J of the transformation is

$$J = \frac{2}{r + \rho}\begin{vmatrix} \dfrac{\partial x}{\partial \rho} & \dfrac{\partial x}{\partial \theta} \\[2mm] \dfrac{\partial y}{\partial \rho} & \dfrac{\partial y}{\partial \theta} \end{vmatrix} = \frac{1}{2}\{(r' + 1) + (r' - 1)\sin\theta\}.$$

Here $r' = dr/d\rho$. Now r' lies between 1/2 and 3/2; so it is readily shown that this Jacobian lies between 1/2 and 3/2.

The right side of (2.2.5) is less than

$$2 \iint | f'_\epsilon(z) |^2 \, d\Sigma$$

where $d\Sigma$ is an element of area and the integration is over an area of G_ϵ which lies near α. A domain in the z-plane bounded by C_{ρ_1} and segments of Γ, Γ_ϵ is mapped by $w = f(z)$ onto a domain whose area is finite and is given by this integral. On the other hand, the left side of (2.2.5) is infinite, which is a contradiction. There is thus a sequence of curves C_ρ, each of which lies in G_ϵ and joins a point z of Γ on the boundary of G_ϵ to the point $z + \epsilon p_\epsilon(z)$ of Γ_ϵ, such that C_ρ lies within the distance 2ρ of α and the length of the map of C_ρ tends to zero as ρ tends to zero. Since $f_\epsilon(z)$ is schlicht in G_ϵ, it follows on letting ρ tend to zero that $f_\epsilon(z)$ is continuous at α and maps α into a single point. Likewise, $f_\epsilon(z)$ is continuous at β and maps β into a single point.

The above statements hold for each fixed ϵ satisfying $| \epsilon | \leqq \epsilon_1$ where ϵ_1 is some positive constant. Now we proceed to find the first variation of the function $f_\epsilon(z)$ as ϵ approaches zero. Well known theorems show that the arc T_ϵ is uniformly bounded as ϵ tends to zero. Indeed if c is large enough so that the boundary of G_ϵ lies inside the circle $| z | \leqq c$, then representation (2.2.2) is valid for $| z | > c$. If τ is a point of T_ϵ, the function

$$\frac{1}{f_\epsilon(1/z) - \tau} = \frac{z}{1 - \tau z + b_1 z^2 + \cdots} = z + \tau z^2 + \cdots$$

is regular and schlicht for $| z | < 1/c$. From (1.1.2) we know that, if a function $\Psi(z) = z + c_2 z^2 + \cdots$ is regular and schlicht in the unit circle $| z | < 1$, then the coefficient c_2 satisfies the inequality $| c_2 | \leqq 2$. This implies, after a change of variable, that $| \tau | \leqq 2c$. It follows that the arc T_ϵ lies in the circle $| w | \leqq 2c$ and that the function $f_\epsilon(z) - z$ is bounded by $3c$ on the boundary of G_ϵ and hence throughout G_ϵ. Thus

$$| f_\epsilon(z) | \leqq 3c + | z |.$$

Let a piecewise analytic Jordan arc Λ be drawn from α to β in D slightly to the left of Γ such that Γ plus Λ forms a closed Jordan curve enclosing a domain K which lies in G_ϵ and D and whose interior angle at α and at β is π. Let Λ' be a piecewise analytic Jordan arc connecting α and β and lying in K except at the end points α and β. Further let Λ' and Γ form a closed Jordan curve enclosing a domain K' whose interior angle at α and at β is $\pi/2$. Clearly K' is a subset of K. There are positive numbers N_1 and ϵ_2, $\epsilon_2 \leqq \epsilon_1$, such that if u is a point of Λ or Λ' and $0 < | u - \alpha | \leqq N_1$, then u is the center of a circle of radius $| u - \alpha | /3$, all of whose points lie in G_ϵ if $| \epsilon | \leqq \epsilon_2$. There are completely analogous statements for points u near β. Every point u of Λ or Λ' which is at a distance greater than N_1 from α and from β is the center of a circle of radius $N_2 > 0$ which lies entirely in G_ϵ. All points of Γ and Γ_ϵ lie outside all these circles. Choose ϵ_2, N_1, and N_2 small in such a way that the circles with centers on Λ do not overlap any of the circles with centers on Λ'. Let λ be the family of circles thus defined whose centers are on Λ, and λ' the family of circles with centers on Λ'.

For reasons which will be clear in the sequel we investigate the modulus of the function

$$\frac{p_\epsilon(u)f'_\epsilon(u)}{f_\epsilon(u) - f_\epsilon(z)} \tag{2.2.6}$$

where u is a point of Λ or Λ' and z belongs to G_ϵ. If u is on Λ or Λ' and $|u - \alpha| \leqq N_1$, then u is the center of a circle of radius $|u - \alpha|/3$ lying in G_ϵ; so the function

$$\frac{f_\epsilon(z) - f_\epsilon(u)}{f'_\epsilon(u)} = (z - u) + \cdots \tag{2.2.7}$$

is a regular schlicht function of z in this circle. A theorem of Koebe asserts that if a function

$$\Psi(s) = s + c_2 s^2 + \cdots$$

is regular and schlicht in the unit circle $|s| < 1$, then the map of $|s| < 1$ in the Ψ-plane covers the circle $|\Psi| < 1/4$. The map of the circle $|z - u| < |u - \alpha|/3$ by the function (2.2.7) therefore covers the interior of a circle of radius $|u - \alpha|/12$ with center at the origin. Since $f_\epsilon(z)$ is schlicht in G_ϵ, it follows that

$$\left|\frac{f_\epsilon(z) - f_\epsilon(u)}{f'_\epsilon(u)}\right| \geqq \frac{|u - \alpha|}{12} \tag{2.2.8}$$

when z lies in the part of G_ϵ which satisfies $|z - u| \geqq |u - \alpha|/3$. Now $p_\epsilon(u)$ is bounded in D and vanishes at α and β; so we have

$$|p_\epsilon(u)| \leqq N_3 |u - \alpha||u - \beta| \tag{2.2.9}$$

for $u \subset D$ and $|\epsilon| \leqq \epsilon_0$. Combining these results, we see that

$$\left|\frac{p_\epsilon(u)f'_\epsilon(u)}{f_\epsilon(u) - f_\epsilon(z)}\right| \leqq N_4 \qquad (|z - u| \geqq |u - \alpha|/3), \tag{2.2.10}$$

when u lies on Λ or Λ' within the distance N_1 of α, and z lies in G_ϵ, $u \neq \alpha$.

If u lies on Λ or Λ' at a distance greater than or equal to N_1 from α and β, we use a similar argument but with circles of radius N_2. We obtain in place of (2.2.8) the inequality

$$\left|\frac{f_\epsilon(z) - f_\epsilon(u)}{f'_\epsilon(u)}\right| \geqq \frac{N_2}{4}$$

when z lies in the part of G_ϵ which satisfies $|z - u| \geqq N_2$. Then since $p_\epsilon(u)$ is bounded in D, it follows that

$$\left|\frac{p_\epsilon(u)f'_\epsilon(u)}{f_\epsilon(u) - f_\epsilon(z)}\right| \leqq N_5 \qquad (|z-u| \geqq N_2) \tag{2.2.11}$$

when u lies on Λ or Λ' at a distance greater than N_1 from α and β, and z lies in G_ϵ.

Now consider the function

$$\Phi_\epsilon(z) = z + \frac{\epsilon}{2\pi i}\int_\alpha^\beta \frac{p_\epsilon(u)f_\epsilon'(u)}{f_\epsilon(u) - f_\epsilon(z)}\,du \tag{2.2.12}$$

where for large z the path of integration is Λ'. If, beginning with large z, the point z crosses the contour of integration, then the residue must be added, giving

$$\Phi_\epsilon(z) = z + \epsilon p_\epsilon(z) + \frac{\epsilon}{2\pi i}\int_\alpha^\beta \frac{p_\epsilon(u)f_\epsilon'(u)}{f_\epsilon(u) - f_\epsilon(z)}\,du. \tag{2.2.13}$$

The path of integration is Λ' but can be moved to Λ. If z belongs to K' or one of the circles of the family λ', then representation (2.2.13) for $\Phi_\epsilon(z)$ is used where the path of integration is along Λ. Then according to inequalities (2.2.10) and (2.2.11), the integrand is uniformly bounded; so $\Phi_\epsilon(z) - z$ is bounded. If z lies outside K' and all circles of the family λ', then representation (2.2.12) is used where the path of integration is Λ'. Since the integrand is again bounded, $\Phi_\epsilon(z) - z$ is bounded. Thus

$$|\Phi_\epsilon(z) - z| \leqq N_6$$

for $|\epsilon| \leqq \epsilon_2$ and for all z in G_ϵ.

If z is a point of Γ on the boundary of G_ϵ, then $z + \epsilon p_\epsilon(z)$ is a point of Γ_ϵ. Now $f_\epsilon(z) = f_\epsilon(z + \epsilon p_\epsilon(z))$; so using representation (2.2.13) for $\Phi_\epsilon(z)$ and (2.2.12) for $\Phi_\epsilon(z + \epsilon p_\epsilon(z))$, we have

$$\Phi_\epsilon(z) = \Phi_\epsilon(z + \epsilon p_\epsilon(z)). \tag{2.2.14}$$

This is the functional relation satisfied by $f_\epsilon(z)$. The function $f_\epsilon(z) - \Phi_\epsilon(z)$, when considered as a function of w under the mapping $w = f_\epsilon(z)$, is single-valued across T_ϵ and is regular and bounded in the exterior of T_ϵ, even at infinity. The function is therefore a constant, and the constant must be zero since $f_\epsilon(z)$ and $\Phi_\epsilon(z)$ each have zero translation at infinity. Thus

$$f_\epsilon(z) = z + \frac{\epsilon}{2\pi i}\int_\alpha^\beta \frac{p_\epsilon(u)f_\epsilon'(u)}{f_\epsilon(u) - f_\epsilon(z)}\,du. \tag{2.2.15}$$

This shows that as ϵ approaches zero, $f_\epsilon(z)$ tends to z. It follows that $f_\epsilon'(u)$ converges to 1 and the function (2.2.6) tends to $p_0(u)/(u - z)$ as ϵ approaches zero.

Let

$$\Psi_\epsilon(z) = z + \frac{\epsilon}{2\pi i}\int_\alpha^\beta \frac{p_0(u)}{u - z}\,du \tag{2.2.16}$$

where for large z the path of integration is Λ'. If z crosses the contour of integration, $\Psi_\epsilon(z)$ can be written

$$\Psi_\epsilon(z) = z + \epsilon p_0(z) + \frac{\epsilon}{2\pi i}\int_\alpha^\beta \frac{p_0(u)}{u - z}\,du. \tag{2.2.17}$$

If z lies in G_ϵ but outside K' and all circles of the family λ', then by using representations (2.2.15) and (2.2.16) where the path of integration is along Λ', it is clear that

$$f_\epsilon(z) = \Psi_\epsilon(z) + o(\epsilon)$$

where the $o(\epsilon)$ holds uniformly since each integrand is uniformly bounded. If z lies in K' or in one of the circles of the family λ', then representations (2.2.13) and (2.2.17) show that

$$f_\epsilon(z) = \Psi_\epsilon(z) + o(\epsilon)$$

where the $o(\epsilon)$ holds uniformly.

Given any z in G_ϵ, there is a path of integration such that $\Psi_\epsilon(z)$ is given by equation (2.2.16). Since $f_\epsilon(z) = \Psi_\epsilon(z) + o(\epsilon)$ uniformly in G_ϵ, the lemma follows.

If z remains in any closed subset of the interior of G_ϵ, then we have a better estimate of the error; namely,

$$f_\epsilon(z) = z + \frac{\epsilon}{2\pi i}\int_\alpha^\beta \frac{p(u)}{u - z}\,du + O(\epsilon^2).$$

For relation (2.2.15) shows that $f_\epsilon(z) = z + O(\epsilon)$; so if T is a polygonal line drawn in D and G_ϵ from α to β and making an angle equal to $\pi/2$ with Γ at α and β, we have on T,

$$| f_\epsilon'(u) - 1 | \leqq N_7 | \epsilon | \{ 1 + | u - \alpha |^{-1} + | u - \beta |^{-1} \}.$$

Then, in view of relations (2.2.1),

$$| f_\epsilon'(u) p_\epsilon(u) - p_0(u) | \leqq N_8 | \epsilon |, \qquad u \subset T,$$

where N_8 is some constant independent of ϵ. It is also clear that if u is on T and z remains at a fixed distance $\delta > 0$ from T, then

$$| \{f_\epsilon(u) - f_\epsilon(z)\}^{-1} - \{u - z\}^{-1} | \leqq | \epsilon | N_9(\delta).$$

Comparing equations (2.2.15) and (2.2.16), we see that the integrands on the right-hand side must differ by less than $N_{10} | \epsilon |$. This shows that the difference between $f_\epsilon(z)$ and $\Psi_\epsilon(z)$ is of order $O(\epsilon^2)$.

Lemma V. *Let Γ with end points at α and β be an analytic arc in $| z | < 1$ not passing through the origin and let $p_\epsilon(z)$ satisfy conditions* (2.2.1). *Let G_ϵ be the one-sheeted or two-sheeted domain bounded by $| z | = 1$, the arc Γ, and the arc Γ_ϵ traced by $z + \epsilon p_\epsilon(z)$ when z traces Γ, such that in moving from α to β along Γ the area to the left lies in G_ϵ. Let $w = f_\epsilon(z)$ be schlicht and regular in G_ϵ, mapping $| z | = 1$ onto $| w | = 1$ and satisfying the relation*

$$f_\epsilon(z + \epsilon p_\epsilon(z)) = f_\epsilon(z)$$

when z belongs to Γ on the boundary of G_ϵ. Further let $f_\epsilon(z)$ have the development

$$f_\epsilon(z) = a_1 z + a_2 z^2 + \cdots, \qquad a_1 > 0,$$

near the origin. Then as ϵ approaches zero,

$$f_\epsilon(z) = z\left\{1 + \frac{\epsilon}{2\pi i}\int_\alpha^\beta \frac{p(u)}{2u^2}\,\frac{u+z}{u-z}\,du - \frac{\bar\epsilon}{2\pi i}\int_\alpha^\beta \frac{\overline{p(u)}}{2\bar u^2}\,\frac{1+\bar u z}{1-\bar u z}\,d\bar u\right\} + o(\epsilon)$$

where the $o(\epsilon)$ holds uniformly in G_ϵ and $p(u) = p_0(u)$. The path of integration depends on z wherever two sheets of G_ϵ overlap.

Proof: According to Lemma IV, for each small ϵ there is a function

$$g_\epsilon(z) = z + b_1/z + b_2/z^2 + \cdots, \qquad |z| \text{ large},$$

such that $\zeta = g_\epsilon(z)$ maps the infinite domain with boundary $\Gamma + \Gamma_\epsilon$ onto the exterior of a Jordan curve T_ϵ in the ζ-plane. Under this mapping the unit circle $|z| = 1$ is mapped onto a Jordan curve C_ϵ in the ζ-plane and T_ϵ lies inside C_ϵ. The interior of C_ϵ can be mapped by a function $w = \Phi_\epsilon(\zeta)$ onto the interior of the unit circle $|w| < 1$ in such a way that $\zeta_0 = g_\epsilon(0)$ goes into $w = 0$ and $(d\Phi/d\zeta)\,(dg/dz)$ is positive at the point $z = 0$. Then

$$f_\epsilon(z) = \Phi_\epsilon(g_\epsilon(z)) = a_1 z + a_2 z^2 + \cdots$$

is the required function.

Since $f_\epsilon(z)$ is of modulus 1 on $|z| = 1$, it may be continued analytically beyond the unit circle by the relation

$$f_\epsilon(z) = \frac{1}{\bar f_\epsilon(1/\bar z)}\,.$$

Then $f_\epsilon(z)$ is schlicht in G_ϵ and in the domain inverse to G_ϵ with respect to the unit circle. Since the mapping $w = f_\epsilon(z)$ carries $z = 0$ into $w = 0$, it must carry $z = \infty$ into $w = \infty$ and therefore $f_\epsilon(z) = z/a_1 + b_0 + b_1/z + \cdots$ near $z = \infty$. Let Γ' and Γ_ϵ' be the respective arcs inverse to Γ and Γ_ϵ with respect to the unit circle. If z is a point of Γ', then the point

$$\frac{1}{1/z + \bar\epsilon\bar p_\epsilon(1/\bar z)} = z - \bar\epsilon\,\frac{z^2\bar p_\epsilon(1/\bar z)}{1 + \bar\epsilon z\bar p_\epsilon(1/\bar z)} \tag{2.2.18}$$

belongs to Γ_ϵ'. The mapping $w = f_\epsilon(z)$ will identify the arcs Γ' and Γ_ϵ', mapping a point z of Γ' and the point (2.2.18) into the same point in the w-plane.

According to Lemma IV, the function $g_\epsilon(z)$ is of the form

$$g_\epsilon(z) = z + \frac{\epsilon}{2\pi i}\int_\alpha^\beta \frac{p_\epsilon(u)g_\epsilon'(u)}{g_\epsilon(u) - g_\epsilon(z)}\,du. \tag{2.2.19}$$

By the remark at the end of the proof of Lemma IV,

$$g_\epsilon(z) = z + \frac{\epsilon}{2\pi i}\int_\alpha^\beta \frac{p(u)}{u - z}\,du + O(\epsilon^2). \tag{2.2.20}$$

Subtracting a constant from $g_\epsilon(z)$ to make $g_\epsilon(0) = 0$, we obtain the mapping

$$w_1 = z\left\{1 + \frac{\epsilon}{2\pi i}\int_\alpha^\beta \frac{p(u)}{u(u-z)}\,du + O(\epsilon^2)\right\}. \tag{2.2.21}$$

Writing

$$Y(z) = U + iV = \frac{\epsilon}{2\pi i}\int_\alpha^\beta \frac{p(u)}{u(u-z)}\,du,$$

we observe that $Y(z)$ is regular in a ring $1 - \delta < |z| < 1 + \delta$, where δ is a positive number independent of ϵ, and that the mapping (2.2.21) transforms $|z| = 1$ into a curve C_ϵ given by

$$w_1 = e^{i\theta}\{1 + U(e^{i\theta}) + O(\epsilon^2)\}.$$

The function

$$\Psi(z) = \frac{1}{2\pi}\int_{-\pi}^{\pi} \frac{e^{i\theta} + z}{e^{i\theta} - z}\,U(e^{i\theta})\,d\theta$$

is regular in $|z| < 1$, and on $|z| = 1$ its real part is equal to $U(e^{i\theta})$. Hence

$$\Psi(z) = \bar{Y}(1/\bar{z}) + ik = \frac{\bar{\epsilon}z}{2\pi i}\int_\alpha^\beta \frac{\overline{p(u)}}{\bar{u}(1 - \bar{u}z)}\,d\bar{u} + ik$$

where k is a real constant. But it is easily verified that $\Psi(0) = (1/2\pi)\int U(e^{i\theta})d\theta = 0$ and so $k = 0$. For all sufficiently small δ_1, $\delta_1 > 0$, the function

$$\Psi(w_1) = \frac{\bar{\epsilon}w_1}{2\pi i}\int_\alpha^\beta \frac{\overline{p(u)}}{\bar{u}(1 - \bar{u}w_1)}\,d\bar{u}$$

is regular in the circle $|w_1| \leqq 1 + \delta_1$. Since the derivative of $\Psi(w_1)$ is of order $O(\epsilon)$ there, the real part of $\Psi(w_1)$ on C_ϵ is $U(w_1) + O(\epsilon^2)$, and it follows that the function

$$w_2 = w_1 \exp(-\Psi(w_1)) = w_1 - w_1\Psi(w_1) \pm \cdots$$

provides a schlicht mapping of the interior of C_ϵ onto a domain which lies inside a circle $|w_2| = 1 + M|\epsilon|^2$ and contains a circle $|w_2| = 1 - M|\epsilon|^2$. The function which maps the latter domain onto the interior of the unit circle with the origin preserved and leading coefficient positive is of the form

$$w_3 = w_2\{1 + O(\epsilon^2)\}$$

where the $O(\epsilon^2)$ holds uniformly in every circle $|w_2| \leqq a < 1$. Hence

$$\begin{aligned} w_3 &= w_1 - \frac{\bar{\epsilon}w_1^2}{2\pi i}\int_\alpha^\beta \frac{\overline{p(u)}}{\bar{u}(1 - \bar{u}w_1)}\,d\bar{u} + o(\epsilon) \\ &= z + \frac{\epsilon z}{2\pi i}\int_\alpha^\beta \frac{p(u)}{u(u-z)}\,du - \bar{\epsilon}\,\frac{z^2}{2\pi i}\int_\alpha^\beta \frac{\overline{p(u)}}{\bar{u}(1 - \bar{u}z)}\,d\bar{u} + o(\epsilon). \end{aligned}$$

Multiplying w_3 by $e^{-i\theta}$, where θ is the amplitude of the first coefficient, we obtain

$$f_\epsilon(z) = w_3 \left\{1 - \frac{\epsilon}{2\pi i}\int_\alpha^\beta \frac{p(u)}{2u^2}\,du - \frac{\bar\epsilon}{2\pi i}\int_\alpha^\beta \overline{\frac{p(u)}{2u^2}}\,d\bar u\right\} + o(\epsilon)$$

$$= z\left\{1 + \frac{\epsilon}{2\pi i}\int_\alpha^\beta \frac{p(u)}{2u^2}\,\frac{u+z}{u-z}\,du - \frac{\bar\epsilon}{2\pi i}\int_\alpha^\beta \overline{\frac{p(u)}{2\bar u^2}}\cdot\frac{1+\bar u z}{1-\bar u z}\,d\bar u\right\} + o(\epsilon).$$

The error $o(\epsilon)$ holds uniformly in the part of G_ϵ that lies in any circle $|z| \leqq b < 1$. By inversion with respect to the unit circle it is seen that the error is of order $o(\epsilon)$ for $|z| = 1/b$ also, and so throughout G_ϵ.

Using the results of Lemma IV and Lemma V, we obtain a variation of a function $w = f(z)$ of class $\mathfrak{S}$.

LEMMA VI. *Let Γ with end points at α and β be an analytic Jordan arc in $|z| < 1$ not passing through the origin $z = 0$, and let $p_\epsilon(z)$ satisfy conditions* (2.2.1). *Let G_ϵ be the one-sheeted or two-sheeted domain bounded by $|z| = 1$, the arc Γ, and the arc Γ_ϵ traced by $z + \epsilon p_\epsilon(z)$ when z traces Γ, such that in moving from α to β along Γ the area to the left lies in G_ϵ. Here we suppose that ϵ is small. Given any function $f(z)$ of class $\mathfrak{S}$, there is a function $f^*(z) = f^*(z, \epsilon)$ of class $\mathfrak{S}$ which for small ϵ is of the form*

$$f^*(z) = f(z) + \frac{\epsilon}{2\pi i}\int_\alpha^\beta \frac{p(u)}{2u^2}$$

$$\cdot\left\{\left(u\frac{f'(u)}{f(u)}\right)^2 \frac{2f(z)^2}{f(u)-f(z)} - zf'(z)\,\frac{u+z}{u-z} + f(z)\right\}du \tag{2.2.22}$$

$$+ \frac{\bar\epsilon}{2\pi i}\int_\alpha^\beta \overline{\frac{p(u)}{2\bar u^2}}\left\{zf'(z)\,\frac{1+\bar u z}{1-\bar u z} - f(z)\right\}d\bar u + o(\epsilon).$$

PROOF: The function $w = f(z)$ maps Γ and Γ_ϵ into Jordan arcs J and J_ϵ in the w-plane, and the points z of Γ and $z + \epsilon p_\epsilon(z)$ of Γ_ϵ are mapped into points $w = f(z)$ of J and $w + \epsilon q_\epsilon(w) = f(z + \epsilon p_\epsilon(z))$ of J_ϵ respectively. Thus

$$\epsilon q_\epsilon(w) = f(z + \epsilon p_\epsilon(z)) - f(z) = \epsilon p_\epsilon(z) f'(z) + o(\epsilon);$$

so

$$q_\epsilon(v) = p(u)f'(u) + o(1) \tag{2.2.23}$$

where $v = f(u)$ and $p(u) = p_0(u)$. Let H_ϵ be the one-sheeted or two-sheeted domain bounded by J and J_ϵ. By Lemma IV there is a function $w_1(w)$ which is of the form

$$w_1 = w + \frac{\epsilon}{2\pi i}\int \frac{q(v)}{v - w}\,dv + o(\epsilon),$$

where $q(v) = q_0(v)$, and which maps H_ϵ into the w_1-plane, the arcs J and J_ϵ mapping into the opposite edges of a slit Λ in such a way that identified points come together. Subtracting a term from w_1 to make the origins in the two planes correspond to one another, we obtain

$$w_2 = w\left\{1 + \frac{\epsilon}{2\pi i}\int \frac{q(v)}{v(v-w)}\,dv\right\} + o(\epsilon).$$

On substituting $v = f(u)$ and using (2.2.23),

$$(2.2.24)\qquad w_2 = w\left\{1 + \frac{\epsilon}{2\pi i}\int_\alpha^\beta \frac{p(u)}{u^2}\left(\frac{uf'(u)}{f(u)}\right)^2 \frac{f(u)}{f(u) - w}\,du\right\} + o(\epsilon).$$

By Lemma V there is a function $z_1(z)$ which maps G_ϵ onto $|z_1| < 1$ cut along a Jordan arc L in such a way that identified points come together and $z_1(z)$ is of the form

$$(2.2.25)\quad z_1(z) = z\left\{1 + \frac{\epsilon}{2\pi i}\int_\alpha^\beta \frac{p(u)}{2u^2}\frac{u+z}{u-z}\,du - \frac{\bar\epsilon}{2\pi i}\int_\alpha^\beta \frac{\overline{p(u)}}{2\bar u^2}\frac{1+\bar u z}{1-\bar u z}\,d\bar u\right\} + o(\epsilon).$$

The chain of mappings, z_1 to z by (2.2.25), z to w by $w = f(z)$, w to w_2 by (2.2.24), defines a mapping $w_2 = f_1(z_1)$ of $|z_1| < 1$ cut along L onto a domain in the w_2-plane. Since $f_1(z_1)$ is single-valued across the slit L in the interior of $|z_1| < 1$, we see that $w_2 = f_1(z)$ is regular and schlicht in $|z_1| < 1$ and it maps $z_1 = 0$ into $w_2 = 0$. From (2.2.24) and (2.2.25) we have

$$w_2 = f_1(z_1) = f(z_1) + \frac{\epsilon}{2\pi i}\int_\alpha^\beta \frac{p(u)}{2u^2}\left\{u^2\frac{f'(u)^2}{f(u)^2}\frac{2f(z_1)f(u)}{f(u)-f(z_1)} - z_1 f'(z_1)\frac{u+z_1}{u-z_1}\right\}du$$
$$+ \frac{\bar\epsilon}{2\pi i}\int_\alpha^\beta \frac{\overline{p(u)}}{2\bar u^2}\, z_1 f'(z_1)\frac{1+\bar u z_1}{1-\bar u z_1}\,d\bar u + o(\epsilon).$$

The leading coefficient of this mapping is seen to be

$$1 + \frac{\epsilon}{2\pi i}\int_\alpha^\beta \frac{p(u)}{2u^2}\left\{2\left(u\frac{f'(u)}{f(u)}\right)^2 - 1\right\}du + \frac{\bar\epsilon}{2\pi i}\int_\alpha^\beta \frac{\overline{p(u)}}{2\bar u^2}\,d\bar u + o(\epsilon).$$

If we divide w_2 by this leading coefficient, we obtain (dropping the subscript 1 from z_1) a function of class $\mathfrak{S}$ which can be written in the form given by (2.2.22).

We remark for future reference that if

$$f(z) = \sum_{\nu=1}^\infty a_\nu z^\nu, \qquad a_1 = 1,$$

and if

$$f^*(z) = \sum_{\nu=1}^\infty a_\nu^* z^\nu, \qquad a_1^* = 1,$$

then from (2.2.22) we have

$$
\begin{aligned}
a_k^* = a_k &+ \frac{\epsilon}{2\pi i}\int_\alpha^\beta \frac{p(u)}{u^2}\left\{\left(u\frac{f'(u)}{f(u)}\right)^2 \sum_{\nu=2}^{k} \frac{a_k^{(\nu)}}{f(u)^{\nu-1}}\right. \\
&\left. - \frac{1}{2}(k-1)a_k - \sum_{\nu=1}^{k-1}\frac{\nu a_\nu}{u^{k-\nu}}\right\} du \qquad (2.2.26) \\
&+ \frac{\bar\epsilon}{2\pi i}\int_\alpha^\beta \frac{\overline{p(u)}}{\bar u^2}\left\{\frac{1}{2}(k-1)a_k + \sum_{\nu=1}^{k-1} \nu a_\nu \bar u^{k-\nu}\right\} d\bar u + o(\epsilon).
\end{aligned}
$$

In Lemma V the function maps the unit circumference into itself. By making a simple change of variable, we obtain a corresponding function which maps $|z| = M$ into $|w| = M$, namely,

$$
(2.2.27)\quad f_\epsilon(z) = z\left\{1 + \frac{\epsilon}{2\pi i}\int_\alpha^\beta \frac{p(u)}{2u^2}\frac{u+z}{u-z}\,du - \frac{\bar\epsilon}{2\pi i}\int_\alpha^\beta \frac{\overline{p(u)}}{2\bar u^2}\frac{M^2+\bar u z}{M^2-\bar u z}\,d\bar u\right\} + o(\epsilon).
$$

By means of this formula we can obtain a variation $v^*(z)$ of any function $v(z)$ which is regular, schlicht, and bounded by M in $|z| < 1$, and the bound M will be preserved. In fact, let G_ϵ and $p_\epsilon(z)$ be as in Lemma V. The function $w = v(z)$ maps G_ϵ onto a domain bounded by the maps J and J_ϵ of Γ and Γ_ϵ and lying in the circle $|w| < M$. If w belongs to J, then $w = v(z)$ where $z \subset \Gamma$, and the point $w + \epsilon q_\epsilon(w) = v(z + \epsilon p_\epsilon(z))$ belongs to J_ϵ. Then

$$
q_\epsilon(w) = p(z)v'(z) + O(\epsilon)
$$

where $w = v(z)$ and $q_\epsilon(w)$ satisfies the conditions (2.2.1). Using (2.2.27) with $M = 1$, we can map G_ϵ onto the circle $|z_1| < 1$ such that corresponding points on Γ and Γ_ϵ become identical. Similarly, using (2.2.27) with z replaced by w, we can transform H_ϵ into a domain lying in $|w_1| < M$, and in this mapping corresponding points on J and J_ϵ go into the same point. It is readily seen that the composite mapping $w_1(z_1)$ is single-valued, and so regular and schlicht, in $|z_1| < 1$. Composing these mappings, we obtain

$$
\begin{aligned}
v^*(z) = v(z) &+ \frac{\epsilon}{2\pi i}\int_\alpha^\beta \frac{p(u)}{2u^2}\left\{\left(u\frac{v'(u)}{v(u)}\right)^2 v(z)\frac{v(u)+v(z)}{v(u)-v(z)} - zv'(z)\frac{u+z}{u-z}\right\} du \\
&\qquad (2.2.28) \\
&- \frac{\bar\epsilon}{2\pi i}\int_\alpha^\beta \frac{\overline{p(u)}}{2\bar u^2}\left\{\left(\bar u\frac{\overline{v'(u)}}{\overline{v(u)}}\right)^2 v(z)\frac{M^2+\overline{v(u)}v(z)}{M^2-\overline{v(u)}v(z)} - zv'(z)\frac{1+\bar u z}{1-\bar u z}\right\} d\bar u + o(\epsilon).
\end{aligned}
$$

We make a remark which, however, will not be used in the sequel. Instead of taking Γ of Lemma IV to be a Jordan arc extending from α to β, we may take Γ to be a closed analytic Jordan curve. Let D be a region containing Γ in its interior and let $p_\epsilon(z)$ be regular in D for $|\epsilon| \leqq \epsilon_0$. We then suppose that the first two conditions (2.2.1) are satisfied but not necessarily the third. If ϵ is sufficiently small, then as z moves completely around Γ, the point $z + \epsilon p_\epsilon(z)$ describes a neighboring Jordan curve Γ_ϵ. Let G_ϵ denote the disconnected set of points composed of the exterior of Γ and the interior of Γ_ϵ. By identifying the point z of Γ with the point $z + \epsilon p_\epsilon(z)$ of $\Gamma_\epsilon(z)$ and introducing local uniformizers τ at pairs of identified points, we obtain from G_ϵ a closed Riemann surface of genus zero. By the uniformization theorem there is a function $w = f_\epsilon(z)$ which maps this Riemann surface onto the closed w-plane in a one-one way and which satisfies the relations of Lemma IV with the path of integration replaced by a closed contour which may depend on z. If z is away from Γ, we may take the path of integration to be Γ traversed clockwise. In particular, if

$$p_\epsilon(z) = p(z) = \frac{1}{z - z_0} \tag{2.2.29}$$

where z_0 is a point lying in the interior of Γ, we obtain by the Cauchy residue theorem from Lemma IV (with the path of integration replaced by a closed contour)

$$f_\epsilon(z) = \begin{cases} z + \dfrac{\epsilon}{z - z_0} + o(\epsilon), & z \text{ outside } \Gamma, \\ z + o(\epsilon), & z \text{ inside } \Gamma_\epsilon. \end{cases}$$

But the function

$$f_\epsilon(z) = \begin{cases} z + \dfrac{\epsilon}{z - z_0}, & z \text{ outside } \Gamma, \\ z, & z \text{ inside } \Gamma_\epsilon, \end{cases} \tag{2.2.30}$$

plainly satisfies the conditions (2.2.2) and (2.2.3) and is therefore unique. Hence the choice (2.2.6) gives the simple function (2.2.7) as, of course, may be seen directly.

If we carry through this program of using a closed Jordan arc for Γ, then the formula (2.2.22) of Lemma VI is valid, but the path of integration is a closed contour which, if z is outside Γ, may be taken to be Γ traversed clockwise. Choosing, in particular,

$$p_\epsilon(z) = p(z) = \frac{z^2}{z - z_0}$$

where z_0 is a point interior to Γ, we obtain from the Cauchy residue theorem the formula

$$\begin{aligned} f^*(z) = f(z) - \frac{\epsilon}{2}\left\{\left(z_0 \frac{f'(z_0)}{f(z_0)}\right)^2 \frac{2f(z)^2}{f(z_0) - f(z)} - zf'(z)\frac{z_0 + z}{z_0 - z} + f(z)\right\} \\ + \frac{\bar{\epsilon}}{2}\left\{zf'(z)\frac{1 + \bar{z}_0 z}{1 - \bar{z}_0 z} - f(z)\right\} + o(\epsilon). \end{aligned} \tag{2.2.31}$$

A formula similar to (2.2.31) was obtained by Schiffer [24c], using a different method.

2.3. Let us now apply the variational method to obtain information concerning functions which belong to boundary points of the nth coefficient region.

If

$$f(z) = \sum_{\nu=1}^{\infty} b_\nu z^\nu$$

is a function of class $\mathfrak{S}$, let

$$b_\nu = x_\nu + iy_\nu$$

where x_ν and y_ν are real. Since

$$x_\nu = \frac{1}{2}(b_\nu + \bar{b}_\nu), \qquad y_\nu = \frac{1}{2i}(b_\nu - \bar{b}_\nu),$$

any real-valued function F of $x_2, y_2, \cdots, x_n, y_n$ can be expressed as a function of $b_2, \bar{b}_2, \cdots, b_n, \bar{b}_n$. Write

$$F_\nu = \frac{1}{2}\left(\frac{\partial F}{\partial x_\nu} - i\frac{\partial F}{\partial y_\nu}\right), \qquad \bar{F}_\nu = \frac{1}{2}\left(\frac{\partial F}{\partial x_\nu} + i\frac{\partial F}{\partial y_\nu}\right).$$

Now let $F = F(b_2, \bar{b}_2, \cdots, b_n, \bar{b}_n)$ be a function which satisfies the following three conditions:

(a) F is defined and takes real values in an open set $\boldsymbol{O}$ containing the closed set $\boldsymbol{V}_n$;

(b) F and its derivatives F_ν are continuous in $\boldsymbol{O}$;

(c) $|\operatorname{grad} F| = \left(\sum_{\nu=2}^{n} |F_\nu|^2\right)^{1/2} > 0$ in $\boldsymbol{O}$.

Condition (c) implies that the maximum value of F in $\boldsymbol{V}_n$ is attained at a point on the boundary of $\boldsymbol{V}_n$.

LEMMA VII. *If F satisfies the conditions* (a), (b), *and* (c), *then every function $f(z)$ of class $\mathfrak{S}$ belonging to a point $(a_2, a_3, \cdots, a_n)$ where F attains its maximum in $\boldsymbol{V}_n$ must satisfy the differential equation*

$$\left(z\frac{f'(z)}{f(z)}\right)^2 \sum_{\nu=1}^{n-1} \frac{A_\nu}{f(z)^\nu} = \sum_{\nu=-(n-1)}^{n-1} \frac{B_\nu}{z^\nu} \tag{2.3.1}$$

where

$$A_\nu = \sum_{k=\nu+1}^{n} a_k^{(\nu+1)} F_k, \qquad B_\nu = \sum_{k=1}^{n-\nu} k a_k F_{k+\nu}, \qquad \nu = 1, 2, \cdots, n-1,$$
$$B_0 = \sum_{k=1}^{n} (k-1) a_k F_k, \qquad B_{-\nu} = \bar{B}_\nu. \tag{2.3.2}$$

Here

$$F_k = \frac{1}{2}\left(\frac{\partial F}{\partial x_k} - i\,\frac{\partial F}{\partial y_k}\right), \qquad x_k = \frac{1}{2}\,(a_k + \bar{a}_k), \qquad y_k = \frac{1}{2i}\,(a_k - \bar{a}_k),$$

$$k = 2, 3, \cdots, n,$$

the derivatives being taken at the point $(a_2, a_3, \cdots, a_n)$. *Moreover,* $B_0 > 0$ *and the right side of* (2.3.1) *is non-negative on* $|z| = 1$ *with at least one zero there.*

Proof: Let the maximum of F in V_n occur at the point $(a_2, a_3, \cdots, a_n)$, and let the function

$$f(z) = \sum_{\nu=1}^{\infty} a_\nu z^\nu$$

of class $\mathfrak{S}$ belong to this point.

Let us consider the neighboring function $f^*(z)$ defined by (2.2.22) whose coefficients are given by (2.2.26). Writing

$$F^* = F(a_2^*, \bar{a}_2^*, \cdots, a_n^*, \bar{a}_n^*)$$

and

$$\Delta a_k^* = a_k^* - a_k, \qquad \Delta F = F^* - F,$$

we have

$$\Delta F = 2\,\mathrm{Re}\left\{\sum_{k=2}^{n} F_k\,\Delta a_k\right\} + o(\epsilon). \tag{2.3.3}$$

By (2.2.26)

$$\Delta F = 2\,\mathrm{Re}\left\{\frac{\epsilon}{2\pi i}\int_\alpha^\beta \frac{p(u)}{u^2}\left[\left(u\,\frac{f'(u)}{f(u)}\right)^2 \sum_{\nu=1}^{n-1} \frac{A_\nu}{f(u)^\nu} - \frac{1}{2}\,B_0 - \sum_{\nu=1}^{n-1} \frac{B_\nu}{u^\nu}\right] du \right.$$
$$\left. + \frac{\bar{\epsilon}}{2\pi i}\int_\alpha^\beta \overline{\frac{p(u)}{\bar{u}^2}}\left[\frac{1}{2}\,B_0 + \sum_{\nu=1}^{n-1} B_\nu \bar{u}^\nu\right] d\bar{u}\right\} + o(\epsilon),$$

that is,

$$\Delta F = 2\,\mathrm{Re}\left\{\frac{\epsilon}{2\pi i}\int_\alpha^\beta \frac{p(u)}{u^2}\left[\left(u\,\frac{f'(u)}{f(u)}\right)^2 \sum_{\nu=1}^{n-1} \frac{A_\nu}{f(u)^\nu} - \frac{1}{2}\,(B_0 + \bar{B}_0) \right.\right.$$
$$\left.\left. - \sum_{\nu=1}^{n-1}\left(\frac{B_\nu}{u^\nu} + \bar{B}_\nu u^\nu\right)\right] du\right\} + o(\epsilon). \tag{2.3.4}$$

Now $\Delta F \leqq 0$ for all small ϵ. Since arg (ϵ) can have any value, we see that the coefficient of ϵ in the expression on the right of (2.3.4) must vanish. Since $p(u)$ is an arbitrary function subject to the conditions stated, we obtain the differential equation (2.3.1) with B_0 replaced by $(B_0 + \bar{B}_0)/2$.

To show that B_0 is real, we note that if $f(z)$ belongs to the point $(a_2, a_3, \cdots, a_n)$ where F attains its maximum in V_n, then the neighboring function

$$f^*(z) = e^{-i\epsilon}f(e^{i\epsilon}z) = z + a_2e^{i\epsilon}z^2 + a_3e^{2i\epsilon}z^3 + \cdots$$

belongs to class $\mathfrak{S}$ when ϵ is real. In this case

$$a_k^* = e^{i(k-1)\epsilon}a_k = a_k + i\epsilon(k-1)a_k + o(\epsilon);$$

so the relation (2.3.3) becomes

$$\Delta F = 2 \operatorname{Re}\left\{i\epsilon \sum_{k=2}^{n} (k-1)a_k F_k\right\} + o(\epsilon) = 2 \operatorname{Re} \{i\epsilon B_0\} + o(\epsilon).$$

Since ϵ can be taken either positive or negative, this shows that B_0 is real. We may thus replace $(B_0 + \bar{B}_0)/2$ by B_0.

To show that the right side of (2.3.1) is non-negative on $|z| = 1$, consider the mapping which carries the unit circle $|\zeta| < 1$ into the unit circle $|z| < 1$ minus a portion of a rectilinear slit orthogonal to the circumference $|z| = 1$ at the point $e^{i\theta}$. This mapping is

$$z(\zeta, t) = f_0^{-1}\{e^{-t}f_0(\zeta)\}, \qquad t > 0, \tag{2.3.5}$$

where

$$f_0(\zeta) = \frac{\zeta}{(1 + e^{-i\theta}\zeta)^2}$$

and f_0^{-1} denotes the inverse function. Differentiating (2.3.5) at $t = 0$, we obtain

$$\left.\frac{\partial z(\zeta, t)}{\partial t}\right|_{t=0} = -\frac{f_0(\zeta)}{f_0'(\zeta)} = -\zeta\frac{1 + e^{-i\theta}\zeta}{1 - e^{-i\theta}\zeta}$$

and so for all small $t = \epsilon$ $(\epsilon > 0)$ we have

$$z(\zeta, \epsilon) = \zeta - \epsilon\zeta\frac{1 + e^{-i\theta}\zeta}{1 - e^{-i\theta}\zeta} + o(\epsilon).$$

The function

$$f(z(\zeta, \epsilon)) = f(\zeta) - \epsilon\zeta f'(\zeta)\frac{1 + e^{-i\theta}\zeta}{1 - e^{-i\theta}\zeta} + o(\epsilon) \tag{2.3.6}$$

is regular and schlicht in $|\zeta| < 1$ and its leading coefficient is

$$1 - \epsilon + o(\epsilon). \tag{2.3.7}$$

Dividing (2.3.6) by (2.3.7), we obtain a function

$$f^*(\zeta) = f(\zeta) + \epsilon\left\{f(\zeta) - \zeta f'(\zeta)\frac{1 + e^{-i\theta}\zeta}{1 - e^{-i\theta}\zeta}\right\} + o(\epsilon) \tag{2.3.6$'$}$$

which belongs to class $\mathfrak{S}$. Writing

$$f^*(\zeta) = \sum_{k=1}^{\infty} a_k^* \zeta^k,$$

we have

$$\Delta a_k = a_k^* - a_k = -\epsilon \left\{ (k-1)a_k + 2 \sum_{\nu=1}^{k-1} \nu a_\nu e^{-i(k-\nu)\theta} \right\} + o(\epsilon),$$

and so if $z = e^{i\theta}$,

$$\Delta F = 2 \operatorname{Re} \left\{ \sum_{k=2}^{n} F_k \Delta a_k \right\} + o(\epsilon) = -\epsilon \operatorname{Re} \left\{ \sum_{\nu=-(n-1)}^{n-1} \frac{B_\nu}{z^\nu} \right\} + o(\epsilon).$$

Since $\Delta F \leqq 0$ and $\epsilon > 0$, we see that the right side of (2.3.1) is non-negative on $|z| = 1$.

The representation

$$B_0 = \frac{1}{2\pi} \int_{-\pi}^{\pi} \left\{ \sum_{\nu=-(n-1)}^{n-1} B_\nu e^{-i\nu\theta} \right\} d\theta \tag{2.3.8}$$

shows that B_0 is greater than or equal to zero and that B_0 can vanish only if all the B_ν vanish. However, the expression

$$B_\nu = \sum_{k=1}^{n-\nu} k a_k F_{k+\nu}$$

shows that all B_ν can vanish only if all the F_ν vanish. But the gradient of F is not zero by hypothesis (c) above, and so

$$B_0 > 0.$$

Finally, if m is the largest value of ν such that $A_\nu \neq 0$, then the left side of (2.3.1) has a pole of precise order m at the origin. Owing to the fact that $w = f(z)$ behaves like z near $z = 0$, the function $Q(z)$ must have a pole of the same order; so $B_\nu = 0$ for $\nu > m$ but $B_m \neq 0$. Since $f(z)$ is schlicht, the function

$$\sum_{\nu=1}^{m} \frac{A_\nu}{f(z)^\nu}$$

has $m - 1$ or fewer zeros as z ranges over $|z| < 1$. The right side of (2.3.1) has exactly $2m$ zeros and if there were no zeros on $|z| = 1$, there would be m zeros in $|z| < 1$ and the same number in $|z| > 1$. The differential equation (2.3.1) shows that this is impossible since $f'(z)$ does not vanish in $|z| < 1$. Thus the right side of (2.3.1) must have at least one zero on $|z| = 1$ and this zero must be of even order. This completes the proof of Lemma VII.

For simplicity, we shall henceforth write

$$P(w) = \sum_{\nu=1}^{n-1} \frac{A_\nu}{w^\nu}, \qquad Q(z) = \sum_{\nu=-(n-1)}^{n-1} \frac{B_\nu}{z^\nu}. \tag{2.3.9}$$

By writing $w = f(z)$, the differential equation (2.3.1) becomes

$$\left(\frac{z}{w}\frac{dw}{dz}\right)^2 P(w) = Q(z). \tag{2.3.10}$$

The formula (2.3.8) or the alternative formula

$$B_0 = -\min_{|z|=1} \left\{ \sum_{\nu=1}^{n-1} \left(\frac{B_\nu}{z^\nu} + \bar{B}_\nu z^\nu \right) \right\} \tag{2.3.8$'$}$$

shows that the value of B_0 is uniquely determined in terms of $B_1, B_2, \cdots, B_{n-1}$. Hence the functions $P(w)$, $Q(z)$ and corresponding equation (2.3.10) are determined if we know the two vectors

$$\mathfrak{A} = (A_1, A_2, \cdots, A_{n-1}), \qquad \mathfrak{B} = (B_1, B_2, \cdots, B_{n-1}). \tag{2.3.11}$$

One of our main problems will be to determine one of the vectors when the other is given.

2.4. We now prove:

LEMMA VIII. *To every point $(a_2, a_3, \cdots, a_n)$ on the boundary of V_n there belongs at least one function $f(z)$ of class $\mathfrak{S}$ and a corresponding pair of vectors $\mathfrak{A}$, $\mathfrak{B}$ such that $w = f(z)$ satisfies the differential equation*

$$\left(\frac{z}{w}\frac{dw}{dz}\right)^2 P(w) = Q(z) \tag{2.4.1}$$

where $Q(z)$ is non-negative on $|z| = 1$ with at least one zero there and where the constant term B_0 in $Q(z)$ is positive.

PROOF: Let the set E in the coefficient space be defined as follows. A point $\mathfrak{p}$ belongs to E if there is a sphere having only the point $\mathfrak{p}$ in common with V_n. It is clear that the set E lies on the boundary of V_n. To show that E is everywhere dense in the boundary of V_n, let $\mathfrak{a}$ be a boundary point. Given $\epsilon > 0$, there is a point $\mathfrak{b}$ not belonging to V_n such that $\|\mathfrak{a} - \mathfrak{b}\| < \epsilon$. Here $\|\mathfrak{a} - \mathfrak{b}\|$ denotes the distance of $\mathfrak{a}$ from $\mathfrak{b}$ in the euclidean space of $2n - 2$ real dimensions. Since V_n is closed by Lemma II, there is an $\epsilon_1 > 0$ such that $\|\mathfrak{b} - \mathfrak{t}\| \geqq \epsilon_1$ for all points $\mathfrak{t}$ of V_n, with equality at one or more points. Let the equality occur at the point $\mathfrak{c}$. Then $\mathfrak{c}$ is on the boundary of V_n and $\|\mathfrak{b} - \mathfrak{c}\| = \epsilon_1$ while $\|\mathfrak{b} - \mathfrak{t}\| \geqq \epsilon_1$ for every point $\mathfrak{t}$ of V_n. Let $\mathfrak{m}$ be the mid-point of the segment of the straight line joining $\mathfrak{b}$ and $\mathfrak{c}$. Then $\|\mathfrak{t} - \mathfrak{m}\| \geqq \epsilon_1/2$ for all points $\mathfrak{t}$ of V_n and the equality occurs if and only if $\mathfrak{t} = \mathfrak{c}$. Since

$$\|\mathfrak{a} - \mathfrak{c}\| \leqq \|\mathfrak{a} - \mathfrak{b}\| + \|\mathfrak{b} - \mathfrak{c}\| \leqq \epsilon + \epsilon_1 \leqq 2\epsilon,$$

we see that the set E lies everywhere dense in the boundary of V_n.

Let $\mathfrak{p}$ be a point of the set E and let $\mathfrak{m} = (\alpha_2, \cdots, \alpha_n)$ be the center of a sphere which has only the point $\mathfrak{p}$ in common with V_n. The function F defined by the equation

$$F(b_2, \bar{b}_2, \cdots, b_n, \bar{b}_n) = \frac{1}{\sum_{\nu=2}^{n} | b_\nu - \alpha_\nu |^2}$$

has the property that it is continuous together with its derivatives in an open set containing V_n and its gradient is not zero in this open set. The maximum of F in V_n occurs at the single point $\mathfrak{p}$. Lemma VII asserts that there is a function $f(z)$ of class $\mathfrak{S}$ belonging to the point $\mathfrak{p}$ which satisfies a differential equation of the form (2.3.1).

Now let $(a_2, a_3, \cdots, a_n)$ be a point on the boundary of V_n. Then it is a limit point of points of E, and to each of these points of E we have shown that there belongs a function $f(z)$ of class $\mathfrak{S}$ which satisfies a differential equation of the form (2.3.1). Now let the right side of (2.3.1) be normalized by the condition

$$(2.4.2) \qquad || \mathfrak{B} || = \left(\sum_{\nu=1}^{n-1} | B_\nu |^2 \right)^{1/2} = 1.$$

This may be accomplished by multiplying the function F of Lemma VII by a suitable constant. Since $a_1 = 1$, the equations

$$B_\nu = \sum_{k=1}^{n-\nu} k a_k F_{k+\nu}, \qquad \nu = 1, 2, \cdots, n - 1,$$

may be solved for the F_ν in terms of the coefficients a_ν and the numbers B_ν, and it follows that the F_ν are bounded. The equations

$$A_\nu = \sum_{k=\nu+1}^{n} a_k^{(\nu+1)} F_k$$

then show that the A_ν are bounded. Thus under the restriction (2.4.2) the functions of class $\mathfrak{S}$ which satisfy a differential equation of the form (2.3.1) constitute a compact set of functions. One such function belongs to the point $(a_2, a_3, \cdots, a_n)$. The relations (2.3.2) of Lemma VII remain true in the limit, provided that we no longer interpret the numbers F_ν as the derivatives of a function F having a maximum at the point but interpret them instead only as parameters.

2.5. Let

$$(2.5.1) \qquad v(z) = b_1 z + b_2 z^2 + b_3 z^3 + \cdots, \qquad 0 < b_1 \leqq 1,$$

be a function which is regular, schlicht, and bounded by 1 in $| z | < 1$. As $v(z)$ ranges over all such functions, the point $(b_1, b_2, \cdots, b_n)$ sweeps out a region $\mathcal{V}_n$ in euclidean space of $2n - 1$ real dimensions. The region $\mathcal{V}_n$ is not a closed set as it does not contain points for which $b_1 = 0$. By inequality (1.1.8) we have $| b_k | < e | b_1 | k$, and so as b_1 tends to zero, the first n coefficients of $v(z)$ tend to zero. The region $\mathfrak{V}_n$ is made into a closed set by adding the point $b_1 = 0, b_2 = 0, \cdots, b_n = 0$.

A variational formula for $v(z)$ is given by (2.2.28) with $M = 1$, namely

$$v^*(z) = v(z) + \frac{\epsilon}{2\pi i}\int_\alpha^\beta \frac{p(u)}{2u^2}\left\{\left(u\,\frac{v'(u)}{v(u)}\right)^2 v(z)\,\frac{v(u)+v(z)}{v(u)-v(z)}\right.$$

$$\left. - zv'(z)\,\frac{u+z}{u-z}\right\} du \tag{2.5.2}$$

$$-\frac{\bar\epsilon}{2\pi i}\int_\alpha^\beta \overline{\frac{p(u)}{2u^2}}\left\{\left(\bar u\,\overline{\frac{v'(u)}{v(u)}}\right)^2 v(z)\,\frac{1+\overline{v(u)}v(z)}{1-\overline{v(u)}v(z)} - zv'(z)\,\frac{1+\bar u z}{1-\bar u z}\right\} d\bar u + o(\epsilon).$$

Writing

$$v^*(z) = \sum_{\nu=1}^{\infty} b_\nu^* z^\nu,$$

we have:

$$b_k^* = b_k + \frac{\epsilon}{2\pi i}\int_\alpha^\beta \frac{p(u)}{2u^2}\left\{\left(u\,\frac{v'(u)}{v(u)}\right)^2\left[b_k + 2\sum_{\nu=2}^{k}\frac{b_k^{(\nu)}}{v(u)^{\nu-1}}\right]\right.$$

$$\left. - \left[kb_k + 2\sum_{\nu=1}^{k-1}\frac{\nu b_\nu}{u^{k-\nu}}\right]\right\} du \tag{2.5.3}$$

$$-\frac{\bar\epsilon}{2\pi i}\int_\alpha^\beta \overline{\frac{p(u)}{2u^2}}\left\{\left(\bar u\,\overline{\frac{v'(u)}{v(u)}}\right)^2\left[b_k + 2\sum_{\nu=2}^{k} b_k^{(\nu)}\overline{v(u)}^{\nu-1}\right]\right.$$

$$\left. - \left[kb_k + 2\sum_{\nu=1}^{k-1}\nu b_\nu \bar u^{k-\nu}\right]\right\} d\bar u + o(\epsilon).$$

As in the preceding paragraphs we find that any function $v(z)$ belonging to a boundary point $(b_1, b_2, \cdots, b_n)$ of $\mathfrak{V}_n$ which is different from the origin $b_1 = 0, b_2 = 0, \cdots, b_n = 0$ satisfies an equation

$$\left(\frac{z}{v}\frac{dv}{dz}\right)^2 R(v) = Q(z) \tag{2.5.4}$$

where

$$R(v) = \sum_{\nu=-(n-1)}^{n-1}\frac{\beta_\nu}{v^\nu}, \qquad Q(z) = \sum_{\nu=-(n-1)}^{n-1}\frac{B_\nu}{z^\nu}. \tag{2.5.5}$$

Here

$$\beta_\nu = \sum_{k=\nu+1}^{n} F_k b_k^{(\nu+1)}, \qquad \beta_{-\nu} = \bar\beta_\nu \qquad (\nu = 1, 2, \cdots, n-1),$$

$$\beta_0 = F_1 b_1 + \operatorname{Re}\sum_{k=2}^{n} b_k F_k,$$

$$B_\nu = \sum_{k=1}^{n-\nu} kb_k F_{k+\nu}, \qquad B_{-\nu} = \bar B_\nu \qquad (\nu = 1, 2, \cdots, n-1), \tag{2.5.6}$$

$$B_0 = F_1 b_1 + \operatorname{Re}\sum_{k=2}^{n} kb_k F_k,$$

where F_1 is a real number and $F_2, F_3, \cdots, F_n$ are certain complex constants. The functions $R(v)$ and $Q(z)$ are non-negative on the unit circumference and it can be shown by methods to be developed in subsequent chapters that each has at least one zero there.

The function $f(z) = v(z)/b_1$ satisfying (2.5.4) is bounded by $1/b_1$, and so by Lemma I it belongs to an interior point of V_n. Conversely, given any interior point $(a_2, a_3, \cdots, a_n)$ of V_n, consider the intersection of the line

$$b_2 = a_2 b_1, \quad b_3 = a_3 b_1, \quad \cdots, \quad b_n = a_n b_1 \tag{2.5.7}$$

with $\mathcal{V}_n$. The point $(b_1, b_2, \cdots, b_n)$ of this line for which b_1 is maximal is certainly a boundary point of $\mathcal{V}_n$. Hence there is a function $f(z)$ belonging to $(a_2, a_3, \cdots, a_n)$ for which the least upper bound in $|z| < 1$ is minimal, and this function satisfies an equation (2.5.4). Thus at any interior point of V_n there is a function $f(z)$ belonging to the point which satisfies an equation (2.5.5).

So far as the region V_n is concerned (n fixed), it is sufficient to consider only functions which satisfy an equation (2.3.1) (boundary functions) or functions which satisfy an equation (2.5.4) (interior functions).

2.6. The variations described in this chapter are interior variations of a schlicht function; that is to say, these variations do not involve the nature of the boundary in the w-plane corresponding to $|z| = 1$ in the mapping $w = f(z)$. It will be shown, however, that in the case of an extremal schlicht function the boundary in the w-plane is composed of analytic arcs. The nature of these slits will be discussed in detail in the following chapters.

Chapter III

THE CRITICAL POINTS OF THE DIFFERENTIAL EQUATION

3.1. In the preceding chapter we have shown that any function $w = f(z)$ belonging to a boundary point of V_n satisfies a differential equation of the form

$$\left(\frac{z}{w}\frac{dw}{dz}\right)^2 P(w) = Q(z) \tag{3.1.1}$$

where

$$P(w) = \sum_{\nu=1}^{n-1} \frac{A_\nu}{w^\nu}, \qquad Q(z) = \sum_{\nu=-(n-1)}^{n-1} \frac{B_\nu}{z^\nu}. \tag{3.1.2}$$

Here $Q(z) \geqq 0$ on $|z| = 1$ with at least one zero there, which must be of even order.

Our purpose is to study solutions of (3.1.1) which are regular in the unit circle $|z| < 1$ and which are normalized by the condition that the Taylor expansion of w about the origin has the form

$$w = z + a_2 z^2 + a_3 z^3 + \cdots. \tag{3.1.3}$$

Because of this normalization we see that

$$A_{n-1} = B_{n-1}, \tag{3.1.4}$$

and there is no loss of generality in assuming that $A_{n-1} = B_{n-1} \neq 0$.

The differential equation (3.1.1) is separable. Integrating, we obtain

$$\int (P(w))^{1/2} \frac{dw}{w} = \int (Q(z))^{1/2} \frac{dz}{z}. \tag{3.1.5}$$

Since $Q(z) \geqq 0$ on $|z| = 1$, we see that the integral

$$\int (Q(z))^{1/2} \frac{dz}{z}$$

has a constant real part on the circumference $|z| = 1$, and by (3.1.5) so does the integral

$$\int (P(w))^{1/2} \frac{dw}{w}.$$

It is thus intuitive, and we shall later prove that the boundary in the w-plane corresponding to $|z| = 1$ in the mapping (3.1.3) consists of loci defined by

$$\operatorname{Re} \int (P(w))^{1/2} \frac{dw}{w} = \text{constant}. \tag{3.1.6}$$

By introducing a suitable real parameter τ we see from (3.1.1) that these loci may be regarded as integral curves of the differential equation

$$\left(\frac{1}{w}\frac{dw}{d\tau}\right)^2 P(w) + 1 = 0. \tag{3.1.6$'$}$$

Writing $w = u + iv$, taking square roots, and separating (3.1.6)$'$ into real and imaginary parts, we see that (3.1.6)$'$ is equivalent to a system of two equations

$$\frac{du}{d\tau} = L(u, v), \qquad \frac{dv}{d\tau} = M(u, v), \tag{3.1.6$''$}$$

where L and M are certain real functions of u and v. A method for investigating the integral curves of such systems of equations has been given by Poincaré but the determination of the structure of the integral curves in the large is generally a difficult problem which has been solved only in certain special cases. In our case the problem is complicated by the fact that the functions L, M are not single-valued, but since the system of two equations arises from one complex equation (3.1.6)$'$, we are able to employ methods based on conformal mapping which are not available in the general Poincaré theory.

We shall investigate the loci (3.1.6) in detail. Inasmuch as we shall also investigate the loci

$$\mathrm{Re} \int (Q(z))^{1/2} \frac{dz}{z} = \text{constant}, \tag{3.1.7}$$

it is sometimes desirable to bring both under a single treatment. Let k, $1 \leqq k \leqq n - 1$, be the smallest value of ν for which $A_\nu \neq 0$. Then

$$\frac{P(w)}{w^2} = A_k w^{-n-1}(w - w_1)(w - w_2) \cdots (w - w_{n-k-1}). \tag{3.1.8}$$

Also

$$\frac{Q(z)}{z^2} = \bar{B}_{n-1} z^{-n-1}(z - z_1)(z - z_2) \cdots (z - z_{2n-2}). \tag{3.1.9}$$

These two rational functions are special cases of the general rational function

$$R(w) = A \frac{(w - q_1)^{\lambda_1}(w - q_2)^{\lambda_2} \cdots (w - q_k)^{\lambda_k}}{(w - p_1)^{\mu_1}(w - p_2)^{\mu_2} \cdots (w - p_m)^{\mu_m}} \tag{3.1.10}$$

where λ_j, μ_j are positive integers. Thus, writing

$$\zeta = \int (R(w))^{1/2}\, dw, \tag{3.1.11}$$

we shall consider the loci defined by

$$\mathrm{Re}(\zeta) = \mathrm{Re} \int (R(w))^{1/2}\, dw = \text{constant}. \tag{3.1.12}$$

In this chapter we consider the behavior of the loci (3.1.12) in the small, and in the following chapter we discuss the behavior in the large.

3.2. It is clear that if w_0 is not one of the points $q_1, q_2, \cdots, q_k, p_1, p_2, \cdots, p_m$ or ∞, then the function $\zeta(w)$ defined by (3.1.11) is regular and schlicht in a neighborhood of w_0. Further there are precisely two arcs emerging from w_0 on which $\mathrm{Re}(\zeta) = c$, where c is a constant, and they form a single analytic arc through w_0. In the neighborhood of w_0 we have

$$\zeta - \zeta_0 = \sum_{\nu=1}^{\infty} b_\nu (w - w_0)^\nu, \qquad b_1 \neq 0,$$

$$w - w_0 = \sum_{\nu=1}^{\infty} c_\nu (\zeta - \zeta_0)^\nu, \qquad c_1 \neq 0.$$

Next, let $R(w)$ have a zero of precise order $\lambda > 0$ at the finite point q. Then locally

$$\zeta - \zeta_0 = \sum_{\nu=1}^{\infty} b_\nu (w - q)^{\lambda/2+\nu}, \qquad b_1 \neq 0,$$

$$w - w_0 = \sum_{\nu=1}^{\infty} c_\nu (\zeta - \zeta_0)^{2\nu/(\lambda+2)}, \qquad c_1 \neq 0,$$

where ζ is defined by (3.1.11). There are exactly $\lambda + 2$ arcs $\mathrm{Re}(\zeta) = c$ emerging from the point q and they are equally spaced at angles of $2\pi/(\lambda + 2)$. These arcs divide the circle $| w - q | < \delta$, where $\delta > 0$ is a sufficiently small positive number, into $\lambda + 2$ domains, each of which is mapped in a one–one manner onto a domain which lies in $\mathrm{Re}(\zeta) > c$ or in $\mathrm{Re}(\zeta) < c$. This is proved by the usual method of local uniformization.

Suppose now that $R(w)$ has a pole of first order at the point p. Then

$$\zeta = \int \left(\frac{F(w)}{w - p}\right)^{1/2} dw$$

where $F(w)$ is regular and non-vanishing at the point p. If we let

$$w - p = s^2,$$

then

$$\zeta = \int \left\{\sum_{\nu=0}^{\infty} c_\nu s^{2\nu}\right\}^{1/2} ds, \qquad c_0 \neq 0.$$

Thus ζ is an odd function of s and there are two arcs emerging from $s = 0$ which form a single analytic arc. Mapping back into the w-plane, we see that there is a single arc emerging from the point p on which $\mathrm{Re}(\zeta) = c$. There is a $\delta > 0$ such that the portion of this arc in the circle $| w - p | \leqq \delta$ together with the circumference of the circle forms the boundary of a domain which is mapped into a domain in the ζ-plane lying in $\mathrm{Re}(\zeta) > c$ or $\mathrm{Re}(\zeta) < c$. The portion of the arc

$\operatorname{Re}(\zeta) = c$ lying in the circle $|w - p| \leqq \delta$ maps into a segment of the line $\operatorname{Re}(\zeta) = c$ which has $\zeta(p)$ as its mid-point.

For poles of order higher than the first the behavior is somewhat more complicated. We consider separately the two cases where the order μ is 2 (Lemma IX) and greater than 2 (Lemma X).

Lemma IX. *Let $R(w)$ have a pole of precise order $\mu = 2$ at a finite point p, and near $w = p$ let*

$$(R(w))^{1/2} = \frac{a + ib}{w - p} + b_0 + b_1(w - p) + \cdots .$$

(1) $b \neq 0$. *There is an $\alpha > 0$ such that no locus $\operatorname{Re}(\zeta) = c$, $-\infty < \operatorname{Im}(\zeta) < +\infty$, remains entirely in $|w - p| < \alpha$. There is a number $\rho > 0$ such that if the locus $\operatorname{Re}(\zeta) = c$, $\operatorname{Im}(\zeta)$ monotonic, enters $|w - p| < \rho$ from $|w - p| \geqq \rho$, then the locus remains in $|w - p| < \rho$ and tends to $w = p$ with $|w - p|$ strictly decreasing. If $a \neq 0$, the locus $\operatorname{Re}(\zeta) = c$, $\operatorname{Im}(\zeta)$ monotonic, spirals around $w = 0$ and behaves asymptotically like a logarithmic spiral while if $a = 0$, the locus has an asymptotic direction at $w = p$ which, however, depends on the constant c, $\operatorname{Re}(\zeta) = c$.*

(2) $b = 0$. *There is a $\beta > 0$ such that if the locus $\operatorname{Re}(\zeta) = c$, $-\infty < \operatorname{Im}(\zeta) < +\infty$, has a point in common with the circle $|w - p| < \beta$, then the locus remains in $|w - p| < 2\beta$ and forms in $|w - p| < 2\beta$ a closed Jordan curve enclosing the point $w = p$ in its interior.*

Proof: Without loss of generality we may suppose that $p = 0$. The locus considered is

$$\zeta = \int (R(w))^{1/2}\, dw = it + c$$

where c is a real constant and t is a real variable, strictly increasing or strictly decreasing. This can be written in the form

$$it + c = (a + ib) \log w + \sum_{\nu=0}^{\infty} \rho_\nu e^{i\alpha_\nu} w^\nu$$

or

$$it + c = (a + ib)(\log r + i\theta) + \sum_{\nu=0}^{\infty} \rho_\nu r^\nu \exp i(\alpha_\nu + \nu\theta)$$

where $w = re^{i\theta}$. Let $u(r, \theta)$ and $v(r, \theta)$ be the real and imaginary parts respectively of the right side,

$$\begin{aligned} u(r, \theta) &= a \log r - b\theta + \sum_{\nu=0}^{\infty} \rho_\nu r^\nu \cos(\alpha_\nu + \nu\theta), \\ v(r, \theta) &= a\theta + b \log r + \sum_{\nu=0}^{\infty} \rho_\nu r^\nu \sin(\alpha_\nu + \nu\theta). \end{aligned} \tag{3.2.1}$$

The locus considered is then the locus

$$u(r, \theta) = c, \qquad v(r, \theta) = t,$$

where c is a constant and t is monotonic. If subscripts denote differentiation with respect to the variable indicated, then

$$ru_r = v_\theta = a + \sum_{\nu=1}^{\infty} \nu\rho_\nu r^\nu \cos(\alpha_\nu + \nu\theta),$$

$$rv_r = -u_\theta = b + \sum_{\nu=1}^{\infty} \nu\rho_\nu r^\nu \sin(\alpha_\nu + \nu\theta).$$

(1) $b \neq 0$. Without loss of generality assume that $b > 0$. Let α, $0 < \alpha < (1/4)(\log 2)^2(a^2 + b^2)^2$, be chosen so small that

$$(|a| + |b|) \sum_{\nu=1}^{\infty} \rho_\nu r^\nu \leqq r^{1/2}$$

for $r \leqq \alpha$. We have

$$\text{(3.2.2)} \quad au + bv = (a^2 + b^2) \log r + \sum_{\nu=0}^{\infty} \rho_\nu r^\nu [a \cos(\alpha_\nu + \nu\theta) + b \sin(\alpha_\nu + \nu\theta)].$$

Let w_1 and w_2 be any two points of the locus $u = c$ lying in $|w| \leqq \alpha$ and let v_1 and v_2 be the corresponding values of v. We suppose that $v_2 > v_1$. Then

$$b(v_2 - v_1) = (a^2 + b^2) \log \frac{r_2}{r_1} + R$$

where $|w_1| = r_1$, $|w_2| = r_2$, and

$$|R| \leqq 2(|a| + |b|) \sum_{\nu=1}^{\infty} \rho_\nu \alpha^\nu < (\log 2)(a^2 + b^2).$$

Hence

$$\text{(3.2.3)} \left|\frac{w_2}{w_1}\right| = \frac{r_2}{r_1} \geqq \exp\left\{\frac{b}{a^2 + b^2}(v_2 - v_1) - \log 2\right\} = \frac{1}{2} \exp\left\{\frac{b}{a^2 + b^2}(v_2 - v_1)\right\}.$$

The locus $u = c$, $-\infty < v < +\infty$, cannot remain entirely in $|w| < \alpha$. For let w_1 be a point of the locus, $0 < |w_1| < \alpha$, and let v_1 be the corresponding value of v. Choose

$$v_2 = v_1 + \frac{a^2 + b^2}{b} \log \tau, \qquad \tau > 1,$$

and let w_2 be the point on the locus when $v = v_2$. Then if $|w_2| \leqq \alpha$, we have, by (3.2.3),

$$\frac{\tau}{2} |w_1| \leqq |w_2| \leqq \alpha,$$

that is,

$$\tau \leqq 2\alpha/|w_1|.$$

This gives a contradiction if τ is chosen large enough.

Next, on the locus we have

$$du = u_r\,dr + u_\theta\,d\theta = 0, \qquad dv = v_r\,dr + v_\theta\,d\theta = dt,$$

that is,

$$u_r\,dr + u_\theta\,d\theta = 0, \qquad -u_\theta\,dr + r^2 u_r\,d\theta = r\,dt.$$

It follows that

$$\frac{dr}{dt} = -\frac{ru_\theta}{r^2u_r^2 + u_\theta^2} = \frac{r^2 v_r}{r^2u_r^2 + u_\theta^2}, \qquad \frac{d\theta}{dt} = \frac{ru_r}{r^2u_r^2 + u_\theta^2}.$$

Let δ, $\delta > 0$, be such that

$$\sum_{\nu=1}^{\infty} \nu\rho_\nu r^\nu \leqq \frac{1}{2}b \tag{3.2.4}$$

for $0 < r \leqq \delta$. Then

$$rv_r = b + \sum_{\nu=1}^{\infty} \nu\rho_\nu r^\nu \sin(\alpha_\nu + \nu\theta) \geqq \frac{1}{2}b > 0.$$

Hence

$$\frac{dr}{dt} = \frac{dr}{dv} > 0.$$

Let the locus enter $|w| < \delta$ from $|w| \geqq \delta$ at the point P on $|w| = \delta$. At P we have $dr < 0$ and so $dv < 0$. That is, $v = t$ is decreasing at P. As $v = t$ continues to decrease, we see that $|w| = r$ is strictly decreasing. Thus the locus remains in $|w| < \delta$. The number ρ in the statement of the lemma is any number satisfying $0 < \rho \leqq \delta$.

To show that the locus tends to $w = 0$, we have from (3.2.2) and (3.2.4):

$$au + bv = (a^2 + b^2)\log r + R_1$$

where

$$|R_1| \leqq (|a| + |b|)\{\rho_0 + b/2\}$$

for $|w| = r \leqq \delta$. If r does not tend to zero, $au + bv = ac + bt$ would remain bounded, which is impossible since $b \neq 0$ and t tends to minus infinity.

If $a \neq 0$, we have

$$\frac{d\theta}{dt} = \frac{a}{a^2 + b^2} + o(1)$$

as r tends to zero, and so θ tends to infinity monotonically as t tends to minus infinity. If $a = 0$, then by (3.2.1)

$$u = -b\theta + \rho_0 + o(1) = c,$$

and so in this case θ tends to $(\rho_0 - c)/b$ as r tends to zero.

(2) $b = 0$. Then $a \neq 0$ and we may assume without loss of generality that $a > 0$. Let β be a positive number such that

$$\sum_{\nu=1}^{\infty} \nu \rho_\nu r^\nu < \frac{1}{4} a$$

for $r \leqq 2\beta$. If the locus has a point w_1 in common with $|w| < \beta$, it cannot have a point w_2 in common with $|w| = 2\beta$. For by (3.2.1) we would then have

$$u\,|_{w_2} - u\,|_{w_1} > a \log 2 - \frac{1}{2} a = a\left(\log 2 - \frac{1}{2}\right) > 0.$$

Hence if the locus has a point in common with $|w| < \beta$, it remains in $|w| < 2\beta$. To show that the locus forms a Jordan curve, we observe that for $|w| = r < 2\beta$ we have

$$\frac{3}{4} a < r u_r < \frac{5}{4} a$$

and so $u_r > 0$. But u is a single-valued function in $0 < |w| < 2\beta$ and u is an increasing function of r along any ray $\theta = \arg w =$ constant. Hence the locus forms a closed Jordan curve.

Lemma X. *Let $R(w)$ have a pole at a finite point p of precise order μ, where $\mu \geqq 3$. There are $\mu - 2$ asymptotic directions at which the arcs* $\mathrm{Re}(\zeta) = c$ *can approach the point p, and these arcs are equally spaced at angles of $2\pi/(\mu - 2)$. For each $\alpha > 0$ there is a number $\rho(\alpha) > 0$ such that if an arc* $\mathrm{Re}(\zeta) = c$, $\mathrm{Im}(\zeta)$ *monotonic, beginning at a point in $|w - p| \geqq \alpha$, enters the circle $|w - p| \leqq \rho(\alpha)$, then it remains in $|w - p| \leqq \rho(\alpha)$ and tends to the point p with the modulus of $w - p$ strictly decreasing as* $|\mathrm{Im}(\zeta)|$ *tends to infinity. The argument of $w - p$ deviates from its asymptotic direction by less than $\pi/[6(\mu - 2)]$ in the circle $|w - p| \leqq \rho(\alpha)$.*

There is a $\beta > 0$ such that if the locus $\mathrm{Re}(\zeta) = c$, $-\infty < \mathrm{Im}(\zeta) < +\infty$, *lies entirely in the circle $|w - p| \leqq \beta$, then it forms, by adjunction of the point p, a closed Jordan curve whose interior angle at p is $2\pi/(\mu - 2)$. The interior of this curve is mapped by* (3.1.11) *in a one–one manner onto a half-plane* $\mathrm{Re}(\zeta) > c$ *or* $\mathrm{Re}(\zeta) < c$.

Proof: There are two cases to consider, μ even and μ odd. Without loss of generality we may suppose that $p = 0$.

Case (1). Let μ be even. Then $\mu = 2m + 2$ where $m \geqq 1$ since $\mu \geqq 3$, and

$$(R(w))^{1/2} = \sum_{\nu=0}^{\infty} b_\nu w^{-m-1+\nu}, \qquad b_0 \neq 0.$$

The locus considered is

$$\int (R(w))^{1/2}\, dw = it + c$$

where c is a real constant and t is a real variable, strictly increasing or strictly decreasing. This can be written in the form

$$it + c = (a + ib) \log w + \sum_{\nu=0}^{\infty} \rho_{m-\nu} e^{i\alpha_{m-\nu}} w^{\nu-m}, \qquad \rho_m \neq 0,$$

or, writing $w = re^{i\theta}$, we have

$$it + c = (a + ib)(\log r + i\theta) + \sum_{\nu=0}^{\infty} \rho_{m-\nu} r^{\nu-m} \exp i[\alpha_{m-\nu} + (\nu - m)\theta].$$

Let $u(r, \theta)$ and $v(r, \theta)$ be the real and imaginary parts respectively of the right side,

$$u(r, \theta) = a \log r - b\theta + \sum_{\nu=0}^{\infty} \rho_{m-\nu} r^{\nu-m} \cos [\alpha_{m-\nu} + (\nu - m)\theta],$$

$$v(r, \theta) = a\theta + b \log r + \sum_{\nu=0}^{\infty} \rho_{m-\nu} r^{\nu-m} \sin [\alpha_{m-\nu} + (\nu - m)\theta].$$

The locus being considered is the locus

$$u(r, \theta) = c, \qquad v(r, \theta) = t,$$

where t is monotonic. Then

$$ru_r = v_\theta = a + \sum_{\nu=0}^{\infty} (\nu - m)\rho_{m-\nu} r^{\nu-m} \cos [\alpha_{m-\nu} + (\nu - m)\theta],$$

$$rv_r = -u_\theta = b + \sum_{\nu=0}^{\infty} (\nu - m)\rho_{m-\nu} r^{\nu-m} \sin [\alpha_{m-\nu} + (\nu - m)\theta].$$

Now let r_0 be such that for $0 < r \leqq r_0$ we have

$$(|\, a\,| + |\, b\,|)(|\log r\,| + 10\pi) + \sum_{\nu=1}^{\infty} \rho_{m-\nu} r^{\nu-m} < \rho_m r^{-m}/10 \tag{3.2.5}$$

and

$$|\, a\,| + |\, b\,| + \sum_{\nu=1}^{\infty} |\, \nu - m\,|\, \rho_{m-\nu} r^{\nu-m} < m\rho_m r^{-m}/10. \tag{3.2.6}$$

Further, let r_0 be so small that the only zero or pole of $R(w)$ in the circle $|\, w\,| \leqq r_0$ is the pole at $w = 0$. Let r_1 be some constant satisfying $0 < r_1 \leqq r_0$ and let the locus $u(r, \theta) = c, v(r, \theta) = t$ enter the circle $| w | \leqq r_1$ at the point (r_1, θ_1). There is no loss of generality in supposing that $| \theta_1 | \leqq \pi$ since the same locus is traced if θ is increased or decreased by an integer multiple of 2π, the constant c being altered to correspond. Moreover, since the same locus is traced whichever branch of the square root $(R(w))^{1/2}$ is chosen in the integrand of (3.1.11), we may also suppose without loss of generality that

$$u(r_1, \theta_1) \leqq 0.$$

This means that the constant c is equal to or less than zero. Now $\cos(\alpha_m - m\theta) = 1$ at points which are at distance $2\pi/m$ apart, and so there are points φ and ψ such that

$$\cos(\alpha_m - m\varphi) = \cos(\alpha_m - m\psi) = 1, \qquad \psi = \varphi + 2\pi/m,$$

and such that θ_1 lies in the interval

$$\varphi \leqq \theta_1 \leqq \psi.$$

Since we have supposed that $| \theta_1 | \leqq \pi$, it follows that $| \psi | \leqq 3\pi$ and likewise $| \varphi | \leqq 3\pi$. Inequality (3.2.5) then implies that

(3.2.7) $$u(r, \varphi) > 0, \qquad u(r, \psi) > 0$$

in the range $0 < r \leqq r_0$. Thus θ_1 cannot be equal to φ or ψ; so

$$\varphi < \theta_1 < \psi.$$

Indeed, since c is a non-positive constant, inequalities (3.2.7) show that so long as the locus $u = c$, $v = t$ remains in the circle $| w | \leqq r_1$, it must remain in the sector

(3.2.8) $$\varphi < \theta < \psi, \qquad 0 < r \leqq r_1 .$$

From (3.2.5) we see that the inequalities

$$\rho_m r^{-m} \left\{| \cos (\alpha_m - m\theta) | - \frac{1}{10}\right\} < | u | < \rho_m r^{-m} \left\{| \cos (\alpha_m - m\theta) | + \frac{1}{10}\right\},$$

$$\rho_m r^{-m} \left\{| \sin (\alpha_m - m\theta) | - \frac{1}{10}\right\} < | v | < \rho_m r^{-m} \left\{| \sin (\alpha_m - m\theta) | + \frac{1}{10}\right\}$$

are valid in the sector (3.2.8). It therefore follows that

(3.2.9) $$\frac{4}{5} \rho_m r^{-m} < | u | + | v | < 2\rho_m r^{-m}$$

in this sector.

As $| v |$ becomes large, the locus must therefore either leave the sector or tend to $w = 0$. If the locus leaves the sector, both the point of entry and the point of exit must lie on the circle $| w | = r_1$, and at each of these points

(3.2.10) $$| u | + | v | < 2\rho_m r_1^{-m}.$$

But u is constant and v is monotonic; so the same bound holds at all points in between. Hence by (3.2.9) and (3.2.10) we have

$$\frac{4}{5}\rho_m r^{-m} < 2\rho_m r_1^{-m}$$

at all points on the portion of the locus between the point of entry and the point of exit. This inequality implies that $r > (1/10)r_1$; so if the locus leaves the sector, it cannot have entered the part $|w| \leqq (1/10)r_1$ of the sector.

Suppose that the locus does enter the circle $|w| \leqq (1/10)r_1$, having begun at a point in $|w| > r_1$. Then

$$\rho_m r^{-m}\left(|\cos(\alpha_m - m\theta)| - \frac{1}{10}\right) < |u| < \rho_m r_1^{-m}\left(1 + \frac{1}{10}\right)$$

on the portion of the locus in $|w| \leqq r_1$; so

$$|\cos(\alpha_m - m\theta)| < \frac{1}{10} + \frac{11}{10}\left(\frac{r}{r_1}\right)^m \leqq \frac{1}{10} + \frac{11}{100} = \frac{21}{100}$$

on the portion of the arc in $|w| \leqq (1/10)r_1$. Then

$$|\sin(\alpha_m - m\theta)| > \left\{1 - \left(\frac{21}{100}\right)^2\right\}^{1/2} > \frac{9}{10}$$

and hence $\sin(\alpha_m - m\theta)$ has a constant sign on this portion of the locus. Solving the equations

$$u_r\,dr + u_\theta\,d\theta = 0, \qquad v_r\,dr + v_\theta\,d\theta = dt$$

where dt is of constant sign, we obtain

$$\frac{dr}{dt} = -\frac{ru_\theta}{r^2u_r^2 + u_\theta^2}.$$

Since $|\sin(\alpha_m - m\theta)| > 9/10$, inequality (3.2.6) shows that $\operatorname{sgn}(-u_\theta) = \operatorname{sgn}[\sin(m\theta - \alpha_m)]$ on the portion of the locus that lies in $|w| \leqq (1/10)r_1$. Thus dr/dt is of constant sign on this portion of the locus and, since r is decreasing at the point of entry, it continues to decrease. Now $|\cos(\alpha_m - m\theta)| < 21/100$ on the portion of the locus that lies in $|w| \leqq (1/10)r_1$ and clearly $\cos(\alpha_m - m\theta)$ tends to zero as r tends to zero. Thus the asymptotic directions exist and θ does not deviate from its limiting value by as much as $\pi/(12m) = \pi/[6(\mu - 2)]$ in $|w| \leqq (1/10)r_1$. Part of the lemma now follows, at least for even μ, if we let $\rho(\alpha) = (1/10)r_0$ when $\alpha > r_0$ and $\rho(\alpha) = (1/10)\alpha$ when $\alpha \leqq r_0$.

Suppose now that a locus $u(r, \theta) = c$, $-\infty < v(r, \theta) < +\infty$, lies entirely in the circle $|w| \leqq r_0$. We suppose without loss of generality that $c \leqq 0$ and, taking some point (r_1, θ) on the locus, that $|\theta_1| \leqq \pi$. There are φ and ψ, each bounded by 3π, such that

$$\cos(\alpha_m - m\varphi) = \cos(\alpha_m - m\psi) = 1, \qquad \psi = \varphi + 2\pi/m, \tag{3.2.11}$$

and such that θ_1 lies in the interval $\varphi \leqq \theta_1 \leqq \psi$. By the preceding argument it is shown that the locus lies in the sector

$$\varphi < \theta < \psi$$

of the circle $| w | \leqq r_0$. Inequality (3.2.9) then shows that r approaches zero as v approaches plus or minus infinity. We have already proved that when the locus tends to zero, it does so along an asymptotic direction. Thus w approaches zero along some asymptotic direction as v approaches plus infinity and w approaches zero along some asymptotic direction as v approaches minus infinity. There are only two asymptotic directions in the sector $\varphi \leqq \theta \leqq \psi$, namely the angles for which $\cos(\alpha_m - m\theta) = 0$.

To show that the asymptotic directions as v approaches plus infinity and as v approaches minus infinity are different, we suppose that they are the same and arrive at a contradiction. Let θ approach θ_2 as v approaches plus or minus infinity where $\cos(\alpha_m - m\theta_2) = 0$ and $\varphi < \theta_2 < \psi$, the case of equality being ruled out by (3.2.11). If $\rho \leqq (1/10)r_1$, then, by what has already been shown, the modulus of the point on the locus tends monotonically to zero in the circle $| w | \leqq \rho$ as v approaches plus infinity and as v approaches minus infinity. Thus the locus touches the circle $| w | = \rho$ only at two points (ρ, θ_3) and (ρ, θ_4), where θ_3 tends to θ_2 and θ_4 tends to θ_2 as ρ tends to zero. Referring to the formula for u_θ and to inequality (3.2.6), we see that

$$| u_\theta | \geqq \rho_m r^{-m} \left(| \sin(\alpha_m - m\theta) | - \frac{1}{10} \right)$$

in the circle $| w | \leqq r_0$. Since $\cos(\alpha_m - m\theta_2) = 0$ and $\theta_3 \to \theta_2$, $\theta_4 \to \theta_2$, it follows that for small ρ, $u_\theta \neq 0$ in the interval (θ_3, θ_4). Then u_θ is of constant sign in this interval and $u(\rho, \theta_3) \neq u(\rho, \theta_4)$. But each is equal to c; so we have arrived at a contradiction. Here we note that $u(r, \theta)$ is single-valued in the sector $\varphi < \theta < \psi$ of the circle $| w | \leqq r_0$.

Finally, the locus $u(r, \theta) = c$, $-\infty < v < +\infty$, lying in this sector is a Jordan arc since any closed curve in the finite part of the plane on which the harmonic function u is a constant must have a singularity of u inside or on it. If the point $w = 0$ is adjoined to the locus, it becomes a closed Jordan curve whose interior angle at $w = 0$ is $\pi/m = 2\pi/(\mu - 2)$. It is readily seen that this domain is mapped onto a half-plane $\text{Re}(\zeta) > c$ or $\text{Re}(\zeta) < c$ with the point $w = 0$ corresponding to $\zeta = \infty$. This completes the proof of the lemma for even μ.

Case (2). Let μ be odd, $\mu = 2m + 1$ where $m \geqq 1$. If we make the substitution $w = s^2$, then essentially the same argument used in case (1) can be used again but the argument will be simpler because the logarithmic term does not appear. In this case ζ is an odd function of s and there are $4m - 2$ asymptotic directions in the s-plane spaced at angles of $\pi/(2m - 1)$. These directions map into $2m - 1 = \mu - 2$ asymptotic directions spaced at angles of $2\pi/(2m - 1) = 2\pi/(\mu - 2)$ in the w-plane. For small α it is seen that $\rho(\alpha)$ may be chosen equal to $(1/100)\alpha$. If an arc $\text{Re}(\zeta) = c$, $\text{Im}(\zeta)$ monotonic, beginning at a point in $| s | \geqq \alpha$, enters the circle $|s| \leqq (1/100)\alpha$, the amplitude of the arc does not deviate from its limit by as much as $\pi/[12(\mu - 2)]$. Since $w = s^2$, the bound for the deviation becomes $\pi/[6(\mu - 2)]$ in the w-plane.

3.3. The behavior of $\zeta(w)$ in the neighborhood of $w = \infty$ can be inferred from the results of the preceding section. From (3.1.10) we see that

$$R = Aw^{\lambda-\mu}\left\{1 + \frac{b_1}{w} + \frac{b_2}{w^2} + \cdots\right\}$$

for large values of w where

$$\lambda = \sum_{j=1}^{k} \lambda_j, \qquad \mu = \sum_{j=1}^{m} \mu_j.$$

Let $w = 1/t$. Then

$$\zeta = \int (R(w))^{1/2}\, dw = B\int \{t^{\mu-\lambda-4}(1 + b_1 t + b_2 t^2 + \cdots)\}^{1/2}\, dt.$$

If $\lambda - \mu + 4 < 0$, then there are $\mu - \lambda - 2$ arcs $\mathrm{Re}(\zeta) = c$ extending to $w = \infty$ and these arcs are spaced at equal angles. On each arc $\mathrm{Im}(\zeta)$ remains bounded as we approach $w = \infty$.

If $\lambda - \mu + 4 = 0$, there are two arcs $\mathrm{Re}(\zeta) = c$ extending to $w = \infty$ and they form an angle π at infinity. In this case the two arcs together form a single analytic arc passing through the point at infinity. As we approach $w = \infty$, $\mathrm{Im}(\zeta)$ remains bounded.

If $\lambda - \mu + 4 = 1$, there is a single analytic arc $\mathrm{Re}(\zeta) = c$ extending to $w = \infty$ and $\mathrm{Im}(\zeta)$ is bounded on it.

If $\lambda - \mu + 4 = 2$, Lemma IX applies with $a + ib = \pm B$.

If $\lambda - \mu + 4 \geqq 3$, Lemma X applies. There are $\lambda - \mu + 2$ asymptotic directions at which the arcs $\mathrm{Re}(\zeta) = c$, $\mathrm{Im}(\zeta)$ monotonic, can approach $w = \infty$. Moreover, $\mathrm{Im}(\zeta)$ becomes infinite as w approaches infinity on an arc $\mathrm{Re}(\zeta) = c$.

Chapter IV

THE Γ-STRUCTURE. BEHAVIOR IN THE LARGE

4.1. In this chapter we shall investigate the behavior of the loci

$$\text{Re}(\zeta) = \text{Re} \int (R(w))^{1/2}\, dw = \text{constant} \tag{4.1.1}$$

in the large. In the preceding chapter we discussed the behavior in the small, and from the results obtained we make the following observations.

The finite critical points of the hyperelliptic integral

$$\zeta = \zeta(w) = \int (R(w))^{1/2}\, dw$$

are the finite zeros q_j $(j = 1, 2, \cdots, k)$ and the finite poles p_j $(j = 1, 2, \cdots, m)$ of $R(w)$. Even though $\zeta(w)$ is regular at a zero of $R(w)$ of even order, we shall still call this zero a critical point; it is a point where the mapping defined by ζ is not conformal.

According to the usual convention, we say that a function $g(w)$ has the order m at a finite point w_0 if, near w_0,

$$g(w) = b_m(w - w_0)^m + b_{m+1}(w - w_0)^{m+1} + \cdots$$

where m is an integer (positive, negative, or zero). Thus a zero has positive order and a pole has negative order. At $w = \infty$ we make the substitution $w = 1/t$; the order at $w = \infty$ is then usually defined to be the order at $t = 0$. For investigating the behavior of ζ at $w = \infty$ we also make the substitution $w = 1/t$. However, since $dw = -dt/t^2$, this introduces a factor $1/t^2$ into the integrand. If we absorb this factor under the radical, the function $R(1/t)$ is multiplied by $1/t^4$.

We define the order of a finite critical point of ζ to be the order of $R(w)$ at the corresponding zero q_j or pole p_j. At $w = \infty$ we define the order to be the same as that of the function $R(1/t)/t^4$ at $t = 0$. With this convention concerning order, we see that $\lambda + 2$ arcs $\text{Re}(\zeta) = \text{constant}$ emerge from a point of order λ, $\lambda \geqq -1$.

At each critical point p_j of order less than -1 let $| w - p_j | < \alpha_j$ be a circle whose closure does not contain any other critical point, and let $\rho_j = \rho(\alpha_j)$ be the number defined in Lemmas IX and X. Let the set of circles $| w - p_j | < \rho_j$ be S_1. Each critical point of order $\lambda \geqq -1$ is the center of an open circle in which the mapping defined by $\zeta(w)$ is known, and we denote the set of these circles by S_2. Each circle of S_2 has a finite number of arcs emerging from its center on which $\text{Re}(\zeta) = \text{constant}$. We suppose that no two circles of the sets S_1 and S_2 have a point in common. Each point of the plane which lies outside all circles S_1 and S_2 is the center of an open circle which is mapped in a one–one way onto a domain in the ζ-plane. A finite subset of these circles together with the finite sets S_1 and S_2 covers the closed w-plane. In particular, the circle with center at infinity belongs to one of these sets and is the circle $|w| > \gamma$, where γ is some constant. It then follows that there is a $\delta > 0$ such that every point w_0 of the plane outside the circles S_1 and S_2 has a neighborhood which is mapped in a one–one manner onto a square in the ζ-plane with center at $\zeta(w_0)$ and sides of length greater than δ.

Let w_0 be some point in the w-plane which is not a critical point of order less than -1. We may then suppose that w_0 lies outside all circles $|w - p_j| < \alpha_j$. A locus $\text{Re}(\zeta) = \text{constant}$, $\text{Im}(\zeta)$ monotonic, emerging from w_0 can be continued indefinitely with $|\text{Im}(\zeta)|$ becoming infinite unless it meets a critical point of order λ, $\lambda \geqq -1$. It will be shown under various further restrictions on the function $R(w)$ that such a locus strikes a critical point of order $\lambda \geqq -1$ for a finite value of $\text{Im}(\zeta)$, or forms a Jordan curve, or tends to a critical point of order less than -1 as $|\text{Im}(\zeta)|$ becomes infinite.

We suppose that $R(w)$ is a not identically vanishing rational function of the form

$$R(w) = \varphi_\lambda(w)/\psi_\mu(w) \tag{4.1.2}$$

where $\varphi_\lambda(w)$ and $\psi_\mu(w)$ are polynomials of precise degrees λ and μ respectively with no common zeros.

Lemma XI. *Let Λ be a simply-connected domain in the closed w-plane bounded by a finite number of arcs on each of which* $\text{Re}(\zeta)$ *is constant and* $\text{Im}(\zeta)$ *is bounded and at most one arc on which* $\text{Im}(\zeta)$ *is constant and* $\text{Re}(\zeta)$ *is bounded. Suppose further that the arc on which* $\text{Im}(\zeta)$ *is constant lies on the frontier of the closure of Λ. Then Λ contains at least one critical point of negative order.*

By an arc $\text{Re}(\zeta) = \text{constant}$ we mean a locus $\text{Re}(\zeta) = \text{constant}$ which can have a critical point on it only at its end points. A similar remark applies to the arc $\text{Im}(\zeta) = \text{constant}$. Two arcs $\text{Re}(\zeta) = \text{constant}$ can intersect only at a cricital point of $R(w)$ which is then an end point of each. If an arc $\text{Re}(\zeta) = \text{constant}$ and an arc $\text{Im}(\zeta) = \text{constant}$ intersect, then the point of intersection is considered to be an end point of each. Thus two arcs can intersect only at their end points.

Since ζ is bounded on each arc of the boundary of Λ, all critical points on the boundary of Λ have orders not less than -1.

We recall that the frontier of a set is the set of all points that are closure points of the set and of its complement. Since Λ is a simply-connected domain in the closed plane, its boundary is connected. Since the arc on which $\text{Im}(\zeta)$ is constant lies on the frontier of the closure of Λ, it is easily seen that the set of arcs on each of which $\text{Re}(\zeta)$ is constant forms a connected set. Then the same constant c suffices on all arcs $\text{Re}(\zeta) = c$. Each arc $\text{Re}(\zeta) = c$ as well as the arc $\text{Im}(\zeta) =$ constant is a Jordan arc, closed or unclosed. This follows from the supposition that the only critical points of $R(w)$ on the arcs are at their end points. The boundary of Λ is then one of the following:

(i) a closed Jordan curve $\mathrm{Im}(\zeta) = \text{constant}$ only;

(ii) a finite number of arcs $\mathrm{Re}(\zeta) = c$ only;

(iii) an arc $\mathrm{Im}(\zeta) = \text{constant}$ which is a closed Jordan curve and one or more arcs $\mathrm{Re}(\zeta) = \text{constant}$;

(iv) one or more arcs $\mathrm{Re}(\zeta) = c$ and a single unclosed Jordan arc $\mathrm{Im}(\zeta) = \text{constant}$ which meets this set at precisely two points.

To illustrate the last three cases, suppose that Λ lies in the annulus $1 < |w| < 2$ and is bounded by the two circles $|w| = 1$, $|w| = 2$ and the straight-line segment from $w = 1$ to $w = 2$. Then the boundary of Λ consists of three or more arcs (depending on the number of critical points on the boundary). In case (ii) the real part of ζ is constant on the entire boundary. In case (iii) $\mathrm{Im}(\zeta) = \text{constant}$ on one of the circles and $\mathrm{Re}(\zeta) = \text{constant}$ on the other and also on the line segment. In case (iv) the arc $\mathrm{Im}(\zeta) = \text{constant}$ is part of one of the circles but it cannot be part of the segment from 1 to 2.

Before proceeding to the proof of Lemma XI we prove a special case of it.

LEMMA XII. *Under the conditions of Lemma* XI *the domain* Λ *contains at least one critical point of* $\zeta(w)$.

PROOF OF LEMMA XII: If there are no arcs $\mathrm{Re}(\zeta) = c$, the boundary of Λ consists of the single closed arc $\mathrm{Im}(\zeta) = c$. Then the function $\mathrm{Im}(\zeta)$ is bounded in a neighborhood of the boundary of Λ and is constant on the boundary itself. Since $\mathrm{Im}(\zeta)$ is harmonic, either it is identically equal to a constant or Λ contains a critical point. If $\mathrm{Im}(\zeta)$ were equal to a constant throughout Λ, ζ would be equal to a constant and so $R(w)$ would vanish identically—contrary to hypothesis. The argument is similar when the arc $\mathrm{Im}(\zeta) = \text{constant}$ is absent.

If an arc $\mathrm{Im}(\zeta) = \text{constant}$ as well as one or more arcs $\mathrm{Re}(\zeta) = \text{constant}$ occurs in the boundary of Λ, then $\mathrm{Re}(\zeta)$ is monotonic on the arc $\mathrm{Im}(\zeta) = \text{constant}$. Thus ζ is not single-valued around the boundary of Λ and so Λ contains a critical point.

The proof gives more than the lemma states, namely that Λ contains a singularity of $\zeta(w)$.

PROOF OF LEMMA XI: We suppose that Λ contains no critical point of negative order and arrive at a contradiction. Among the simply-connected domains which lie in Λ and satisfy all the conditions imposed on Λ in the statement of Lemma XI, let Λ_0 be one which contains as few critical points of positive order as possible. There are no critical points of negative order in Λ_0 since Λ_0 is part of Λ, but Lemma XII shows that there is at least one critical point of positive order in Λ_0.

Consider all possible arcs $\mathrm{Re}(\zeta) = \text{constant}$, $\mathrm{Im}(\zeta)$ bounded, which lie entirely in Λ_0 and have their end points at critical points in Λ_0. We exclude any arc which lies in Λ_0 but has one or both end points on the boundary of Λ_0. The arcs, if any exist, and the critical points which lie in Λ_0 form a closed set which is the sum of a finite number of closed connected sets $S_1, S_2, \cdots, S_k$, no pair of sets having a point in common.

There are at least three arcs $\text{Re}(\zeta) = \text{constant}$, $\text{Im}(\zeta)$ monotonic, emerging from the same or different points of S_1, none of these arcs belonging to S_1. For if S_1 consists of k arcs and j critical points, we know that no subset of the arcs of S_1 can form a closed Jordan curve because of the minimizing property of Λ_0. Thus $j = k + 1$ where $k \geqq 0$. Now at least three arcs $\text{Re}(\zeta) = \text{constant}$, $\text{Im}(\zeta)$ monotonic, emerge from each critical point. Hence there are at least $3j$ arcs emerging from the j critical points. Each arc belonging to S_1 is counted twice; so there are at least $3j - 2k = k + 3$ of them which do not belong to S_1. Since $k \geqq 0$, there are at least three arcs not belonging to S_1 which emerge from S_1; let these three arcs be denoted by A, B, and C.

If any one of the arcs A, B, or C meets the boundary of Λ_0, let it terminate there. No more than one of them can meet the boundary. For suppose that A and B both meet the boundary of Λ_0 for a finite value of $\text{Im}(\zeta)$. By considering the several cases we see that there is a domain Λ_1 which satisfies the conditions imposed on Λ in Lemma XI and which is a subdomain of Λ_0. But the boundary of Λ_1 lies in the set which is the sum of the boundary of Λ_0 plus the set S_1 plus the arcs A and B. At least one critical point interior to Λ_0 lies on the boundary of Λ_1 and this violates the minimizing property of Λ_0.

We thus suppose without loss of generality that the arcs A and B do not meet the boundary of Λ_0 no matter how far they are prolonged. Also, neither A nor B can meet another critical point in Λ_0, for if either did, the zero would be a part of S_1.

The arc A has at least one limit point in the closure of Λ_0. Since $\text{Im}(\zeta)$ is bounded in the neighborhood of each point of Λ_0 and its boundary, there is more than one limit point. Then there are infinitely many; so there is one limit point p in the closure of Λ_0 at which $R(w)$ is regular and not zero. A neighborhood N of p is mapped into a square in the ζ-plane having its center at $\zeta(p)$. The portions of A near p are mapped into a series of vertical lines in the ζ-plane which have $\zeta(p)$ as a limit point. Since A does not meet the boundary Λ_0, it is clear that the point p does not lie on the arc $\text{Im}(\zeta) = \text{constant}$ which may form part of the boundary of Λ_0. If p lies on one of the arcs $\text{Re}(\zeta) = \text{constant}$ on the boundary of Λ_0, then this arc is mapped into a vertical line in the ζ-plane through the point $\zeta(p)$. It follows that an arc of A plus an arc α of the locus $\text{Im}(\zeta) = \text{constant}$ drawn in N through the point p forms a closed Jordan curve J_A lying in Λ_0. Likewise, if p lies in the open set Λ_0, then an arc of A plus an arc α of the locus $\text{Im}(\zeta) = \text{constant}$ drawn through p forms a closed Jordan curve lying in Λ_0. We observe that the arc α does not necessarily contain the point p. The interior of J_A is a domain of type Λ; so because of the minimizing property of Λ_0 it follows that the curve J_A contains in its interior all zeros of $R(w)$ which lie in Λ_0.

If the arc B meets the arc A at a point different from the point of S_1 from which they may both emerge, then an arc of A plus an arc of B plus a part of S_1 forms a closed Jordan curve which encloses a domain Λ_2. But the boundary of Λ_2 contains some of the critical points which lie in Λ_0 and this violates the minimizing property of Λ_0. Thus if B meets the curve J_A, it meets it at a point of α. But then an arc of B plus a part of S_1 plus an arc of A forms a curve on which $\text{Re}(\zeta)$ is constant; this curve plus an arc of α forms a closed Jordan curve enclosing a domain of type Λ, in violation of the minimizing property of Λ_0. Hence B does not meet the curve J_A. Since J_A contains in its interior all zeros of $R(w)$ which lie in Λ_0 and B issues from one of these zeros, it follows that B lies inside J_A. In particular, the arc B does not meet the boundary of Λ_0.

In the same way we can show that there is a closed Jordan curve J_B which consists of an arc of B plus an arc β on which $\text{Im}(\zeta)$ is constant, $R(w)$ being regular and not zero on β. We remark that the arc B has infinitely many limit points inside or on J_A and that these limit points cannot all lie on the arc α of J_A; if they did, B would cross J_A. Choose a limit point q not on α. Since β is drawn in a neighborhood of this limit point, β can be chosen in such a way that it has no point in common with J_A. Thus J_B contains A in its interior, and J_A contains B in its interior. The closed Jordan curves J_A and J_B cannot have a point in common, for J_A is α plus a part of A and J_B is β plus a part of B. By considering the four possibilities the statement follows that J_A and J_B have no point in common. Thus J_A contains J_B in its interior and J_B contains J_A in its interior which is a contradiction.

The above method shows that if $R(w)$ is the square of a rational function, then the locus $\text{Re}(\zeta) = \text{constant}$, $\text{Im}(\zeta)$ monotonic, when prolonged must either meet a point where $R(w)$ has positive order for a finite value of $\text{Im}(\zeta)$, tend to a unique limit which is a point where $R(w)$ has negative order as $\text{Im}(\zeta)$ becomes infinite, or form a closed Jordan curve enclosing at least one pole of $R(w)$.

4.2. Let

$$P(w) = \frac{A_k}{w^k} + \frac{A_{k+1}}{w^{k+1}} + \cdots + \frac{A_{n-1}}{w^{n-1}} \tag{4.2.1}$$

where

$$A_k \neq 0, \quad A_{n-1} \neq 0, \qquad 1 \leqq k \leqq n - 1; \tag{4.2.1$'$}$$

$$Q(z) = \sum_{\nu=-(n-1)}^{n-1} \frac{B_\nu}{z^\nu} \tag{4.2.2}$$

where

$$B_{-\nu} = \bar{B}_\nu, \qquad B_{n-1} \neq 0, \text{ and } B_0 > 0; \tag{4.2.2$'$}$$

and $Q(z) \geqq 0$ on $|z| = 1$ with at least one zero there. Finally let

$$\Phi(v) = A\frac{2v - 1}{v^2(v - 1)^2}. \tag{4.2.3}$$

We shall now consider in detail the three special cases where the rational function has one of the forms

$$\frac{P(w)}{w^2}, \quad \frac{Q(z)}{z^2}, \quad \text{or} \quad \Phi(v).$$

The results concerning $\Phi(v)$ (Lemmas XV and XX) are included because they are required in Chapter XIV.

Lemma XIII. *Let*

(4.2.4)
$$\zeta = \int (P(w))^{1/2} \frac{dw}{w}$$

where $P(w)$ *is defined by* (4.2.1). *Beginning at any point in the plane other than the origin, prolong the locus* $\mathrm{Re}(\zeta) = $ *constant,* $\mathrm{Im}(\zeta)$ *monotonic. Then the locus* (i) *strikes* $w = \infty$ *or some other zero of* $P(w)$ *for a finite value of* $\mathrm{Im}(\zeta)$ *or* (ii) *tends to the unique limit* $w = 0$ *as* $\mathrm{Im}(\zeta)$ *becomes infinite.*

Proof: We make the substitution

$$w = 1/s^2$$

and obtain

$$\zeta = \int \{4A_k s^{2k-2} + \cdots + 4A_{n-1} s^{2n-4}\}^{1/2} ds = \int (F(s))^{1/2} ds.$$

Beginning at any point in the s-plane other than $s = \infty$ and prolonging the locus $\mathrm{Re}(\zeta) =$ constant, $\mathrm{Im}(\zeta)$ monotonic, we see that the locus must (a) tend to the unique limit $s = \infty$; (b) strike a zero of $F(s)$ for a finite value of $\mathrm{Im}(\zeta)$; or (c) do neither (a) nor (b). In case (c) Lemma X shows that the entire locus must lie in some circle $|s| \leqq a$. Then it has at least one limit point in this circle. Since $\mathrm{Im}(\zeta)$ is bounded in a neighborhood of each point of $|s| \leqq a$, the locus has at least two limit points in $|s| \leqq a$ and therefore infinitely many. Thus there is a limit point p where $F(s)$ is regular and not zero. It follows that a locus $\mathrm{Im}(\zeta) =$ constant drawn through p and lying in a neighborhood of p plus an arc of the locus $\mathrm{Re}(\zeta) =$ constant forms a closed Jordan curve bounding a domain of the type described in Lemma XI. The interior of this Jordan curve can contain no critical point of negative order, which is a contradiction. Hence either case (a) or case (b) must hold. But then Lemma XIII follows.

Lemma XIV. *Let*

(4.2.5)
$$\zeta = \int (Q(z))^{1/2} \frac{dz}{z}$$

where $Q(z)$ *is defined by* (4.2.2). *Beginning at any point in the plane other than* $z = 0$ *or* $z = \infty$, *prolong the locus* $\mathrm{Re}(\zeta) = $ *constant,* $\mathrm{Im}(\zeta)$ *monotonic. Then the locus either* (i) *strikes a zero of* $Q(z)$ *for a finite value of* $\mathrm{Im}(\zeta)$ *or* (ii) *tends to a unique limit* $z = 0$ *or* $z = \infty$ *as* $\mathrm{Im}(\zeta)$ *becomes infinite.*

PROOF: Making the substitution $z = s^2$, we obtain

$$\zeta = \int (F(s))^{1/2}\, ds$$

where

$$F(s) = \frac{4}{s^2} \sum_{\nu=-(n-1)}^{n-1} \frac{B_\nu}{s^{2\nu}}.$$

The function $s^2F(s)$ is non-negative on the unit circumference $|s| = 1$, and so

$$\zeta = \int (F(s))^{1/2}\, ds = \int (s^2 F(s))^{1/2} \frac{ds}{s}$$

has a constant real part on $|s| = 1$. Moreover, $F(s)$ has at least two distinct zeros on $|s| = 1$, each of order 2 or higher.

Beginning at a point in the s-plane other than $s = 0$ or $s = \infty$ and prolonging the locus $\mathrm{Re}(\zeta) =$ constant, $\mathrm{Im}(\zeta)$ monotonic, we see that the locus must (a) strike a zero of $F(s)$ for a finite value of $\mathrm{Im}(\zeta)$; (b) tend to a unique limit $s = 0$ or $s = \infty$ as $\mathrm{Im}(\zeta)$ becomes infinite; or (c) do neither (a) nor (b).

In case (c), Lemma X shows that the locus lies in $\delta \leqq |s| \leqq 1/\delta$ where δ is some positive number. Then it has a limit point p at which $F(s)$ is regular and not zero, and a closed Jordan curve J can be formed of an arc $\mathrm{Re}(\zeta) =$ constant (part of the locus) and an arc $\mathrm{Im}(\zeta) =$ constant. Moreover, there will be no zeros or poles of $F(s)$ on the Jordan curve J. The curve J is not the unit circumference $|s| = 1$ since that has zeros of $F(s)$ on it. If J lies partly in $|s| > 1$ and partly in $|s| < 1$, then it crosses the unit circle at two distinct points. There is then a closed Jordan curve consisting of part of J and part of $|s| = 1$ which bounds a domain not containing a pole of $F(s)$. This violates Lemma XI; so J lies in $|s| \leqq 1$ or $|s| \geqq 1$. It is clear that there is no loss of generality in assuming the latter case. The interior of J must contain the origin $s = 0$ by Lemma XI. If J meets $|s| = 1$ in one or more points, there would be a domain whose boundary consists of parts of $|s| = 1$ and parts of J lying in $|s| > 1$ and not containing the origin. Lemma XI would then be violated; so J lies entirely in $|s| > 1$.

Now let J denote any closed Jordan curve lying in $|s| > 1$ whose boundary consists of a finite number of arcs $\mathrm{Re}(\zeta) =$ constant, $\mathrm{Im}(\zeta)$ bounded, and at most one arc $\mathrm{Im}(\zeta) =$ constant, $\mathrm{Re}(\zeta)$ bounded. We suppose that each of these arcs has zeros of $F(s)$ only at its end points. Further let J contain the unit circumference $|s| = 1$ in its interior. Among all such Jordan curves J there is one, J_0 say, which contains as few zeros of $F(s)$ as possible.

Let Σ be the set of zeros of $F(s)$ which are interior to J_0 and lie in $|s| \geqq 1$. Construct all possible arcs $\mathrm{Re}(\zeta) =$ constant, $\mathrm{Im}(\zeta)$ bounded, which lie in the interior of J_0, which join two zeros of the set Σ, and which lie in $|s| > 1$ except perhaps for their end points. These arcs, if any exist, together with the set Σ of zeros form a set which is the sum of a finite number of closed connected sets $S_1, S_2, \cdots, S_m$. Each set S_j is closed, lies in the interior of J_0, and except for a finite number of points lies in $|s| > 1$.

We remark that none of the sets S_j contains a subset forming a closed Jordan curve. For this curve would have some zero of $F(s)$ on it and, since it lies in $|s| \geqq 1$, it would therefore either violate the minimizing property of J_0 or have a point in common with $|s| = 1$. In the latter case a part of the curve plus a part of $|s| = 1$ would bound a simply-connected domain not containing $s = 0$, contradicting Lemma XI.

Now there are at least two zeros s_1 and s_2 of the set Σ which lie on $|s| = 1$. No pair of zeros on $|s| = 1$ can belong to the same connected set. For if s_1 and s_2 belong to S_1, say, then a part of S_1 plus a part of the circumference $|s| = 1$ forms the boundary of a simply-connected domain in $|s| > 1$ on which $\text{Re}(\zeta)$ is constant, in violation of Lemma XI. Thus let s_1 belong to S_1, s_2 to S_2.

Since s_1 and s_2 are zeros of order 2 or more, there are arcs $\text{Re}(\zeta) = \text{constant}$ emerging from s_1 and s_2 which lie in $|s| > 1$ except for their end points. These arcs may or may not belong to S_1 and S_2. If S_1 consists of k arcs and j zeros of the set Σ, then $j = k + 1$. Each zero except the zero s_1 on $|s| = 1$ has at least three arcs $\text{Re}(\zeta) = \text{constant}$ emerging from it into $|s| > 1$. Since s_1 is the only zero on $|s| = 1$ which belongs to S_1, we see that from the j zeros there are at least $3(j - 1) + 1 = 3j - 2$ arcs $\text{Re}(\zeta) = \text{constant}$ emerging. Since each of the k arcs is counted twice, there are at least $3j - 2 - 2k = k + 1$ arcs $\text{Re}(\zeta) = \text{constant}$ emerging from the zeros of S_1 and not belonging to S_1. Since $k \geqq 0$, at least one arc A not belonging to S_1 emerges from it. The arc A lies except perhaps for one end point in $|s| > 1$. If it meets J_0, we terminate A there. Likewise there is an arc B emerging from a zero of S_2 and lying except perhaps for one end point in $|s| > 1$; B is terminated wherever it meets J_0.

Neither A nor B can meet a zero of $F(s)$ in the interior of J_0, no matter how far they are extended. In particular, they cannot terminate on $|s| = 1$. Moreover they cannot intersect in the interior of J_0, for they can do so only at a zero of $F(s)$. They cannot both meet J_0, for if they did, then J_0 plus A plus B plus S_1 plus S_2 plus $|s| = 1$ would contain a subset which forms a closed Jordan curve bounding a domain of the type described in Lemma XI but not containing $s = 0$.

We therefore suppose without loss of generality that the arc A does not meet J_0 no matter how far it is extended. By a previous argument it then has a limit point at which $F(s)$ is regular and not zero. Then a subarc of A plus a piece α of a locus $\text{Im}(\zeta) = \text{constant}$ forms a closed Jordan curve U which lies in J_0 and in $|s| > 1$, and U does not have any zero or pole of $F(s)$ on it. By the minimizing property of J_0 it follows that U contains in its interior all zeros of $F(s)$ which lie inside J_0.

Since B has one end point at one of these zeros, it follows that at least part of B is inside U. If B meets U, it must do so at a point of the arc α. In this case A plus B plus S_1 plus S_2 plus α plus $|s| = 1$ contains a subset which bounds a simply-connected domain not containing the origin. For beginning at the point where B meets α, trace back along B until B meets S_2, then over a part of S_2 to the point s_2, then along $|s| = 1$ from s_2 to s_1 and from s_1 along A until α is encountered. Since there are two arcs of $|s| = 1$ joining s_1 and s_2, we obtain in this way two Jordan arcs extending from α back again to α and these two Jordan arcs differ only in the path taken along $|s| = 1$. Adding a piece of α, we thus obtain two closed Jordan curves, one of which does not contain the origin $s = 0$ in its interior. This is impossible by Lemma XI. Thus B does not meet the Jordan curve U, and so it lies entirely inside U.

Since B does not meet U, it has a limit point q which is inside or on U. Since B cannot cross α, it cannot have an interior point of α as limit point. But B has infinitely many limit points, and so we may suppose that q is not on α. Then a subarc of B plus an arc $\text{Im}(\zeta) = \text{constant}$ in the neighborhood of q forms a closed Jordan curve V. It is easily seen that U and V do not meet.

Thus V lies inside U. But by symmetry U lies inside V, and this contradiction shows that the case (c) is impossible. Hence either (a) or (b) occurs. Lemma XIV then follows.

LEMMA XV. *Let*

$$\zeta = \int (\Phi(v))^{1/2}\, dv \tag{4.2.6}$$

where $\Phi(v)$ is defined by (4.2.3). *Starting at any point of the plane other than $v = 0$ or $v = 1$, prolong the locus* $\text{Re}(\zeta) =$ *constant,* $\text{Im}(\zeta)$ *monotonic. Then the locus* (i) *strikes $v = 1/2$ or $v = \infty$ for a finite value of* $\text{Im}(\zeta)$; (ii) *tends to one of the two points $v = 0$ or $v = 1$ as a unique limit as* $\text{Im}(\zeta)$ *becomes infinite*; *or* (iii) *forms a Jordan curve or cycle on which there are no zeros or poles of $\Phi(v)$.*

PROOF: Let

$$A^{1/2} = a + ib \tag{4.2.7}$$

(according to one or the other determination of sign). Making the substitution $u^2 = 2v - 1$, we obtain

$$\zeta = \int (\Phi(v))^{1/2}\, dv = \pm(a + ib) \int \frac{4u^2}{u^4 - 1}\, du.$$

The variation of ζ around any simple closed curve in the u-plane which does not pass through one of the four points $1, -1, i, -i$ is, according to the residue theorem, equal to

$$2\pi i(a + ib)\,(k + im)$$

where k, m can have only the values $-1, 0, 1$. Thus there are at most nine possible values for the variation of ζ around such a Jordan curve.

Starting from any point of the u-plane other than 1, -1, i, or $-i$, prolong the locus $\text{Re}(\zeta) = \text{constant}$, $\text{Im}(\zeta)$ monotonic. The locus must (a) strike $u = 0$ or $u = \infty$ for a finite value of $\text{Im}(\zeta)$; (b) tend to one of the points 1, -1, i, or $-i$ as a unique limit; (c) form a closed Jordan curve not passing through $u = 0$, ∞, 1, -1, i, $-i$; or (d) exhibit none of these three types of behavior. In case (d), Lemma IX shows that the locus remains at a distance not less than δ from the points 1, -1, i, $-i$, where δ is some positive number. Then the locus will have infinitely many limit points of which one, say p, will be different from $u = 0$ or $u = \infty$. A closed Jordan curve J can then be formed from an arc of the locus $\text{Re}(\zeta) = \text{constant}$ plus an arbitrarily small arc α through p on which $\text{Im}(\zeta) = \text{constant}$. In fact, given $\epsilon > 0$, the Jordan curve may be formed such that the variation of $\text{Re}(\zeta)$ on the arc α is less than ϵ in value. Since the variation of $\text{Re}(\zeta)$ on α cannot be zero, we obtain an infinite sequence of Jordan curves J_1, J_2, $\cdots$ on which the variation of $\text{Re}(\zeta)$ has values tending to zero, no two of which are equal. This contradicts the fact that the variation of ζ around any closed Jordan curve has one of nine possible values.

The above proof can clearly be applied to any rational function having no more than two points of odd order.

4.3. If

$$\zeta = \int (P(w))^{1/2} \frac{dw}{w} \tag{4.3.1}$$

where $P(w)$ is defined by (4.2.1), let Γ_w be the set of all arcs $\text{Re}(\zeta) = \text{constant}$ which have one or both end points at a zero of $P(w)$ or at $w = \infty$.

If

$$\zeta = \int (Q(z))^{1/2} \frac{dz}{z} \tag{4.3.2}$$

where $Q(z)$ is defined by (4.2.2), let Γ_z be the set of all arcs $\text{Re}(\zeta) = \text{constant}$ which have one or both end points at a zero of $Q(z)$.

If

$$\zeta = \int (\Phi(v))^{1/2}\, dv \tag{4.3.3}$$

where $\Phi(v)$ is defined by (4.2.3), let Γ_v be the set of all arcs $\text{Re}(\zeta) = \text{constant}$ which have one or both end points at the zero of $\Phi(v)$ or at $v = \infty$.

In all three cases we see that Γ is the set of all loci of the form $\text{Re}(\zeta) = \text{constant}$ which have one or both end points at critical points of orders not less than -1. The set Γ, or the Γ-structure as we shall sometimes call it, plays an important role in the following pages.

Let $w_1, w_2, \cdots, w_j$ be the finite zeros of $P(w)$. Then Γ_w is a finite set of arcs since there is only a finite number of arcs $\text{Re}(\zeta) = \text{constant}$ emerging from each zero and from the point at infinity. Each arc is analytic except perhaps at its end points, and the end points which do not lie at a zero of $P(w)$ or at $w = \infty$ must lie at the origin $w = 0$. If an arc has one end point at $w = 0$, then ζ is unbounded on the arc; otherwise ζ is bounded on the arc. Likewise, if $z_1, z_2, \cdots, z_m$ are the zeros of $Q(z)$, $\mathbf{\Gamma}_z$ is a finite set of arcs, each of which has its end points in the set $0, \infty, z_1, z_2, \cdots, z_m$, and each arc is analytic except perhaps at its end points. If an arc has one end point at $z = 0$ or $z = \infty$, then ζ is unbounded on the arc; otherwise ζ is bounded on the arc. The unit circumference $|z| = 1$ is part of $\mathbf{\Gamma}_z$ since $Q(z) \geqq 0$ on $|z| = 1$ and dz/z is pure imaginary there.

Lemma XVI. *If ζ is defined by* (4.3.1), *then any locus* $\text{Re}(\zeta) = $ *constant which forms a Jordan curve in the closed plane (sphere) must pass through the point* $w = 0$.

Proof: For if the curve does not pass through $w = 0$, then by Lemma X it must lie in some domain $|w| > \delta > 0$. It follows from Lemma XIII that the locus is the sum of finitely many arcs $\text{Re}(\zeta) = $ constant having zeros of $P(w)$ only at their end points and that $\text{Im}(\zeta)$ is bounded on each arc. Let this curve be J. Then J is part of $\mathbf{\Gamma}_w$ and, by Lemma XI, one of the two parts into which J divides the plane must contain the origin and the other must contain the point at infinity. Furthermore, the critical point at $w = \infty$ must have a negative order. Since the critical point at infinity has an order not less than -1, we see that the order at infinity must be equal to -1. Then only one arc $\text{Re}(\zeta) = $ constant emerges from the point at infinity.

If $w = \infty$ is connected to J by arcs of $\mathbf{\Gamma}_w$, then the exterior of J cut from some point of J to infinity along a subset of $\mathbf{\Gamma}_w$ forms a domain Λ of the type described in Lemma XI, and this domain contains no critical point of negative order, in violation of Lemma XI. Hence $w = \infty$ is not connected to J by arcs of $\mathbf{\Gamma}_w$. Thus, if a point begins at infinity and moves along an arc of $\mathbf{\Gamma}_w$ emerging from $w = \infty$, it must meet a zero of $P(w)$ in the exterior of J. Continuing along another arc of $\mathbf{\Gamma}_w$ which emerges from this zero, and so on, the path of the moving point must ultimately return to a point previously encountered. Thus there is a closed Jordan curve J' and a set of arcs connecting J' to infinity on which the real part of ζ is constant. Lemma XI shows that the interior of J' contains the origin, but the exterior of J' cut along arcs of $\mathbf{\Gamma}_w$ to $w = \infty$ does not contain a pole of $P(w)$; so we have a contradiction.

Lemma XVII. *The set $\mathbf{\Gamma}_w$ is connected if the point $w = 0$ is adjoined to it.*

Proof: For if a point begins somewhere on $\mathbf{\Gamma}_w$ and moves along $\mathbf{\Gamma}_w$, the moving point cannot return to a point previously met without first passing through $w = 0$ (Lemma XVI). There are only a finite number of arcs belonging to $\mathbf{\Gamma}_w$; so the moving point must ultimately tend to $w = 0$. Thus if $w = 0$ is added to $\mathbf{\Gamma}_w$, every point of $\mathbf{\Gamma}_w$ is connected to the origin.

LEMMA XVIII. *If ζ is defined by* (4.3.2), *then any locus* $\text{Re}(\zeta) =$ *constant which forms a Jordan curve in the closed plane must be the unit circumference* $|z| = 1$ *or pass through* $z = 0$ *or* $z = \infty$.

PROOF: Let the locus be J, and suppose that J is not $|z| = 1$ and that it does not pass through $z = 0$ or $z = \infty$. Lemma XIV shows that J is the sum of a set of arcs of Γ_z with end points at the zeros of $Q(z)/z^2$, and ζ is bounded on J. The exterior of J contains the point at infinity, and the interior contains the origin $z = 0$ (Lemma XI). Since J is not $|z| = 1$, it has a point in $|z| < 1$ or in $|z| > 1$. Loci $\text{Re}(\zeta) =$ constant are symmetric with respect to $|z| = 1$; so we may suppose without loss of generality that J has a point in $|z| > 1$.

Then J must lie entirely in $|z| > 1$. For if J has one point in $|z| \leqq 1$, then it has at least one point in common with $|z| = 1$. In this case a portion of J plus a portion of $|z| = 1$ forms a closed Jordan curve bounding a simply-connected domain containing no pole of $Q(z)/z^2$. Since $\text{Re}(\zeta)$ is constant on the boundary, Lemma XI shows that this is impossible. Thus J contains the closed unit circle $|z| \leqq 1$ in its interior.

Each closed Jordan curve which (i) lies in $|z| \geqq 1$, (ii) is a subset of Γ_z, (iii) does not pass through $z = \infty$, (iv) has at least one point in $|z| > 1$, must lie in $|z| > 1$ and contain $|z| \leqq 1$ in its interior. There is one such curve, J' say, which contains in its interior as few zeros of $Q(z)$ as possible. Since $Q(z)$ has a zero of second or higher order at some point z_0 of $|z| = 1$, there is an arc of Γ_z extending from z_0 into $|z| > 1$. Let a point begin at z_0 and move along Γ_z into $|z| > 1$ until it meets another zero of $Q(z)$, then along another arc of Γ_z emerging from this zero, and so on. The moving point cannot trace a closed Jordan curve because of the minimizing property of J' and for the same reason it must remain in $|z| > 1$. Thus it ultimately meets a point of J'. Hence there is a Jordan arc joining J' and $|z| = 1$ along which the real part of ζ is constant. This Jordan arc plus $|z| = 1$ plus J' forms the boundary of a simply-connected domain on which $\text{Re}(\zeta)$ is constant, and the domain contains no poles of $Q(z)/z^2$. Lemma XI shows that this is impossible.

LEMMA XIX. *The set* Γ_z *is connected if the points* $z = 0$ *and* $z = \infty$ *are adjoined to it.*

As in the proof of Lemma XVII, we show that each point of Γ_z in $|z| \leqq 1$ is connected to $z = 0$ and that each point of Γ_z in $|z| \geqq 1$ is connected to $z = \infty$.

4.4. The set Γ_w defined above is closed and connected if the point $w = 0$ is adjoined. The complement is therefore the sum of a finite number of disjoint simply-connected domains,

$$O_w^{(1)}, O_w^{(2)}, \cdots, O_w^{(m)}.$$

Each domain $O_w^{(\nu)}$ has $w = 0$ as a boundary point. Otherwise ζ would be bounded in $O_w^{(\nu)}$ with constant real part on the boundary, and so would be a constant throughout the w-plane.

The domain $O_w^{(\nu)}$ is called a "strip domain" if it is mapped by (4.3.1) in a one–one manner onto a strip

$$-\infty < \mathrm{Im}(\zeta) < \infty, \qquad a < \mathrm{Re}(\zeta) < b$$

in the ζ-plane. The domain $O_w^{(\nu)}$ is called an "end domain" if it is mapped by (4.3.1) in a one–one manner onto a half-plane

$$-\infty < \mathrm{Im}(\zeta) < \infty, \qquad \mathrm{Re}(\zeta) < a \text{ or } \mathrm{Re}(\zeta) > a$$

in the ζ-plane.

The set $\mathbf{\Gamma}_z$ is also closed and connected if the points $z = 0$ and $z = \infty$ are added to it. The complement of $\mathbf{\Gamma}_z$ is the sum of a finite number of simply-connected domains

$$O_z^{(1)}, O_z^{(2)}, \cdots, O_z^{(m)}, E_z^{(1)}, E_z^{(2)}, \cdots, E_z^{(m)}.$$

Since $Q(z)$ is non-negative on $|z| = 1$, we see that $\mathbf{\Gamma}_z$ is symmetrical with respect to the unit circumference $|z| = 1$. In other words, if z belongs to $\mathbf{\Gamma}_z$, so does the point $1/\bar{z}$. We arrange the domains $O_z^{(1)}, \cdots, E_z^{(m)}$ such that the domains $O_z^{(\nu)}$, $\nu = 1, 2, \cdots, m$, lie in $|z| < 1$ while the domains $E_z^{(\nu)}$, $\nu = 1, 2, \cdots, m$, lie in $|z| > 1$. Also, as z moves over $O_z^{(\nu)}$, we suppose that the point $1/\bar{z}$ moves over $E_z^{(\nu)}$. We define strip domains and end domains in the case of the $O_z^{(\nu)}$, $E_z^{(\nu)}$ in the same way as above for the $O_w^{(\nu)}$.

THEOREM I. *Each domain $O_w^{(\nu)}$ is either a strip or an end domain. There are precisely $n - 1$ end domains, each subtending an angle $2\pi/(n - 1)$ at the origin. If the point $w = 0$ is removed from $\mathbf{\Gamma}_w$, the remainder of $\mathbf{\Gamma}_w$ falls into components $\gamma_1, \gamma_2, \cdots, \gamma_k$, each of which has $w = 0$ as a limit point. If $O_w^{(\nu)}$ is an end domain, its boundary lies in a single set γ_i plus the point $w = 0$. If $O_w^{(\nu)}$ is a strip domain, its boundary lies in two distinct sets γ_i and γ_j; and every locus* $\mathrm{Re}(\zeta) =$ *constant, $-\infty < \mathrm{Im}(\zeta) < \infty$, in $O_w^{(\nu)}$ forms, with the adjunction of $w = 0$, a Jordan curve whose interior contains one of the sets, say γ_i, and whose exterior contains the other, γ_j.*

THEOREM II. *Each of the domains $O_z^{(1)}, O_z^{(2)}, \cdots, O_z^{(m)}, E_z^{(1)}, E_z^{(2)}, \cdots, E_z^{(m)}$ is either a strip or end domain. There are precisely $n - 1$ end domains $O_z^{(\nu)}$ in $|z| < 1$ and each subtends an angle $2\pi/(n - 1)$ at the origin. If the points $z = 0$ and $z = \infty$ are removed from $\mathbf{\Gamma}_z$, the remainder of $\mathbf{\Gamma}_z$ falls into components $\gamma_1, \gamma_2, \cdots, \gamma_k$. If $O_z^{(\nu)}$ is an end domain, its boundary lies in a single set γ_i plus the point $z = 0$. If $O_z^{(\nu)}$ is a strip domain, its boundary lies in two distinct sets γ_i and γ_j; and every locus* $\mathrm{Re}(\zeta) =$ *constant, $-\infty < \mathrm{Im}(\zeta) < \infty$, in $O_z^{(\nu)}$ forms, with the adjunction of $z = 0$, a Jordan curve whose interior contains one of the sets, say γ_i, and whose exterior contains the other, γ_j.*

Since z belongs to $O_z^{(\nu)}$ if and only if $1/\bar{z}$ belongs to $E_z^{(\nu)}$, the mapping of the domains $E_z^{(\nu)}$ is immediately inferred from Theorem II.

As the proofs of Theorems I and II are similar, although the proof of Theorem II is slightly more complicated, we shall prove Theorem II only.

PROOF OF THEOREM II: Since $Q(z)$ has no singularities in the simply-connected domain $O_z^{(\nu)}$, the function $\zeta(z)$ is single-valued in $O_z^{(\nu)}$. Beginning at some point z_1 on the boundary of $O_z^{(\nu)}$, $z_1 \neq 0$, prolong an arc $\text{Im}(\zeta) = \text{constant}$, $\text{Re}(\zeta)$ monotonic, which extends into $O_z^{(\nu)}$, and let the arc be terminated only if it meets the boundary of $O_z^{(\nu)}$ at some point z_2. Let this arc be K. Suppose without loss of generality that $\text{Re}(\zeta)$ is increasing as the point moves away from z_1. Lemma XI shows that K is a Jordan arc. It also shows that K must either strike the boundary of $O_z^{(\nu)}$ for a finite value of $\text{Re}(\zeta)$ or tend to the unique limit $z = 0$ as $\text{Re}(\zeta)$ becomes infinite. For otherwise K would have a limit point in the closure of $O_z^{(\nu)}$ but not at the origin. The limit point cannot be on the boundary of $O_z^{(\nu)}$ since in that case K would meet the boundary. Then the limit point would have to be in the interior of $O_z^{(\nu)}$ and in this case an arc of K plus a small arc on which $\text{Re}(\zeta) = \text{constant}$ through this limit point would form a closed Jordan curve lying in $O_z^{(\nu)}$ and therefore containing no pole of $Q(z)$. Thus K is either an arc $\text{Im}(\zeta) = \text{constant}$, $a < \text{Re}(\zeta) < b$, or an arc $\text{Im}(\zeta) = \text{constant}$, $a < \text{Re}(\zeta) < \infty$, and K lies except for its end points in the domain $O_z^{(\nu)}$. It is to be shown that in the first case $O_z^{(\nu)}$ is a strip domain mapping onto a strip

$$-\infty < \text{Im}(\zeta) < \infty, \qquad a < \text{Re}(\zeta) < b,$$

in the ζ-plane, and in the second case $O_z^{(\nu)}$ is an end domain mapping onto a half-plane

$$-\infty < \text{Im}(\zeta) < \infty, \qquad \text{Re}(\zeta) > a,$$

in the ζ-plane.

Through any interior point of K prolong the arc

$$\text{Re}(\zeta) = c, \qquad -\infty < \text{Im}(\zeta) < \infty.$$

We call this arc $L(c)$, and we denote the point where $L(c)$ intersects K by $z(c)$. The arc $L(c)$ lies in $O_z^{(\nu)}$ and cuts K in precisely one point (we do not regard the origin as a point belonging to K or $L(c)$). If the point $z = 0$ is adjoined to $L(c)$, we obtain a closed Jordan curve $L_0(c)$.

A neighborhood of each point of $L(c)$ is mapped in a one–one manner onto a domain in the ζ-plane. For each $M > 0$, a neighborhood of the portion of $L(c)$ for which $-M \leqq \text{Im}(\zeta) \leqq M$ is mapped in a one–one manner onto a neighborhood of the straight-line segment $\text{Re}(\zeta) = c$, $-M \leqq \text{Im}(\zeta) \leqq M$ in the ζ-plane. It is therefore clear that for each positive M the portion of the arc $L(c)$ for which $-M \leqq \text{Im}(\zeta) \leqq M$ is a continuous function of c. Using Lemma X, we see that the asymptotic directions of $L(c)$ depend continuously on c. Since they form a discrete set, the asymptotic directions are the same for each c, that is, they are independent of c.

The set of points in the z-plane covered by the arcs $L(c)$ when c traverses all values on K is an open set lying in $\mathbf{O}_z^{(\nu)}$. Let this open set be D. Then D is mapped in a one–one manner onto a strip $a < \text{Re}(\zeta) < b$, $-\infty < \text{Im}(\zeta) < \infty$, in case K is the arc $a < \text{Re}(\zeta) < b$, and onto the half-plane $a < \text{Re}(\zeta) < \infty$, $-\infty < \text{Im}(\zeta) < \infty$, in case K is the arc $a < \text{Re}(\zeta) < \infty$. It remains to show that D is the entire domain $\mathbf{O}_z^{(\nu)}$.

If $c \neq c'$, then $L_0(c)$ and $L_0(c')$ are closed Jordan curves having only the point $z = 0$ in common. Hence either their interiors are disjoint, or the interior of one lies in the interior of the other. Since $L(c)$ varies continuously with c, it follows that for any c, $a < c < b$ or $a < c < \infty$ as the case may be, the interior of one of $L_0(c)$, $L_0(c')$ lies in the interior of the other.

As c approaches a, the arc $L(c)$ approaches a set of arcs of $\mathbf{\Gamma}_z$ which form part of the boundary of $\mathbf{O}_z^{(\nu)}$, and these arcs lie in a component γ_i of $\mathbf{\Gamma}_z$. In case K corresponds to the interval $a < \text{Re}(\zeta) < b$, then as c approaches b, the arc $L(c)$ approaches a set of arcs of $\mathbf{\Gamma}_z$ which lie on the boundary of $\mathbf{O}_z^{(\nu)}$ and which belong to a component γ_j of $\mathbf{\Gamma}_z$. The components γ_i and γ_j are distinct since one has points inside $L_0(c)$ and the other has points outside $L_0(c)$ for all c, $a < c < b$. It follows that $\mathbf{O}_z^{(\nu)}$ contains no points outside the domain D.

Finally, suppose that K corresponds to the interval $a < \text{Re}(\zeta) < \infty$. Then if β is the number mentioned in Lemma X, the function $\zeta(z)$ is bounded by some constant in the part of $\mathbf{O}_z^{(\nu)}$ which lies in $|z| \geqq \beta$. Hence if c is sufficiently large, $c \geqq c'$ say, then the Jordan arc $L(c')$ lies in $|z| < \beta$. The domain covered by the arcs $L(c)$ for $a < c < c'$ maps into a strip $a < \text{Re}(\zeta) < c'$, $-\infty < \text{Im}(\zeta) < \infty$, while the interior of $L(c')$ maps into the half-plane $\text{Re}(\zeta) > c'$ by Lemma X. Since the asymptotic directions of $L(c)$ are independent of c and the asymptotic directions of $L(c')$ differ by $2\pi/(n-1)$, we see that the domain D subtends an angle of $2\pi/(n-1)$ at the origin: D is the domain interior to $L(c')$ plus the set covered by the arcs $L(c)$ for $a < c \leqq c'$; D includes the interior of $L(c)$ for each c satisfying $c > a$; so it includes the interior of the arcs of $\mathbf{\Gamma}_z$ to which $L(c)$ converges as c approaches a. Thus $\mathbf{O}_z^{(\nu)}$ is precisely the domain D.

Each strip domain subtends zero angular space at the origin since the asymptotic directions of $L(c)$ are independent of c, and each end domain subtends an angle of $2\pi/(n-1)$ at the origin. It follows that there are precisely $n-1$ end domains in $|z| < 1$.

4.5. We consider finally the case where ζ is defined by

$$\zeta = \int (\Phi(v))^{1/2}\, dv \tag{4.5.1}$$

where

$$\Phi(v) = A\,\frac{2v-1}{v^2(v-1)^2}, \qquad A = (a+ib)^2. \tag{4.5.2}$$

Since this is a very special case, we state the result as a lemma.

By reflection on the line $\mathrm{Re}(v) = 1/2$ or reflection on the real axis $\mathrm{Im}(v) = 0$, it is clear that we may suppose that $0 \leqq a \leqq b$. The case $0 \leqq b \leqq a$ is obtained by reflection on the line $\mathrm{Re}(v) = 1/2$, whereas the case where a and b have opposite signs is obtained by reflecting on the real axis $\mathrm{Im}(v) = 0$.

LEMMA XX. *If ζ is defined by (4.3.3), $\mathbf{\Gamma}_v$ is one of the following configurations:*

(1) $0 < a < b$. *Two of the loci* $\mathrm{Re}(\zeta) = $ *constant issuing from* $v = 1/2$ *tend to* $v = 1$. *These two loci plus the point* $v = 1$ *form a closed Jordan curve containing* $v = 0$ *in its interior. The third locus issuing from* $v = 1/2$ *lies (apart from the end point at* $v = 1/2$*) entirely in the interior of the Jordan curve formed by the other two and tends to* $v = 0$. *The locus* $\mathrm{Re}(\zeta) = $ *constant issuing from* $v = \infty$ *tends to* $v = 1$. *The structure* $\mathbf{\Gamma}_v$, *which is composed of these four loci, divides the plane into two strip domains.*

(2) $0 < a = b$. *The structure* $\mathbf{\Gamma}_v$ *is composed of the three loci* $\mathrm{Re}(\zeta) = $ *constant issuing from* $v = 1/2$. *One is the locus* $\mathrm{Re}(v) = 1/2$, $\mathrm{Im}(v) \leqq 0$, *extending from* $v = 1/2$ *to* $v = \infty$. *The other two loci are symmetrical about the line* $\mathrm{Re}(v) = 1/2$; *one tends to* $v = 0$, *the other to* $v = 1$. *The complement of* $\mathbf{\Gamma}_v$ *is a strip domain.*

(3) $0 = a < b$. *Two of the loci* $\mathrm{Re}(\zeta) = $ *constant issuing from* $v = 1/2$ *form a closed loop symmetrical about the real axis and enclosing* $v = 0$ *in its interior. The third locus issuing from* $v = 1/2$ *is the segment of the real axis from* $v = 1/2$ *to* $v = 1$ *while the arc extending to* $v = \infty$ *is the segment of the real axis from* $v = 1$ *to* $v = \infty$. *The structure* $\mathbf{\Gamma}_v$ *divides the plane into two simply-connected domains. One is the interior of the loop, and the other is the exterior of the loop cut along the positive axis from* $v = 1/2$ *to* $v = \infty$. *The function* $e^{\zeta/b}$ *is single-valued in the interior of the loop and maps the interior of the loop onto the interior or exterior of a circle as the case may be. Any locus* $\mathrm{Re}(\zeta) = $ *constant prolonged from a point interior to the loop forms a closed Jordan curve. The other domain is a strip domain.*

A more detailed discussion of these cases with diagrams is to be found in Chapter XV.

PROOF: Let the three arcs $\mathrm{Re}(\zeta) = $ constant emerging from $v = 1/2$ be L, M, N. If one of them when prolonged returns to the point $v = 1/2$ for a finite value of $\mathrm{Im}(\zeta)$, then two of them are equal, say $M = N$. Then M plus N is a closed Jordan arc J passing through $v = 1/2$ but not passing through any of the points $0, 1, \infty$. According to Lemma XI, J must enclose at least one of the points 0, 1. If it enclosed both $v = 0$ and $v = 1$, then the arc $\mathrm{Re}(\zeta) = $ constant prolonged from $v = \infty$ would meet the point $1/2$ and therefore be the arc L. Then the exterior of J cut along L would be a domain of the type described in Lemma XI, but would contain no critical point of $\Phi(v)$. Thus J encloses precisely one of the points $v = 0$ and $v = 1$, say J encloses $v = 0$. Then the value of the integral (4.5.1) taken around J is pure imaginary. According to the Cauchy residue theorem, it is equal to

$$\pm 2\pi i(a + ib)i,$$

and this implies that $a = 0$. It immediately follows that $\operatorname{Re}(\zeta)$ is constant on the arc of the positive real axis from $v = 1/2$ to $v = 1$ and also on the arc of the positive real axis from $v = 1$ to $v = \infty$. The first of them is the arc L and the second is the arc $\operatorname{Re}(\zeta) =$ constant prolonged from $v = \infty$. The function $e^{\zeta/b}$ is single-valued inside J and maps the interior of J onto the interior or exterior of a circle as the case may be. The exterior of J, cut along the positive real axis from $v = 1/2$ to $v = 1$ and from $v = 1$ to $v = \infty$, is a strip domain, mapping onto a strip of width πb. The Jordan curve formed by M and N is symmetrical about the real axis.

If two of the three arcs L, M, N form a closed Jordan curve enclosing the point $v = 1$, we should have $b = 0$; so this case is impossible under the restriction $0 \leqq a \leqq b, b > 0$.

We thus suppose that none of the three arcs L, M, N returns to the point $v = 1/2$ for a finite value of $\operatorname{Im}(\zeta)$. If the three tend to different points $0, 1, \infty$, let L meet $v = \infty$ for a finite value of $\operatorname{Im}(\zeta)$, M tend to $v = 0$ as $\operatorname{Im}(\zeta)$ becomes infinite, and N tend to $v = 1$ as $\operatorname{Im}(\zeta)$ becomes infinite. Then part of M plus an arc α_0 on which $\operatorname{Im}(\zeta)$ is constant forms a closed Jordan curve J_0 lying in a small neighborhood of $v = 0$ and enclosing $v = 0$. The value of (4.5.1) taken around J_0 is equal to

$$\pm 2\pi(a + ib)$$

and the value of (4.5.1) taken over α_0 is the real part of this, namely

$$\pm 2\pi a.$$

Likewise, a part of N plus an arc α_1 on which $\operatorname{Im}(\zeta)$ is constant forms a closed Jordan curve J_1 lying in a small neighborhood of $v = 1$ and enclosing $v = 1$. The value of (4.5.1) when the integration is over J_1 is

$$\pm 2\pi i\,(a + ib)$$

and the value of the integral over α_1 is

$$\pm 2\pi b.$$

The exterior of J_0 and J_1, cut by part of M from J_0 to $1/2$ and by N from J_1 to $1/2$ and by L from $1/2$ to infinity is a simply-connected domain in which $(\Phi(v))^{1/2}$ is regular. Then the integral (4.5.1) around the entire boundary is zero; so taking real parts, we have

$$\pm 2\pi a \pm 2\pi b = 0.$$

In the case $0 \leqq a \leqq b, b > 0$, this becomes

$$0 < a = b.$$

Then L is the line $\operatorname{Re}(v) = 1/2$, $\operatorname{Im}(v) \leqq 0$, while M is the image of N in the line $\operatorname{Re}(v) = 1/2$. In this case, the complement of $\mathbf{\Gamma}_v$ is a single strip domain mapping onto a strip of width $2\pi a$.

Now suppose that two of the arcs tend to the same limit, which must be $v = 0$

or $v = 1$. Suppose then that M and N each tend to $v = 1$ as $\text{Im}(\zeta)$ becomes infinite. They form, together with $v = 1$, a closed Jordan curve J_1 which passes through the points $v = 1/2$ and $v = 1$. A part of J_1 plus an arc $\text{Im}(\zeta) = \text{constant}$ near $v = 1$ forms a closed Jordan curve enclosing a domain of the type described in Lemma XI but not enclosing the point $v = 1$. This Jordan curve must therefore enclose $v = 0$, and it then follows that J_1 must enclose $v = 0$. We now show that L cannot have $v = 1$ as limit. For if L has $v = 1$ as limit, then L plus M plus $v = 1$ is a closed Jordan curve J_2, and L plus N plus $v = 1$ is a closed Jordan curve J_3. The curves J_1, J_2, J_3 each enclose $v = 0$ and no point except $v = 1/2$ and $v = 1$ belongs to every pair of them. The interiors of two of these curves are disjoint and lie inside the third curve. Only one of the smaller domains can contain $v = 0$. The contradiction shows that L does not have $v = 1$ as limit. If L were outside J_1, it would meet $v = \infty$. Then a small arc $\text{Im}(\zeta) = \text{constant}$ near $v = 1$ plus a part of J_1 would form a closed Jordan curve J_1' enclosing both $v = 0$ and $v = 1$. The exterior of J_1' cut by L from 1/2 to infinity would be a domain of the type described in Lemma VII but not containing any pole of $\Phi(v)$. This contradiction shows that L does not lie outside J_1.

Thus M and N together with $v = 1$ form a closed Jordan curve J_1 through $v = 1/2$ and $v = 1$ enclosing $v = 0$, and L lies except for its end point inside J_1. Since L does not have $v = 1$ as limit, it tends to $v = 0$ as $\text{Im}(\zeta)$ becomes infinite. Then the arc $\text{Re}(\zeta) = \text{constant}$ prolonged from $v = \infty$ must tend to $v = 1$ as $\text{Im}(\zeta)$ becomes infinite. An arc α_0 on which $\text{Im}(\zeta)$ is constant plus a part of L forms a closed Jordan curve J_4 in the neighborhood of $v = 0$ and enclosing $v = 0$. The arc α_0 lies entirely inside J_1. Then the value of (4.5.1) taken over J_4 is, according to the residue theorem,

$$\pm 2\pi(a + ib);$$

the value of (4.5.1) over α_0 is the real part of this, or

$$\pm 2\pi a.$$

Likewise an arc α_1 on which $\text{Im}(\zeta)$ is constant plus an arc of M forms a closed Jordan curve J_5 in the neighborhood of $v = 1$ and enclosing $v = 1$. Only part of α_1 lies inside J_1. The value of (4.5.1) over J_5 is

$$\pm 2\pi i(a + ib)$$

and the value of (4.5.1) over α_1 is

$$\pm 2\pi b.$$

Then the part of α_1 that lies inside J_1 plus parts of M and N form a Jordan curve J^* which can be deformed into J_4 without crossing any singularity of $(\phi(v))^{1/2}$. Since $(\phi(v))^{1/2}$ is single-valued inside J^*, we have from Cauchy's theorem

$$0 < |2\pi a| < |2\pi b|.$$

Since $0 \leqq a \leqq b$, $b > 0$, this shows that

$$0 < a < b.$$

The interior of J_1 is mapped onto a strip of width $2\pi a$ while the exterior is mapped onto a strip of width $\pi(b - a)$.

The case in which two of the three arcs L, M, N have $v = 0$ as limit would imply that $|b| > |a|$, and is therefore excluded under the hypothesis $0 \leqq a \leqq b, b > 0$.

The structure $\mathbf{\Gamma}_v$ has been shown to fall under three cases $0 \leqq a \leqq b, b > 0$, and these cases imply respectively that

$$\text{(i) } a = 0, b > 0,$$

$$\text{(ii) } 0 < a = b,$$

$$\text{(iii) } 0 < a < b.$$

These three possibilities are mutually distinct and include all possible values of a, b when $0 \leqq a \leqq b, b > 0$. Thus conditions (i), (ii), (iii) imply in turn which configuration the structure $\mathbf{\Gamma}_v$ must have.

CHAPTER V

GEODESICS. CONTINUITY THEOREM

5.1. We now consider the metrics defined by

$$(5.1.1) \qquad |d\zeta|^2 = \left|\frac{P(w)}{w^2}\right| |dw|^2,$$

$$(5.1.2) \qquad |d\zeta|^2 = \left|\frac{Q(z)}{z^2}\right| |dz|^2.$$

In the case of (5.1.1) we shall show that any two points w_1 and w_2, $w_1 \neq 0, w_2 \neq 0$, are connected by a unique geodesic. In the case of (5.1.2) there is a geodesic connecting any two points z_1 and z_2, $0 < |z_1| < \infty, 0 < |z_2| < \infty$, but if, for example, both z_1 and z_2 lie on $|z| = 1$, then the two arcs into which z_1 and z_2 divide this circumference may both be geodesics connecting these two points. Hence the geodesics are not always unique in the metric (5.1.2).

Similarly, any two points v_1, v_2 in the v-plane which are both different from 0 and 1 are connected by at least one geodesic in the metric

$$(5.1.3) \qquad |d\zeta|^2 = |\Phi(v)| \, |dv|^2,$$

but the geodesics are not necessarily unique.

Because the proof of the existence of geodesics for the metric (5.1.2) is almost entirely similar to that for (5.1.1), we shall consider only the case (5.1.1) in detail. Since

$$P(w) = \frac{A_1}{w} + \frac{A_2}{w^2} + \cdots + \frac{A_{n-1}}{w^{n-1}},$$

we obtain, making the substitution $w = 1/s^2$:

$$\zeta = \int (P(w))^{1/2} \frac{dw}{w} = -\int \{4A_1 + 4A_2 s^2 + \cdots + 4A_{n-1} s^{2n-4}\}^{1/2} \, ds.$$

Thus let

$$(5.1.4) \qquad N(s) = a_0 + a_1 s + \cdots + a_m s^m, \qquad a_m \neq 0, \; m \geqq 1,$$

and consider the metric $|d\zeta|^2 = |N(s)| \, |ds|^2$ where

$$(5.1.5) \qquad \zeta = \int (N(s))^{1/2} \, ds.$$

The set of all loci $\mathrm{Re}(\zeta) =$ constant issuing from zeros of $N(s)$ and from $s = 0$ will be denoted by $\mathbf{\Gamma}_s$ while the set of all loci $\mathrm{Im}(\zeta) =$ constant issuing from zeros of $N(s)$ and from $s = 0$ will be called $\mathbf{\Lambda}_s$. We note that $\mathbf{\Lambda}_s$ is a set of arcs on which the real part of

$$\int \{-a_0 - a_1 s - \cdots - a_m s^m\}^{1/2}\, ds$$

is constant. The general properties of $\boldsymbol{\Gamma}_s$ and $\boldsymbol{\Lambda}_s$ are described in Theorem I (with $s = \infty$ corresponding to the pole at the origin).

If α and β are two distinct finite points of the s-plane, let an admissible path from α to β be one which is rectifiable and continuous and which has piecewise continuous first derivatives with respect to the distance along the arc. If there is an admissible path from α to β such that

$$\int_\alpha^\beta |\, d\zeta \,| = \int_\alpha^\beta |\, N(s)\, ds \,| \tag{5.1.6}$$

is not larger for this path than for any other admissible path connecting α and β, this path is called a geodesic from α to β or from β to α.

5.2. LEMMA XXI. *Any two finite points of the s-plane are connected by one and only one geodesic.*

PROOF: In a circle $|\, s \,| \leqq s_0$ it is clear that $(N(s))^{1/2}$ is bounded by some number N_0, and so the integral (5.1.6) extended along any chord of this circle does not exceed $2s_0N_0$. If α and β lie in $|\, s \,| \leqq s_0$, it follows that the geodesic distance from α to β is not larger than $2s_0N_0$. On the other hand, there is an s_1 such that $|\, N(s) \,| \geqq 1$ for $|\, s \,| \geqq s_1$. Thus there is an s_2 (depending on s_0 and s_1) such that if the integral along an admissible path from α to β is less than $4s_0N_0$, then the admissible path lies in $|\, s \,| \leqq s_2$.

Let $\boldsymbol{O}_s^{(\nu)}$ $(\nu = 1, 2, \cdots, r)$ denote the simply-connected domains into which $\boldsymbol{\Gamma}_s$ dissects the plane. Each $\boldsymbol{O}_s^{(\nu)}$ has $s = \infty$ as a boundary point and is mapped in a one–one way onto a strip or half-plane in the ζ-plane by the function (5.1.5). Let L be an admissible path joining two finite points α and β for which the integral

$$\int_\alpha^\beta |\, (N(s))^{1/2}\, ds \,| \tag{5.2.1}$$

has a finite value.

Let α lie in a domain $\boldsymbol{O}_s^{(\mu)}$, and let a variable point moving along L from α towards β first leave the closure of $\boldsymbol{O}_s^{(\mu)}$ at the point A. In the ζ-plane construct a straight line from $\zeta(\alpha)$ to $\zeta(A)$ and map this line back into an arc in the s-plane which lies except for one end point in the domain $\boldsymbol{O}_s^{(\mu)}$. Let L_1 consist of this new arc plus the portion of L from A to β. Then (5.2.1) is not increased, and it is decreased unless L and L_1 coincide. If α does not belong to any domain $\boldsymbol{O}_s^{(\mu)}$, then α belongs to $\boldsymbol{\Gamma}_s$ and $L = L_1$. Make a similar construction near β, obtaining a new curve L_2 which coincides with L_1 except for an arc (B, β) lying in a single domain $\boldsymbol{O}_s^{(\nu)}$ except for one end point.

As we traverse L_2 starting from α, we obtain three paths (α, A), (A, B),

(B, β). Suppose that (A, B) is not a point-path. If (A, B) enters an end domain $\boldsymbol{O}_s^{(i)}$ at p and leaves it at q, then by mapping $\boldsymbol{O}_s^{(i)}$ onto a half-plane in the ζ-plane it is seen that an integral along the boundary of $\boldsymbol{O}_s^{(i)}$ from p to q gives a smaller value than does the integral from p to q along (A, B). In the same way, if (A, B) enters a strip domain and leaves without crossing it, then the value of (5.2.1) can be decreased by moving part of (A, B) over onto the boundary of the strip domain. Thus L can be deformed until it consists of segments (α, A) and (B, β) plus possibly an arc (A, B). The arc (A, B) has no point in common with any end domain and enters a strip domain only to cross it. Let this path from α to β be L_3. If L_3 is not a Jordan arc, then some subset is a Jordan arc, and we integrate only over this subset, thereby decreasing the value of (5.2.1). Each arc of L_3 that crosses a strip domain is deformed if necessary in such a way that it maps into the ζ-plane as a straight-line segment. Moreover, L_3 is a Jordan arc with each point traced only once in moving from α to β.

It is to be shown that there is a geodesic G joining any two finite points α and β and that it has the following properties:

(a) G consists of finitely many arcs, each of which maps into the ζ-plane as a straight-line segment.

(b) The only points of intersection of these arcs are at the zeros of $N(s)$.

(c) G is a Jordan arc joining α and β, and each point of G is traced only once in moving from α to β.

(d) If G passes through a zero s_i of $N(s)$ of precise order λ and the two arcs of G meeting at s_i form an angle φ, $0 \leqq \varphi \leqq 2\pi$, then

$$\frac{2\pi}{\lambda + 2} \leqq \varphi \leqq 2\pi - \frac{2\pi}{\lambda + 2}.$$

(e) G does not contain the point $s = \infty$.

(*f*) If the point $s = \infty$ is removed from $\boldsymbol{\Gamma}_s$, $\boldsymbol{\Gamma}_s$ is divided into a finite number of components. Each such component intersects G in an empty set or a connected set.

(g) Each strip domain which intersects G does so in exactly one interval of G; the interval has one or both end points at α, β or it crosses the strip domain.

(h) If $\alpha \subset \boldsymbol{O}_s^{(\mu)}$ and $\beta \subset \boldsymbol{O}_s^{(\nu)}$, then G contains one or two intervals, each with an end point at α or β. Except for these two possible intervals, G lies entirely on $\boldsymbol{\Gamma}_s$ and in strip domains.

The arc L can be deformed into an arc L_3 which has properties (a), (c), (e), (h) and the value of (5.2.1) is decreased unless L itself has these properties. The reason that the deformed arc L_3 satisfies condition (a) is that L_3 consists of two arcs (α, A) and (B, β) which map into straight-line segments in the ζ-plane and, in case $A \neq B$, a path (A, B). The path (A, B) lies on $\boldsymbol{\Gamma}_s$ and in strip domains, and a point moving along (A, B) can enter a strip domain only to cross it. Thus (A, B) is the sum of arcs that lie on certain subsets of $\boldsymbol{\Gamma}_s$ and arcs that lie in strip domains, and these two types of arcs alternate along (A, B). Each of the arcs maps into the ζ-plane as a straight-line segment, and the arcs that cross strip domains are of length equal to or greater than the width of the narrowest strip domain. Hence L_3 has only a finite number of arcs in all. In fact, if the value of (5.2.1), where the path of integration is along L_3, is not larger than τ and if each strip domain is of width γ or more, then L_3 is the sum of $2[\tau/\gamma] + 3$ arcs at most, each of which maps into the ζ-plane as a straight-line segment.

The locus $\mathrm{Re}(\zeta) = \text{constant}$, $-\infty < \mathrm{Im}(\zeta) < \infty$, prolonged from an interior point of a strip domain lies in that strip domain and forms, by adjoining the point at $s = \infty$, a closed Jordan curve on the Riemann sphere. Let $J_1, J_2, \cdots, J_k$ be the set of such Jordan curves, one in each strip domain, and let none of them pass through α or β. Of the two simply-connected domains complementary to J_i, we shall call that one the exterior which contains the point $s = 0$. The point $s = \infty$ lies on each Jordan curve, but no other point of the plane belongs to more than one curve. Thus for every pair of these curves, one lies inside the other or they are mutually exterior.

In order to prove that the geodesic has property (g), we suppose that the arc L_3 crosses some curve J_i more than once and show that the value of (5.2.1) can then be diminished. We remark that L_3 has only a finite number of points in common with any of the curves J_i. Since L_3 crosses J_i more than once, there are points p and q on a curve J_σ such that the subarc (p, q) of L_3 has no points in common with any curve J_ν ($\nu = 1, 2, \cdots, k$) except the two end points p and q which lie on J_σ. Then there are points p' and q' of (p, q) such that the arcs (p, p') and (q, q') lie in the strip domain $\boldsymbol{O}_s^{(\sigma)}$ in which J_σ lies while the points p' and q' lie on the boundary of $\boldsymbol{O}_s^{(\sigma)}$. The arc (p', q') of L_3 does not cross any strip domain; so it has no point in common with any strip domain, and it must therefore lie entirely on $\boldsymbol{\Gamma}_s$. If there were two arcs of $\boldsymbol{\Gamma}_s$ extending from p' to q' without passing through $s = \infty$, then one of the domains into which $\boldsymbol{\Gamma}_s$ divides the plane would not have $s = \infty$ as a boundary point. Hence there is only one path on $\boldsymbol{\Gamma}_s$ which extends from p' to q' and does not pass through $s = \infty$. The portion of the boundary of $\boldsymbol{O}_s^{(\nu)}$ from p' to q' which does not contain $s = \infty$ is one such path, and it is the only one. Thus the arc (p', q') of L_3 lies on the boundary of $\boldsymbol{O}_s^{(\sigma)}$. It is then clear that L_3 can be deformed in such a way that the value of (5.2.1) is decreased. For, beginning at p, let a point move along L_3 toward α until it meets the boundary of $\boldsymbol{O}_s^{(\sigma)}$ or the point α, whichever is met first, and let this point be p^*. Then, beginning at q, let a point move toward β until it meets the boundary of $\boldsymbol{O}_s^{(\sigma)}$ or β, and let this point be q^*. Then the arc (p^*, q^*) of L_3 lies in the closure of $\boldsymbol{O}_s^{(\sigma)}$ and contains the arc (p, q) of L_3. A straight-line segment in the ζ-plane from $\zeta(p^*)$ to $\zeta(q^*)$ does not cross the map of J_σ, and the image of this segment in the s-plane plus the portions (α, p^*) and (q^*, β) of L_3 is an admissible path from α to β for which the value of (5.2.1) is smaller. Also the number of points at which the path crosses J_σ is reduced by two.

Thus the path of integration can be deformed until the path crosses each J_ν at most once, and the value of (5.2.1) will be decreased. If α lies in a strip domain $\boldsymbol{O}_s^{(i)}$, then α may lie inside or outside J_i, but in either case the path of integration has at most one point in common with J_i. Likewise if β lies in a strip domain

$O_s^{(j)}$, the path of integration has at most one point in common with J_j. It follows that the path of integration may be deformed until it satisfies (g). Let the path of integration thus obtained be L_4.

If $s = \infty$ is removed from $\mathbf{\Gamma}_s$, then by Theorem I $\mathbf{\Gamma}_s$ is divided into components, each of which contains all its limit points except $s = \infty$. To show that L_4 has property (f), let L_4 intersect a component of $\mathbf{\Gamma}_s$ at two points p and q but suppose that the arc (p, q) of L_4 does not lie entirely on $\mathbf{\Gamma}_s$. Now the arc (p, q) has no point in common with any end domain; so this arc must have some points in common with a strip domain and it must then cross one of the closed Jordan curves J_i. Since L_4 has only one point in common with J_i, one of the points p, q must lie inside J_i and the other outside. But $\mathbf{\Gamma}_s$ does not cross J_i except at $s = \infty$; so the arc of $\mathbf{\Gamma}_s$ joining p and q must go to infinity. Thus p and q do not belong to the same component of $\mathbf{\Gamma}_s$, and this contradiction shows that L_4 has property (f). The path L_4 therefore has the properties (a), (c), (e), (f), (g), and (h).

Now let $G_1, G_2, G_3, \cdots$ be a sequence of admissible paths joining α and β for which the value of (5.2.1) tends toward its lower bound. Without loss of generality let each G_ν have the properties of the arc L_4. Then a subsequence of the arcs G_ν tends to a limit path G which is an admissible path joining α and β. The value of the integral (5.2.1) over G is equal to its lower bound for all admissible paths joining α and β. The path G has the properties of L_4, and it is a geodesic.

By a mapping into the ζ-plane it is seen that G must have property (b). Consider property (d). There are $\lambda + 2$ arcs $\mathrm{Re}(\zeta) =$ constant issuing from a zero s_i of order λ and the angle between adjacent arcs is $2\pi/(\lambda + 2)$. These form $\lambda + 2$ angular openings at s_i, each of which maps into a set in the ζ-plane which covers a semi-circle $\mathrm{Re}(\zeta) \geqq 0$, $|\zeta| \leqq \tau$, or $\mathrm{Re}(\zeta) \leqq 0$, $|\zeta| \leqq \tau$, under the mapping

$$\zeta = (s - s_i)^{(\lambda+2)/2} \{ b_0 + b_1 (s - s_i) + \cdots \}, \qquad b_0 \neq 0.$$

If two arcs of G meeting at s_i make an angle φ with each other at s_i, where $0 \leqq \varphi < 2\pi/(\lambda + 2)$, then the portions of the two arcs of G which lie in a neighborhood of s_i must lie in an angular opening of width $4\pi/(\lambda + 2)$ formed by putting together two adjacent angular openings, each of width $2\pi/(\lambda + 2)$. These two angular openings map into a whole neighborhood of $\zeta = 0$ cut along the positive or the negative imaginary axis. In this mapping, angles are multiplied by $(\lambda + 2)/2$; so the two arcs of G map into two rectilinear segments meeting at $\zeta = 0$ and forming an angle which is less than π. This angle does not contain the cut, and it follows that the integral (5.2.1.) can be decreased. Thus the geodesic G has the property (d).

It will now be shown that there is only one path with properties (a) to (h) inclusive; that is, a geodesic between two points is unique. Let G_1 and G_2 have properties (a) to (h) inclusive. If they intersect at some point different from α and β, there will be subarcs G_1' and G_2' which join two points α' and β' and these subarcs will have properties (a) to (h) inclusive. Moreover, the subarcs can be chosen in such a way that together they form a Jordan curve. Thus we may suppose without loss of generality that $G_1 + G_2$ is a closed Jordan curve.

If $G_1 + G_2$ encloses a zero of $N(s)$, then there will be at least three arcs of $\mathbf{\Gamma}_s$ issuing from this zero and each arc must lead to $s = \infty$ along different paths. Otherwise there would be a domain $\boldsymbol{O}_s^{(\nu)}$ not having $s = \infty$ as a boundary point. Thus if $N(s)$ has a zero inside $G_1 + G_2$, there are two arcs of $\mathbf{\Gamma}_s$ which issue from the zero of $N(s)$ and meet the same geodesic, say G_1, without first passing through $s = \infty$. Let the two points of intersection be p and q. These points lie on the same component of $\mathbf{\Gamma}_s$ since each is connected to the zero of $N(s)$ inside $G_1 + G_2$. But we have shown above that two points belonging to the same component of $\mathbf{\Gamma}_s$ are joined by only one path on $\mathbf{\Gamma}_s$ which does not pass through $s = \infty$. This path must therefore contain the zero of $R(s)$ lying inside $G_1 + G_2$. Hence the intersection of G_1 with this component of $\mathbf{\Gamma}_s$ is not connected, contradicting property (f). The contradiction shows that there is no zero of $N(s)$ inside $G_1 + G_2$.

Map the interior of $G_1 + G_2$ onto the interior of the unit circle $|z| < 1$. If $s_1, s_2, \cdots, s_j$ are the zeros of $N(s)$ on $G_1 + G_2$, let the points on $|z| = 1$ which correspond to $\alpha, \beta, s_1, s_2, \cdots, s_j$ be called the vertices. The vertices divide $|z| = 1$ into arcs, each of which maps into a rectilinear segment in the ζ-plane. Furthermore, $\zeta(z) = \zeta(s(z))$ is regular and single-valued in $|z| < 1$. Thus, the function

$$z\frac{d\zeta}{dz}$$

has a constant argument on each arc and is regular in $|z| < 1$.

Now let the point z describe the unit circumference counterclockwise, indenting slightly in the neighborhood of each vertex. At each vertex which does not correspond to α or β, the argument of $z(d\zeta/dz)$ is not increased. At the vertices corresponding to α and β, the argument of $z(d\zeta/dz)$ is decreased or is increased by less than π, and this is true whether or not α, β are zeros of $N(s)$. Hence the total increase in the argument of $z(d\zeta/dz)$ is less than 2π. But this is impossible since $z(d\zeta/dz)$ has a zero at $z = 0$. This completes the proof of the lemma.

We remark that in the case of the metric (5.1.2) this argument breaks down because $G_1 + G_2$ can contain a pole in its interior. In this metric geodesics exist, but two points are not always joined by a unique geodesic as pointed out at the beginning of this chapter.

5.3. Given a domain $\boldsymbol{O}_s^{(\nu)}$, we say that $\boldsymbol{\Omega}_s^{(\nu)}$ is the "extended domain" belonging to $\boldsymbol{O}_s^{(\nu)}$ if $\boldsymbol{\Omega}_s^{(\nu)}$ is a simply-connected domain containing $\boldsymbol{O}_s^{(\nu)}$ which is mapped in a one-one manner onto the entire ζ-plane cut along finitely many straight lines

(i) $$\operatorname{Im}(\zeta) = c_k, \qquad -\infty < \operatorname{Re}(\zeta) \leqq a_k,$$

(ii) $$\operatorname{Im}(\zeta) = c_k, \qquad b_k \leqq \operatorname{Re}(\zeta) < \infty,$$

each rectilinear cut having its finite end point $a_k + ic_k$ or $b_k + ic_k$ at the image of a zero of $N(s)$. If $\boldsymbol{O}_s^{(\nu)}$ is a strip domain mapping onto the strip $a < \operatorname{Re}(\zeta) < b$, $-\infty < \operatorname{Im}(\zeta) < \infty$, then $\boldsymbol{\Omega}_s^{(\nu)}$ maps onto the ζ-plane cut along lines (i) and (ii) where $a_k \leqq a$ and $b_k \geqq b$ and both of the points $a_k + ic_k$, $b_k + ic_k$ are images of zeros of $N(s)$. If $\mathbf{\Omega}_s^{(\nu)}$ is an end domain mapping onto the half-plane $\mathrm{Re}(\zeta) > a$, $-\infty < \mathrm{Im}(\zeta) < \infty$, then $\mathbf{\Omega}_s^{(\nu)}$ maps onto the ζ-plane cut by lines of type (i) only and $a_k \leqq a$ while $a_k + ic_k$ is the image of a zero of $N(s)$. Similarly if $\mathbf{\Omega}_s^{(\nu)}$ is an end domain mapping onto the half-plane $\mathrm{Re}(\zeta) < b$, $-\infty < \mathrm{Im}(\zeta) < \infty$, $\mathbf{\Omega}_s^{(\nu)}$ maps onto the ζ-plane cut by lines of type (ii) only and $b \leqq b_k$ while $b_k + ic_k$ is the image of a zero of $N(s)$; see Figures 3 and 4, pages 96, 97.

To show the existence of these extended domains we make use of the set of arcs $\mathbf{\Lambda}_s$ defined above. Let $\mathbf{P}_s^{(\nu)}$ be the simply-connected domains complementary to $\mathbf{\Lambda}_s$. Each $\mathbf{P}_s^{(\nu)}$ has $s = \infty$ as a boundary point. For each μ and ν the intersection of $\mathbf{O}_s^{(\mu)}$ and $\mathbf{P}_s^{(\nu)}$ is connected. In fact, if p and q are points that lie in $\mathbf{O}_s^{(\mu)}$ and in $\mathbf{P}_s^{(\nu)}$, then the geodesic connecting them clearly lies in both $\mathbf{O}_s^{(\mu)}$ and $\mathbf{P}_s^{(\nu)}$ and so p and q are joined by an arc which lies in the intersection of $\mathbf{O}_s^{(\mu)}$ and $\mathbf{P}_s^{(\nu)}$. Since the intersection of $\mathbf{O}_s^{(\mu)}$ and $\mathbf{P}_s^{(\nu)}$ is connected, it is simply-connected.

There are no zeros of $N(s)$ in any domain $\mathbf{O}_s^{(\mu)}$ and $s = 0$ does not lie in $\mathbf{O}_s^{(\mu)}$. Hence if an arc of $\mathbf{\Lambda}_s$ has a point in common with $\mathbf{O}_s^{(\mu)}$, the arc crosses $\mathbf{O}_s^{(\mu)}$. If $\mathbf{O}_s^{(\mu)}$ is a strip domain, any arc of $\mathbf{\Lambda}_s$ which has a point in common with $\mathbf{O}_s^{(\mu)}$ must cross $\mathbf{O}_s^{(\mu)}$, forming a cross-cut with neither end point at $s = \infty$. If $\mathbf{O}_s^{(\mu)}$ is an end domain, then any arc of $\mathbf{\Lambda}_s$ which has a point in common with $\mathbf{O}_s^{(\mu)}$ must cross $\mathbf{O}_s^{(\mu)}$, one end point being at $s = \infty$. Each domain $\mathbf{O}_s^{(\mu)}$ has $s = 0$ or at least one zero of $N(s)$ on its boundary and each such point has one arc of $\mathbf{\Lambda}_s$ emerging from it which enters $\mathbf{O}_s^{(\mu)}$. The domain $\mathbf{O}_s^{(\mu)}$ is therefore crossed by at least one arc of $\mathbf{\Lambda}_s$ and therefore intersects more than one domain $\mathbf{P}_s^{(\nu)}$. For in the s-plane no arc of $\mathbf{\Lambda}_s$ forms a slit whose two edges make up part of the boundary of a domain $\mathbf{P}_s^{(\nu)}$; this is a consequence of the fact that no critical point in the s-plane is of order -1.

As a point s moves along an arc $\mathrm{Re}(\zeta) = $ constant in $\mathbf{O}_s^{(\nu)}$ with $\mathrm{Im}(\zeta)$ increasing from $-\infty$ to $+\infty$, the point s lies successively in the domains

$$\mathbf{P}_s^{(1)}, \mathbf{P}_s^{(2)}, \cdots, \mathbf{P}_s^{(k)}, \qquad k \geqq 2.$$

Since these domains are disjoint, we see that $\mathbf{O}_s^{(\mu)}$ is the sum of the disjoint simply-connected domains

$$\mathbf{P}_s^{(1)} \cap \mathbf{O}_s^{(\nu)}, \mathbf{P}_s^{(2)} \cap \mathbf{O}_s^{(\nu)}, \cdots, \mathbf{P}_s^{(k)} \cap \mathbf{O}_s^{(\nu)}$$

plus the portion of $\mathbf{\Lambda}_s$ that lies in $\mathbf{O}_s^{(\nu)}$. The domain $\mathbf{\Omega}_s^{(\nu)}$ is the sum of the disjoint domains

$$\mathbf{P}_s^{(1)}, \mathbf{P}_s^{(2)}, \cdots, \mathbf{P}_s^{(k)}$$

plus those arcs of $\mathbf{\Lambda}_s$ which intersect $\mathbf{O}_s^{(\nu)}$. It readily follows that $\zeta(s)$ is single-valued in $\mathbf{\Omega}_s^{(\nu)}$ and maps it in a one-one way onto the whole ζ-plane cut along horizontal lines as described above.

5.4. Given $\mathfrak{A} = (A_1, A_2, \cdots, A_{n-1})$, the function

$$P(w) = \sum_{\nu=1}^{n-1} \frac{A_\nu}{w^\nu}$$

is determined and so is the set $\mathbf{\Gamma}_w$. Thus $\mathbf{\Gamma}_w = \mathbf{\Gamma}_w(\mathfrak{A})$ depends upon $\mathfrak{A}$. Similarly, given $\mathfrak{B} = (B_1, B_2, \cdots, B_{n-1})$, the function

$$Q(z) = \sum_{\nu=-(n-1)}^{n-1} \frac{B_\nu}{z^\nu}$$

is determined and therefore $\mathbf{\Gamma}_z = \mathbf{\Gamma}_z(\mathfrak{B})$. It is clear that $\mathbf{\Gamma}_w$ and $\mathbf{\Gamma}_z$ depend only upon the directions of the vectors $\mathfrak{A}$ and $\mathfrak{B}$ and not on their magnitudes or lengths. For we may multiply $P(w)$ or $Q(z)$ by any positive number without affecting $\mathbf{\Gamma}_w$ or $\mathbf{\Gamma}_z$.

Let

$$\text{(5.4.1)} \qquad \| \mathfrak{A}^* - \mathfrak{A} \| = \left(\sum_{\nu=1}^{n-1} | A_\nu^* - A_\nu |^2 \right)^{1/2}.$$

We say that $\mathbf{\Gamma}_w(\mathfrak{A}^*)$ tends to $\mathbf{\Gamma}_w(\mathfrak{A})$ if the set of limit points of $\mathbf{\Gamma}_w(\mathfrak{A}^*)$ as $\| \mathfrak{A}^* - \mathfrak{A} \|$ tends to zero coincides with $\mathbf{\Gamma}_w(\mathfrak{A})$. If $\mathbf{\Gamma}_w(\mathfrak{A}^*)$ tends to $\mathbf{\Gamma}_w(\mathfrak{A})$ no matter how $\mathfrak{A}^*$ tends to $\mathfrak{A}$, we say that $\mathbf{\Gamma}_w$ is continuous at $\mathfrak{A}$. If the set of limit points of $\mathbf{\Gamma}_w(\mathfrak{A}^*)$ includes $\mathbf{\Gamma}_w(\mathfrak{A})$ no matter how $\mathfrak{A}^*$ tends to $\mathfrak{A}$, then we say that $\mathbf{\Gamma}_w$ is lower semi-continuous at $\mathfrak{A}$. If the set of limit points of $\mathbf{\Gamma}_w(\mathfrak{A}^*)$ is contained in $\mathbf{\Gamma}_w(\mathfrak{A})$ no matter how $\mathfrak{A}^*$ tends to $\mathfrak{A}$, then $\mathbf{\Gamma}_w$ is upper semi-continuous at $\mathfrak{A}$.

Let $N_w(\mathfrak{A})$ be the sum of all subcontinua of $\mathbf{\Gamma}_w(\mathfrak{A})$ which are of interior mapping radius unity and contain the point at infinity. Then $N_w(\mathfrak{A})$ is a subcontinuum of $\mathbf{\Gamma}_w(\mathfrak{A})$ containing the point at infinity. Its mapping radius is equal to or less than 1 and is greater than 1/4 since $N_w(\mathfrak{A})$ lies in $| w | \geqq 1/4$ and does not contain the whole circumference $| w | = 1/4$. Its mapping radius is 1 if and only if $A_1 \neq 0$ and no subcontinuum of $\mathbf{\Gamma}_w(\mathfrak{A})$ containing the point at infinity and of mapping radius greater than 1 contains a zero of $P(w)$. Continuity as well as lower and upper semi-continuity are defined in a way analogous to that for $\mathbf{\Gamma}_w(\mathfrak{A})$.

We suppose that n is a fixed integer and that not all components of the vector $\mathfrak{A}$ are zero. We prove the following lemma (the continuity theorem).

LEMMA XXII. *The set* $\mathbf{\Gamma}_w(\mathfrak{A})$ *is lower semi-continuous and its points of continuity are those for which* $A_{n-1} \neq 0$. *The set* $N_w(\mathfrak{A})$ *is upper semi-continuous and its points of continuity are those for which its mapping radius is* 1.

The set $\mathbf{\Gamma}_z(\mathfrak{B})$ *is lower semi-continuous and continuous if* $B_{n-1} \neq 0$. *The set of loci* $\mathbf{\Gamma}_v$ *described in Lemma* XX *depends continuously on* arg (A), *where*

$$\Phi(v) = A \frac{2v - 1}{v^2(v - 1)^2},$$

unless A *is real. If* A *is real, the set of limit points of* $\mathbf{\Gamma}_v(A^*)$ *as* A^* *tends to* A *is* $\mathbf{\Gamma}_v(A)$ *plus every point interior to the loop. Inside the loop* $\mathbf{\Gamma}_v(A^*)$ *resembles a logarithmic spiral*

$$\text{Re}(A^* \log w) = \text{constant}.$$

As A^* *tends to the real number* A, *the spirals approach closed loops.*

PROOF: We write

$$P(w) = P_{\mathfrak{A}}(w) = \frac{A_1}{w} + \cdots + \frac{A_{n-1}}{w^{n-1}},$$

$$P^*(w) = P_{\mathfrak{A}^*}(w) = \frac{A_1^*}{w} + \cdots + \frac{A_{n-1}^*}{w^{n-1}},$$

$$\zeta = \int (P(w))^{1/2} \frac{dw}{w}, \qquad \zeta^* = \int (P^*(w))^{1/2} \frac{dw}{w}.$$

We prove first the following statement. Each arc of $\mathbf{\Gamma}_w(\mathfrak{A})$ is crossed at some point w_1, $P(w_1) \neq 0$, by an arc I on which $\mathrm{Im}(\zeta)$ is constant; and, if $I(\delta)$ is the part of I that maps into a segment of length $2\delta > 0$ in the ζ-plane with center at $\zeta(w_1)$, then $I(\delta)$ intersects $\mathbf{\Gamma}_w(\mathfrak{A}^*)$, provided that $\| \mathfrak{A}^* - \mathfrak{A} \| \leqq \eta(\delta)$.

Let $q_1, q_2, \cdots, q_k$ be the distinct zeros of $P(w)$ where $q_k = \infty$. Our notion of the order of a zero at infinity will here be the usual one rather than that adopted in Chapter IV. Let $C_\nu(\epsilon_1)$ be the circle $| w - q_\nu | < \epsilon_1$ for $\nu = 1, 2, \cdots, k - 1$, $C_k(\epsilon_1)$ the circle $| w | > 1/\epsilon_1$, and $C_0(\epsilon_1)$ the circle $| w | < \epsilon_1$. We suppose that ϵ_1 is so small that the circles $C_0(\epsilon_1), C_1(\epsilon_1), \cdots, C_k(\epsilon_1)$ are disjoint. Choose $\epsilon_2 = \epsilon_2(\epsilon_1) > 0$ so small that if $P(w)$ has a zero of precise order λ_ν at q_ν, then $P^*(w)$ has precisely λ_ν zeros in $C_\nu(\epsilon_1)$, $\nu = 1, 2, \cdots, k$, provided that $\| \mathfrak{A}^* - \mathfrak{A} \| \leqq \epsilon_2$ (multiple zeros being counted multiply). Given ϵ_1, the existence of ϵ_2 follows from Rouché's theorem. If $A_{n-1} \neq 0$, then $P^*(w)$ has no further zeros, whereas if $A_{n-1} = 0$, there may be additional zeros of $P^*(w)$; but we choose ϵ_2 so small that all additional zeros lie in $C_0(\epsilon_1)$.

Let there be μ_ν arcs $\mathrm{Re}(\zeta) = $ constant emerging from the point q_ν, $1 \leqq \nu \leqq k - 1$, and let ϵ_1 be chosen so small that each of these arcs cuts the circumference of $C_\nu(\epsilon_1)$ at precisely one point. Then $C_\nu(\epsilon_1)$ is divided into μ_ν domains, each of which maps into a domain in the ζ-plane bounded by a straight-line segment and a curved (approximately semi-circular) arc. In $C_\nu(\epsilon_1)$ cut along one of the arcs $\mathrm{Re}(\zeta) = $ constant issuing from q_ν, the function $\zeta(w)$ is single-valued and maps adjacent domains onto half-neighborhoods of $\zeta(q_\nu)$ lying on opposite sides of the line $\mathrm{Re}(\zeta) = $ constant through $\zeta(q_\nu)$. Now construct a regular hexagon with center at $\zeta(q_\nu)$ and with two opposite sides a and a' perpendicular to the line $\mathrm{Re}(\zeta) = $ constant through $\zeta(q_\nu)$. The line $\mathrm{Re}(\zeta) = $ constant through $\zeta(q_\nu)$ divides the hexagon into two halves and, if the hexagon is chosen small enough, each one of the μ_ν half-neighborhoods at $\zeta(q_\nu)$ contains one or the other half of the hexagon. Mapped back into the w-plane, the images of the portions of the half-neighborhoods interior to the hexagon together form a neighborhood of q_ν which is interior to $C_\nu(\epsilon_1)$ and which is bounded by a Jordan curve J. Given any one of the arcs $\mathrm{Re}(\zeta) = $ constant issuing from q_ν, it is cut at some point w_1 by the arc I of J which is the image of side a or side a' of the hexagon, as the case may be. Let $I(\delta)$ be the part of I that maps into a segment of length $2\delta > 0$ in the ζ-plane with center at $\zeta(w_1)$, and let p and p' be the end points of $I(\delta)$.

Let J_p and $J_{p'}$ be arcs $\mathrm{Re}(\zeta) = $ constant extending from p and p' into the interior of J and let J_p and $J_{p'}$ terminate when they again meet J. Then J_p plus $J_{p'}$ plus $I(\delta)$ plus part of J complementary to $I(\delta)$ forms a closed Jordan curve enclosing q_ν. Let ϵ_1', $0 < \epsilon_1' < \epsilon_1$, be such that the closure of $C_\nu(\epsilon_1')$ lies inside this curve. Let D be a domain which includes J and the arcs J_p and $J_{p'}$ but lies at a positive distance from $C_\nu(\epsilon_1')$. There is an $\alpha > 0$ such that all points in the ζ-plane within a distance α of the images of J_p and $J_{p'}$ lie in the map of D. There is an ϵ_2', $0 < \epsilon_2' < \epsilon_2$, such that if $\| \mathfrak{A}^* - \mathfrak{A} \| \leqq \epsilon_2'$, then all zeros of $P^*(w)$ which lie in $C_\nu(\epsilon_1)$ are contained in $C_\nu(\epsilon_1')$ and the arcs $\mathrm{Re}(\zeta^*) = $ constant drawn from p and p' into the interior of J and prolonged until they again meet J lie in D, meeting J on the same arcs as do J_p and $J_{p'}$ respectively. Let these arcs be J_p^* and $J_{p'}^*$. Then J_p^* plus $J_{p'}^*$ plus $I(\delta)$ plus part of J complementary to $I(\delta)$ forms a closed Jordan curve containing $C_\nu(\epsilon_1')$ in its interior.

From any point w_0 of $I(\delta)$ prolong the locus $\mathrm{Re}(\zeta^*) = $ constant into the interior of J and extend it until it again meets J. Let this locus be $J_{w_0}^*$. If $J_{w_0}^*$ does not pass through any zero of $P^*(w)$, then it depends continuously on w_0. But as w_0 moves from p to p', $J_{w_0}^*$ would, if it met no zero of $P^*(w)$, move continuously from J_p^* to $J_{p'}^*$ and so it would sweep out all of $C_\nu(\epsilon_1')$. This is impossible; so there is an arc of $\mathbf{\Gamma}_w(\mathfrak{A}^*)$ which meets $I(\delta)$.

If $A_1 = 0$, the same construction may be used in the neighborhood of infinity. If $A_1 \neq 0$, a modification of the above construction may be used or the result can be obtained directly from the power series expansion of $\zeta(w)$ in powers of $1/w^{1/2}$. Since each arc of $\mathbf{\Gamma}_w(\mathfrak{A})$ has at least one end point at a zero of $P(w)$, the statement which we set out to prove follows. It is readily shown from this statement that each point w_2, $w_2 \neq 0$, of $\mathbf{\Gamma}_w(\mathfrak{A})$ is a limit point of points of $\mathbf{\Gamma}_w(\mathfrak{A}^*)$; so $\mathbf{\Gamma}_w(\mathfrak{A})$ is lower semi-continuous.

We suppose next that $A_{n-1} \neq 0$ and show that $\mathbf{\Gamma}_w$ is then continuous at $\mathfrak{A}$. Let all zeros of $P(w)$ lie at a distance λ or more from $w = 0$. Choose ϵ_1, $\epsilon_1 > 0$, so small that the circles $C_\nu(\epsilon_1)$ are disjoint and $\epsilon_1 < \lambda/2$. Let η be a given positive number which is less than $\rho(\lambda/2)$, where $\rho(\alpha)$ is the function defined in Lemma X. Let the domains complementary to $\mathbf{\Gamma}_w(\mathfrak{A})$ be $\boldsymbol{O}_1, \boldsymbol{O}_2, \cdots, \boldsymbol{O}_m$. If $\boldsymbol{O}_\nu$ is a strip domain mapping onto the strip $a_\nu < \mathrm{Re}(\zeta) < b_\nu$, let subdomains of $\boldsymbol{O}_\nu$ be defined as follows:

$$\boldsymbol{O}_\nu(\rho) \text{ maps onto } a_\nu < \mathrm{Re}(\zeta) < b_\nu,\ |\mathrm{Im}(\zeta)| < 1/\rho,$$

$$\boldsymbol{O}_\nu(\rho, \delta) \text{ maps onto } a_\nu + \delta < \mathrm{Re}(\zeta) < b_\nu - \delta,\ |\mathrm{Im}(\zeta)| < 1/\rho.$$

If $\boldsymbol{O}_\nu$ is an end domain mapping onto a half-plane $\mathrm{Re}(\zeta) > a_\nu$,

$$\boldsymbol{O}_\nu(\rho) \text{ maps onto } a_\nu < \mathrm{Re}(\zeta) < a_\nu + 1/\rho,\ |\mathrm{Im}(\zeta)| < 1/\rho,$$

$$\boldsymbol{O}_\nu(\rho, \delta) \text{ maps onto } a_\nu + \delta < \mathrm{Re}(\zeta) < a_\nu + 1/\rho,\ |\mathrm{Im}(\zeta)| < 1/\rho$$

Here δ is some positive number which is less than unity and less than half the width of the narrowest strip domain and $0 < \rho < 1$. Choose ρ so small that the domains $\boldsymbol{O}_\nu(\rho)$ together with $\mathbf{\Gamma}_w(\mathfrak{A})$ cover the set $|w| \geqq \eta$. Then let δ be so small that all points of the ring $\eta \leqq |w| \leqq 1/\eta$ lie either in one of the domains $\boldsymbol{O}_\nu(\rho, \delta)$

or within a distance η of some point of $\mathbf{\Gamma}_w(\mathfrak{A})$. This determines ρ and δ. Let ϵ_1', $0 < \epsilon_1' < \epsilon_1$, be so small that the circles $C_\nu(\epsilon_1')$ lie outside the domains $\boldsymbol{O}_\nu(\rho, \delta/4)$. In particular, since $\epsilon_1' < \epsilon_1$, the circles $C_\nu(\epsilon_1')$ lie at a distance greater than $\lambda/2$ from $w = 0$ for $1 \leqq \nu \leqq k$. Then choose $\epsilon_2' > 0$ such that for $\| \mathfrak{A}^* - \mathfrak{A} \| \leqq \epsilon_2'$ all zeros of $P^*(w)$ lie in the circles $C_\nu(\epsilon_1')$,

$$\left| \frac{d\zeta^*}{dw} - \frac{d\zeta}{dw} \right| = | \{(P^*(w))^{1/2} - (P(w))^{1/2}\}/w | \leqq \frac{\delta}{4L} \tag{5.4.2}$$

in the domains $\boldsymbol{O}_\nu$ $(\rho, \delta/2)$, and any arc $\operatorname{Re}(\zeta^*) =$ constant which enters $| w | \leqq \eta$ from a point in $| w | \geqq \lambda/2$ must tend to $w = 0$ with the modulus of the moving point strictly decreasing. The proof of Lemma X shows that an $\epsilon_2' > 0$ exists which satisfies the last condition. Here L is the length in the w-plane of the longest arc $\operatorname{Re}(\zeta) =$ constant or $\operatorname{Im}(\zeta) =$ constant lying in any domain $\boldsymbol{O}_\nu(\rho, \delta/2)$.

If an arc $\operatorname{Re}(\zeta^*) =$ constant has a point in $\boldsymbol{O}_\nu(\rho, \delta)$, then (5.4.2) shows that it lies in $\boldsymbol{O}_\nu(\rho, \delta/2)$ and crosses $\boldsymbol{O}_\nu(\rho, \delta/2)$, leaving that domain only on an arc $\operatorname{Im}(\zeta) = \pm 1/\rho$. Since these arcs lie in $| w | < \eta$, it follows that no arc of $\mathbf{\Gamma}_w(\mathfrak{A}^*)$ can have a point in $\boldsymbol{O}_\nu(\rho, \delta)$. Thus if $\| \mathfrak{A}^* - \mathfrak{A} \| \leqq \epsilon_2'$, each point of $\mathbf{\Gamma}_w(\mathfrak{A}^*)$ lies in $| w | \leqq \eta$ or in $| w | \geqq 1/\eta$ or within a distance η of some point of $\mathbf{\Gamma}_w(\mathfrak{A})$. Thus as $\| \mathfrak{A}^* - \mathfrak{A} \|$ tends to zero, the set of limit points of $\mathbf{\Gamma}_w(\mathfrak{A}^*)$ as $\| \mathfrak{A}^* - \mathfrak{A} \|$ tends to zero lies in $\mathbf{\Gamma}_w(\mathfrak{A})$. Since the set of limit points includes $\mathbf{\Gamma}_w(\mathfrak{A})$, it must coincide with $\mathbf{\Gamma}_w(\mathfrak{A})$.

On the other hand, suppose that $A_{n-1} = 0$. We show that if w_1 is any point in the plane, there is a sequence of $\mathfrak{A}^*$ such that $\| \mathfrak{A}^* - \mathfrak{A} \|$ tends to zero and w_1 lies in $\mathbf{\Gamma}_w(\mathfrak{A}^*)$. Suppose without loss of generality that $w_1 \neq 0$ and that w_1 is not a point of $\mathbf{\Gamma}_w(\mathfrak{A})$. Let

$$P^*(w) = P(w) + \epsilon/w^m$$

where $A_{m-1} \neq 0$, $A_\nu = 0$ for $m \leqq \nu \leqq n - 1$. There are positive numbers c and ϵ_0 such that the function $P^*(w)$ has a zero $w(\epsilon)$ which satisfies the inequality $c | \epsilon | \leqq | w(\epsilon) | \leqq 2c | \epsilon |$ for $| \epsilon | \leqq \epsilon_0$. The function $w(\epsilon)$ is single-valued and continuous for $| \epsilon | \leqq \epsilon_0$ and, as ϵ rotates through an angle 2π with constant modulus, the value $w(\epsilon)$ rotates through an angle 2π in the same direction. Let $J(\epsilon)$ be the locus $\operatorname{Re}(\zeta^*) =$ constant, $-\infty < \operatorname{Im}(\zeta^*) < \infty$, passing through w_1. If it passes through a zero of $P^*(w)$, then w_1 lies on $\mathbf{\Gamma}_w(\mathfrak{A}^*)$; so suppose that it does not pass through a zero of $P^*(w)$. It then forms a closed Jordan curve passing through $w = 0$.

Now let $\epsilon \neq 0$ increase in argument by 2π, the modulus being held constant. Then $w(\epsilon)$ and $J(\epsilon)$ move continuously and each returns to the starting position. Moreover, $w(\epsilon)$ must lie always inside or always outside $J(\epsilon)$. Suppose that $w(\epsilon)$ lies inside $J(\epsilon)$. Then join w_1 and $w(\epsilon)$ by a path γ which lies inside $J(\epsilon)$ except by the one end point at w_1. As ϵ increases in argument by 2π, the Jordan curve $J(\epsilon)$ never passes through $w = \infty$ (a zero of $P^*(w)$) and so the path γ may be continuously deformed in such a way that it remains inside $J(\epsilon)$. When the argument of ϵ has been increased by 2π, we obtain another path γ' joining w_1 and $w(\epsilon)$ and lying inside $J(\epsilon)$, except for the end point w_1. The two paths γ and γ' together form a Jordan curve whose interior is contained in the interior of $J(\epsilon)$. As we trace out $\gamma + \gamma'$ with internal area to the left, the argument of the moving point increases by 2π which shows that $w = 0$ is interior to $\gamma + \gamma'$ and therefore a fortiori interior to $J(\epsilon)$, a contradiction. If $w(\epsilon)$ remains outside $J(\epsilon)$, then it is readily seen that some point interior to $J(\epsilon)$ must increase its argument by 2π as ϵ rotates through an angle 2π, and we may repeat the above reasoning, obtaining another contradiction. Thus there is some argument of ϵ such that $J(\epsilon)$ passes through a zero of $P^*(w)$. The result follows, and it is seen that the limit of $\mathbf{\Gamma}_w(\mathfrak{A}^*)$ as $\|\mathfrak{A}^* - \mathfrak{A}\|$ approaches zero through all possible sequences is the entire w-plane.

Finally consider the sets $N_w(\mathfrak{A})$. If $\mathfrak{A}^* = \mathfrak{A}(m)$, $m = 1, 2, \cdots$, is any sequence of vectors such that $\|\mathfrak{A}^* - \mathfrak{A}\|$ approaches zero, the limit points of $N_w(\mathfrak{A}^*)$ lie in $\mathbf{\Gamma}_w(\mathfrak{A})$. This may be proved by using a method essentially the same as that used to prove that the limit points of $\mathbf{\Gamma}_w(\mathfrak{A}^*)$ lie in $\mathbf{\Gamma}_w(\mathfrak{A})$ when $A_{n-1} \neq 0$, for $N_w(\mathfrak{A}^*)$ lies in $|w| \geqq 1/4$. We now show that if $\mathfrak{A}^* = \mathfrak{A}(m)$, $m = 1, 2, \cdots$, is a sequence of vectors such that $\|\mathfrak{A}^* - \mathfrak{A}\|$ approaches zero and if $w(m)$ is a point of $N_w(\mathfrak{A}(m))$, then every limit point of $w(m)$ lies in $N_w(\mathfrak{A})$. The point $w(m)$ lies in some subcontinuum of $\mathbf{\Gamma}_w(\mathfrak{A}^*)$ of mapping radius unity and containing the point at infinity. Thus $w(m)$ is connected to infinity by a chain $\gamma(m)$ composed of finitely many arcs of $\mathbf{\Gamma}_w(\mathfrak{A}^*)$ which have their end points at $w(m)$ and zeros of $P^*(w)$. Then for some subsequence of integers m the zeros of $P^*(w)$ approach limits which are zeros of $P(w)$ and $w(m)$ approaches a limit w' which lies in $\mathbf{\Gamma}_w(\mathfrak{A})$. Then the chain of arcs $\gamma(m)$ connecting $w(m)$ to infinity converges to a limit γ' which lies in $\mathbf{\Gamma}_w(\mathfrak{A})$ and connects w' to infinity. The chains $\gamma(m)$ and therefore γ' are continua of mapping radii not less than unity. It follows that w' lies in $N_w(\mathfrak{A})$. Thus $N_w(\mathfrak{A})$ is upper semi-continuous at $\mathfrak{A}$.

Suppose that $N_w(\mathfrak{A})$ is of mapping radius unity. Then $A_1 \neq 0$ and $N_w(\mathfrak{A})$ is a single arc of $\mathbf{\Gamma}_w(\mathfrak{A})$ extending to infinity and having no zero of $P(w)$ on it except perhaps at its end points. Remove an arc of length $\rho > 0$ from the finite end of $N_w(\mathfrak{A})$ and denote the remainder of $N_w(\mathfrak{A})$ by γ_ρ. The method used to prove that $\mathbf{\Gamma}_w$ is lower semi-continuous shows that every point of γ_ρ is a limit point of points of $N_w(\mathfrak{A}^*)$ as $\|\mathfrak{A}^* - \mathfrak{A}\|$ tends to zero. Since ρ is arbitrary, it follows that for every sequence $\mathfrak{A}^*$ such that $\|\mathfrak{A}^* - \mathfrak{A}\|$ tends to zero, every point of $N_w(\mathfrak{A})$ is a limit point of points of $N_w(\mathfrak{A}^*)$. Thus $N_w(\mathfrak{A})$ is continuous at points where it has mapping radius unity.

Suppose that $N_w(\mathfrak{A})$ has a mapping radius less than unity. Consider the function

$$P_1(w) = P(w) + \epsilon/w^{n-1}$$

where $\epsilon > 0$. There are finitely many geodesics in the metric defined by $P_1(w)$ which join all possible pairs of zeros of $P_1(w)$, and each geodesic is the sum of finitely many arcs on each of which $d\zeta_1 = (P_1(w))^{1/2}\, dw/w$ has a constant argument. Thus, for any small ϵ not equal to zero there is an arbitrarily small value of

θ such that there are no arcs $\mathrm{Re}(\zeta^*) =$ constant connecting a pair of zeros of $P^*(w)$ where

$$P^*(w) = e^{i\theta} P_1(w), \qquad \zeta^* = \int (P^*(w))^{1/2} \frac{dw}{w}.$$

Hence all arcs of $\mathbf{\Gamma}_w(\mathfrak{A}^*)$ join a zero of $P^*(w)$ to the point $w = 0$. It follows that $N_w(\mathfrak{A}^*)$ has a mapping radius equal to unity and that $N_w(\mathfrak{A}^*)$ is composed of a single arc not containing any finite zero of $P^*(w)$. A subsequence can be found such that $N_w(\mathfrak{A}^*)$ converges to a limit as ϵ and θ approach zero; the limit is a subcontinuum of $N_w(\mathfrak{A})$ having mapping radius unity and therefore the subcontinuum is not all of $N_w(\mathfrak{A})$. Therefore N_w is not continuous at $\mathfrak{A}$.

The investigation of the set $\mathbf{\Gamma}_z(\mathfrak{B})$ is similar and we shall therefore omit it.

CHAPTER VI

FUNCTIONS WHICH ARE REGULAR IN $|z| < 1$ AND SATISFY THE DIFFERENTIAL EQUATION

6.1. Lemma VIII shows that to every point on the boundary of the nth coefficient region V_n there belongs a function $w = f(z)$ which satisfies a differential equation of the form

$$\left(\frac{z}{w}\frac{dw}{dz}\right)^2 P(w) = Q(z) \tag{6.1.1}$$

where

$$P(w) = P_n(w) = \sum_{\nu=1}^{n-1} \frac{A_\nu}{w^\nu}, \qquad Q(z) = Q_n(z) = \sum_{\nu=-(n-1)}^{n-1} \frac{B_\nu}{z^\nu}. \tag{6.1.2}$$

Here $Q(z)$ is non-negative on $|z| = 1$ with at least one zero there and the coefficient B_0 is positive. Since

$$w = f(z) = z + a_2 z^2 + \cdots$$

near $z = 0$, it is clear that if k is the largest integer such that $B_k \neq 0$, then $A_k = B_k$ and all $A_\nu = 0$ for $\nu > k$ (and conversely). A differential equation of the form (6.1.1) which has these additional properties will be called a $\mathfrak{D}_n$-equation. A $\mathfrak{D}_n$-equation is determined by a pair of vectors $\mathfrak{A} = (A_1, A_2, \cdots, A_{n-1})$, $\mathfrak{B} = (B_1, B_2, \cdots, B_{n-1})$.

Since $Q(z)$ has a zero on $|z| = 1$, it follows that not all B_ν can vanish and therefore not all A_ν are zero. If $w = f(z)$ satisfies a $\mathfrak{D}_n$-equation and is regular in a neighborhood of the origin $z = 0$, it is clear that $f(0) = 0$ and that

$$f'(0) = e^{i(2\pi q/k)}$$

where k and q are integers, $1 \leqq k \leqq n - 1$. We shall make the normalizing assumption that $f'(0) = 1$.

A function $w = f(z)$ will be called a $\mathfrak{D}_n$-function if it is regular in $|z| < 1$, normalized by the condition that $f'(0) = 1$, and if it satisfies some $\mathfrak{D}_n$-equation.

6.2. We shall prove in this and the following chapter that there is a one-one correspondence between $\mathfrak{D}_n$-functions and boundary points of the region V_n. Thus the $\mathfrak{D}_n$-functions are just the extremal functions associated with the coefficients $a_2, a_3, \cdots, a_n$. In this chapter we shall prove only the following theorem:

THEOREM III. *If $w = f(z)$ is a $\mathfrak{D}_n$-function, then it is schlicht in $|z| < 1$ and maps $|z| < 1$ onto the w-plane minus a subcontinuum of Γ_w containing $w = \infty$. By continuation from within $|z| < 1$ the function $f(z)$ is regular on the unit circumference $|z| = 1$ with the exception of finitely many points where it has the local development*

$$f(z) = \sum_{\nu} \alpha_\nu (z - e^{i\varphi})^{\nu/m}$$

where m is a positive integer and only finitely many negative values of ν occur.

PROOF: We suppose without loss of generality that $w = f(z)$ satisfies a $\mathfrak{D}_n$-equation (6.1.1) where $A_{n-1} = B_{n-1} \neq 0$. If $n = 2$, explicit integration of (6.1.1) shows that if w is regular in $|z| < 1$, then

$$w = \frac{z}{(1 - e^{i\varphi} z)^2}.$$

Thus we may suppose that $n \geqq 3$.

Let

$$Q(z) = \frac{B_{n-1}}{z^{n-1}} (z - z_1)\left(z - \frac{1}{\bar{z}_1}\right)(z - z_2)\left(z - \frac{1}{\bar{z}_2}\right) \cdots (z - z_{n-1})\left(z - \frac{1}{\bar{z}_{n-1}}\right),$$

$$P(w) = \frac{A_{k-1}}{w^{n-1}} (w - w_1)(w - w_2) \cdots (w - w_{n-k}), \qquad k \geqq 2.$$

Integrating both sides of (6.1.1), we see that

$$\int (P(w))^{1/2} \frac{dw}{w} = \int (Q(z))^{1/2} \frac{dz}{z}. \tag{6.2.1}$$

The common value of these two integrals will be denoted by ζ.

We prove first that $w = f(z)$ has a unique limit as z approaches a point z_0 on $|z| = 1$ from the interior.

Let z_0 be a point on $|z| = 1$ and suppose first that $Q(z_0) \neq 0$. Then there is an open set g which contains all points of $|z| < 1$ within a distance δ, $\delta > 0$, of z_0 and which maps by

$$\zeta = \int (Q(z))^{1/2} \frac{dz}{z}$$

into a semi-circle γ in the ζ-plane. Without loss of generality we may suppose that γ is the semi-circle $\mathrm{Re}(\zeta) > 0$, $|\zeta| < \tau$. Take δ so small that the radius τ of γ is less than the width of any strip in the ζ-plane which is the image of a strip domain belonging either to $\mathbf{\Gamma}_w$ or to $\mathbf{\Gamma}_z$. Choosing δ still smaller if necessary, we may suppose that $Q(z)$ has no zero inside or on the boundary of g. Then g lies entirely in a single domain $\boldsymbol{O}_z^{(\nu)}$ in the z-plane. The branch of $\zeta(z)$ which has been selected vanishes at $z = z_0$ and is regular, single-valued, and schlicht in g. The inverse function $z(\zeta)$ is then regular and schlicht in γ and by reflection and inversion is regular and schlicht throughout $|\zeta| < \tau$. It is clear from the $\mathfrak{D}_n$-equation that $w = f(z)$ has no zero in $|z| < 1$ except the simple zero at $z = 0$ and so $f(z)$ has no zero in g.

When ζ lies in γ, the function $z(\zeta)$ is regular and its values lie in g. Hence the function $w(z(\zeta)) = f(z(\zeta))$ is regular in γ. The relation

$$\zeta = \int (P(w))^{1/2} \frac{dw}{w} \tag{6.2.2}$$

then shows that $P(w)$ does not vanish for any w corresponding to a point ζ of γ, and so $\zeta(w)$ is a regular function of w. Since the radius τ of γ is less than the width of any strip domain belonging to $\mathbf{\Gamma}_w$, it follows that γ lies in the map of at most two strip domains in the w-plane. By making τ small, the semi-circle γ will then lie in the map of a single strip or end domain in the ζ-plane. Since γ is a bounded domain, $w(\zeta)$ does not have the limit zero in γ. Thus $1/w(\zeta)$ is continuous in the closure of γ. Let the point $\zeta = 0$, which is the center of the semi-circle γ, map into the point w_0 in the w-plane, where $w_0 \neq 0$.

Hence if z_0 is a point on $|z| = 1$ for which $Q(z_0) \neq 0$ and if w_0 is not one of the points $\infty, w_1, w_2, \cdots, w_{n-k}$, then $w = f(z)$ is regular and schlicht in a neighborhood of z_0, and in this neighborhood we have

$$w - w_0 = \sum_{\nu=1}^{\infty} c_\nu (z - z_0)^\nu, \qquad c_1 \neq 0. \tag{6.2.3}$$

On the other hand, if w_0 is one of the points w_i and if $P(w)$ has a zero of precise order λ at w_i, then

$$\zeta - \zeta_0 = \sum_{\nu=1}^{\infty} \alpha_\nu (z - z_0)^\nu, \qquad \alpha_1 \neq 0,$$

$$w - w_i = \sum_{\nu=1}^{\infty} \beta_\nu (\zeta - \zeta_0)^{2\nu/(\lambda+2)}, \qquad \beta_1 \neq 0,$$

and so

$$w - w_i = \sum_{\nu=2}^{\infty} c_\nu (z - z_0)^{\nu/(\lambda+2)}, \qquad c_2 \neq 0. \tag{6.2.4}$$

If $w_0 = \infty$, then

$$\zeta - \zeta_0 = \sum_{\nu=1}^{\infty} \alpha_\nu (z - z_0)^\nu,$$

$$w = \sum_{\nu=-1}^{\infty} \beta_\nu (\zeta - \zeta_0)^{2\nu/(k-1)},$$

and so

$$w = \sum_{\nu=-2}^{\infty} c_\nu (z - z_0)^{\nu/(k-1)}, \qquad c_{-2} \neq 0. \tag{6.2.5}$$

Suppose now that $Q(z)$ has a zero of precise order λ, $\lambda > 0$, at the point z_0 on $|z| = 1$. Then λ is even and there are $\lambda + 2$ arcs $\mathrm{Re}(\zeta) = $ constant emerging from the point z_0. Two of these are arcs of $|z| = 1$ and $\lambda/2$ of them enter into $|z| < 1$. These $\lambda/2 + 2$ arcs form $\lambda/2 + 1$ angular openings at z_0, each angle being equal to $2\pi/(\lambda + 2)$. Let g be the open set consisting of all points in one of these openings within a distance $\tau > 0$ of z_0, distance being measured in the ζ-plane along geodesics. The image of g by the mapping

$$\zeta = \int (Q(z))^{1/2} \frac{dz}{z}$$

will be a semi-circle γ in the ζ-plane of radius τ. If δ is taken sufficiently small, g will lie in a single domain $O_z^{(\nu)}$ and the semi-circle γ will lie entirely inside the image of a single domain $O_w^{(\mu)}$ in the w-plane. The point z_0 maps into the center of the semi-circle γ and the function $w(\zeta)$ is continuous in the closure of γ. Let w_0 be the point in the w-plane corresponding to the center ζ_0 of γ.

If w_0 is a zero of $P(w)$ of order μ, $\mu \geqq 0$, then

$$\zeta - \zeta_0 = \sum_{\nu=0}^{\infty} \alpha_\nu (z - z_0)^{((\lambda+2)/2)+\nu}, \qquad \alpha_0 \neq 0,$$

$$w - w_0 = \sum_{\nu=1}^{\infty} \beta_\nu (\zeta - \zeta_0)^{2\nu/(\mu+2)}, \qquad \beta_1 \neq 0,$$

and so

$$w - w_0 = \sum_{\nu=\lambda+2}^{\infty} c_\nu (z - z_0)^{\nu/(\mu+2)}, \qquad c_{\lambda+2} \neq 0. \tag{6.2.6}$$

If $w_0 = \infty$, then

$$\zeta - \zeta_0 = \sum_{\nu=0}^{\infty} \alpha_\nu (z - z_0)^{((\lambda+2)/2)+\nu}, \qquad \alpha_0 \neq 0,$$

$$w = \sum_{\nu=-1}^{\infty} \beta_\nu (\zeta - \zeta_0)^{2\nu/(k-1)}, \qquad \beta_{-1} \neq 0,$$

and so

$$w = \sum_{\nu=-(\lambda+2)}^{\infty} c_\nu (z - z_0)^{\nu/(k-1)}, \qquad c_{-\lambda-2} \neq 0. \tag{6.2.7}$$

Formulas (6.2.3)–(6.2.7), proved for z in a domain g near z_0, hold in a complete neighborhood of z_0. A finite number of circles, in each of which one of these series converges, covers the unit circumference $|z| = 1$. Thus $f(z)$ is regular on $|z| = 1$ except at finitely many points and $|f(z)|$ has a positive lower bound in some ring $1 - \epsilon \leqq |z| \leqq 1$. From the $\mathfrak{D}_n$-equation it is seen that the only zero of $f(z)$ in $|z| < 1$ is at the origin $z = 0$.

As z describes the unit circumference $|z| = 1$, the point $\zeta(z)$ moves along a locus $\mathrm{Re}(\zeta) =$ constant and $w = f(z)$ moves on a locus L in the w-plane where the values of $f(z)$ on $|z| = 1$ are those obtained by continuation from within $|z| < 1$. The locus L does not contain the point $w = 0$. If w_0 is any point of $\mathbf{\Gamma}_w$ other than the origin $w = 0$, then $\mathbf{\Gamma}_w$ contains all arcs $\mathrm{Re}(\zeta) =$ constant emerging from w_0. Thus if L has one point in common with $\mathbf{\Gamma}_w$, it is a subset of $\mathbf{\Gamma}_w$.

If L has no point in common with $\mathbf{\Gamma}_w$, it lies in a single domain $\mathbf{O}_w^{(\nu)}$ and is a Jordan arc which may be traced several times as z describes the unit circumference. In either case L is a connected set of arcs not containing $w = 0$; so Lemma XVI shows that no subset of L can form a closed Jordan curve (on the Riemann sphere) and the complement of L is therefore simply-connected. The closure of the complement of L is the entire w-plane.

The function

$$g(z) = \frac{1}{f(z)} = \frac{1}{z} + \sum_{\nu=0}^{\infty} b_\nu z^\nu \tag{6.2.8}$$

is continuous in the closed unit circle except in a neighborhood of the origin, where it has a simple pole with residue 1. As z describes $|z| = 1$, the point $s = g(z)$ moves on a locus L' whose complement D is a simply-connected domain. As in the proof of Lemma III it can be shown that $g(z)$ maps $|z| < 1$ onto D in a one–one manner. Then $w = f(z)$ is schlicht in $|z| < 1$. Since $f(z)$ is regular there, L must contain the point $w = \infty$ and is therefore a subset of $\mathbf{\Gamma}_w$. This completes the proof of Theorem III.

CHAPTER VII

THE LENGTH-AREA PRINCIPLE. TEICHMÜLLER'S METHOD

7.1. Lemma VIII shows that to every point on the boundary of V_n there belongs a $\mathfrak{D}_n$-function. We shall now prove the converse of this result, namely that every $\mathfrak{D}_n$-function belongs to a boundary point of V_n. This will follow from a method due to Teichmüller [27]. Teichmüller's method turns on an important principle in conformal mapping which we may call the "length-area principle." This principle involves an application of the Schwarz inequality to the derivative of a mapping function in such a way that an upper bound is obtained for the square of a length (or of an average length) in terms of an area or product of areas. That is, an inequality of the form

$$\left(\iint |g'| \, dx \, dy\right)^2 \leqq \iint |g'|^2 \, dx \, dy \cdot \iint dx \, dy$$

is always involved implicitly or explicitly, where $g(z)$, $z = x + iy$, is the mapping function. This type of inequality is of fundamental importance because it is an expression in the large of the defining property of a conformal map that the local distortion in length is independent of the direction at the point. An application of the inequality has already been made in **2.2.** This principle in various forms has been used effectively by several writers including Hurwitz, Courant, Faber, Grötzsch, Ahlfors, and Beurling.

7.2. Let

$$P(w) = \frac{A_k}{w^k} + \frac{A_{k+1}}{w^{k+1}} + \cdots + \frac{A_{n-1}}{w^{n-1}} \tag{7.2.1}$$

where $A_{n-1} \neq 0$, $A_k \neq 0$, $n \geqq 3$, $1 \leqq k \leqq n - 1$, and let the set of arcs $\mathbf{\Gamma}_w$ be defined as in Chapter IV.

LEMMA XXIII. *Let $\mathbf{C}_w$ be a subcontinuum of $\mathbf{\Gamma}_w$ containing the point $w = \infty$ and having mapping radius unity. If*

$$\psi(w) = w + \alpha w^n + \beta w^{n+1} + \cdots \tag{7.2.2}$$

is regular and schlicht in the complement of $\mathbf{C}_w$, then

$$\operatorname{Re}(\alpha A_{n-1}) \leqq 0 \tag{7.2.3}$$

with equality if and only if $\psi(w) = w$.

PROOF: Let $w = 1/s^2$. Then the system of arcs $\mathbf{\Gamma}_w$ is mapped into a system $\mathbf{\Gamma}_s$ in the s-plane and the set $\mathbf{C}_w$ is mapped into a set $\mathbf{C}_s$. Apart from an irrelevant factor -2 the function

$$\zeta = \int (P(w))^{1/2} \frac{dw}{w}$$

becomes

$$\zeta = \int (N(s))^{1/2}\, ds \tag{7.2.4}$$

where

$$N(s) = A_k s^{2k-2} + A_{k+1} s^{2k} + \cdots + A_{n-1} s^{2n-4}. \tag{7.2.5}$$

The set $\mathbf{C}_s$ is a subcontinuum of $\mathbf{\Gamma}_s$ containing $s = 0$ and of exterior mapping radius unity. The function

$$S = [\psi(s^{-2})]^{-1/2} = (s^{-2} + \alpha s^{-2n} + \cdots)^{-1/2} = s - \frac{\alpha}{2} s^{-2n+3} + \cdots \tag{7.2.6}$$

is regular and schlicht in the complement of $\mathbf{C}_s$ except at $s = \infty$, where it has a simple pole of residue 1. For large s

$$\pm\zeta = \frac{(A_{n-1})^{1/2}}{n-1} s^{n-1} \left\{1 + \frac{\tau_1}{s^2} + \cdots\right\} + (a + ib) \log s + c. \tag{7.2.7}$$

We choose one of the arcs of $\mathbf{\Gamma}_s$ that extends to $s = \infty$ and designate it by λ. If σ_0 is sufficiently large, then $\zeta(s)$ is regular and single-valued in the simply-connected domain formed by cutting the circular domain $|s| > \sigma_0$ along the arc λ. Beginning at a point t_1 on the arc λ, $|s_1|$ large, we trace the locus

$$|\zeta| = \left| \int (N(s))^{1/2}\, ds \right| = \rho = \text{constant}$$

with argument of ζ increasing. This locus can be continued until it returns to a point t_2 on the arc λ. Then the subarc (t_1, t_2) of λ plus the locus $|\zeta| = \rho$ with end points at t_1 and t_2 forms a closed Jordan curve J_ρ in the s-plane which lies in $|s| > \sigma_0$ if $|t_1|$ is sufficiently large. Let γ_ρ be the portion of $\mathbf{\Gamma}_s$ that lies in the interior of J_ρ. For large ρ, $\mathbf{C}_s$ is a subset of γ_ρ. Cut each strip domain $\mathbf{O}_s^{(\nu)}$ into two subdomains by a cross-cut $\operatorname{Im}(\zeta) = \text{constant}$ and let I_ν be the cross-cut in $\mathbf{O}_s^{(\nu)}$.

The arcs I_ν and the set $\mathbf{\Gamma}_s$ together form a connected set in the plane minus the point $s = \infty$. For let $\tau_1, \tau_2, \cdots, \tau_k$ be the arcs of $\mathbf{\Gamma}_s$ that tend to $s = \infty$ and let them be arranged in order of increasing argument in a neighborhood of $s = \infty$. The opening between τ_i and τ_{i+1} belongs to a strip or end domain. In the latter case the arcs τ_i and τ_{i+1} belong to the same component of $\mathbf{\Gamma}_s$. In the former τ_i and τ_{i+1} lie on the boundary of a strip domain and the two bounding edges of the strip domain are connected by an arc I_ν; so τ_i and τ_{i+1} are connected in the set $\mathbf{\Gamma}_s + \Sigma I_\nu$. Since each point of $\mathbf{\Gamma}_s$ is connected in $\mathbf{\Gamma}_s$ to one of the arcs τ_i, it follows that the set $\mathbf{\Gamma}_s + \Sigma I_\nu$ is connected.

Let ρ be so large that J_ρ contains all zeros of $N(s)$ and all arcs I_ν in its interior. Then the set $\gamma_\rho + \lambda + \Sigma I_\nu$ is connected and its complement is therefore a simply-connected domain D_ρ in the s-plane. The function $\zeta(s)$ is regular and single-valued in D_ρ. As s moves from t_1 to t_2 around the portion of J_ρ that does

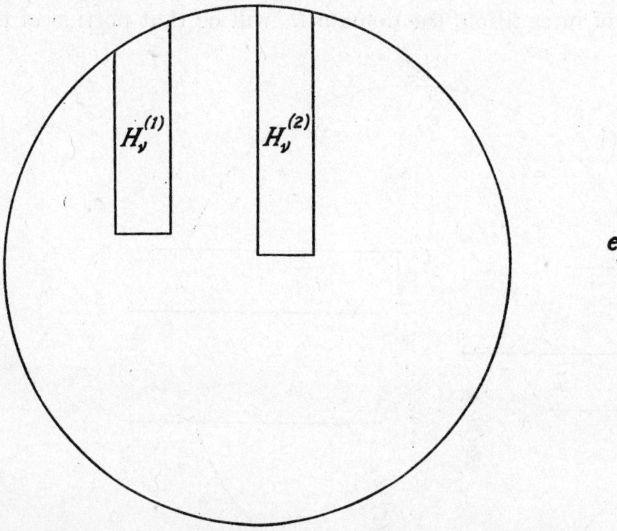

FIG. 1

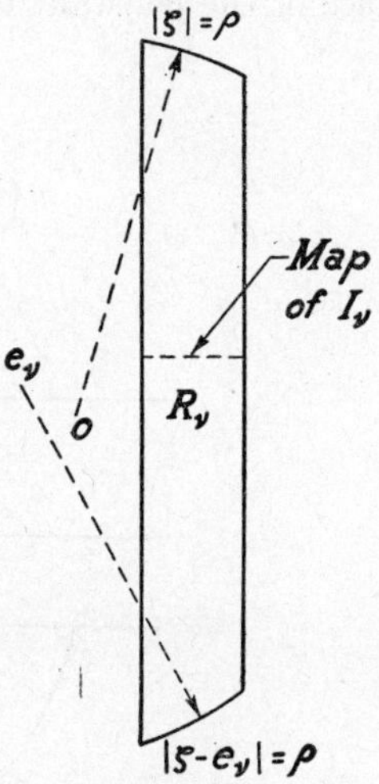

FIG. 2

not lie on λ, the point $\zeta(s)$ moves several times around the circle $|\zeta| = \rho$, although it does not necessarily return to its starting point. If $O_s^{(\nu)}$ is an end domain, the part of $O_s^{(\nu)}$ that lies inside J_ρ is mapped in a one–one way onto a domain

$$|\zeta| < \rho, \mathrm{Re}(\zeta) > c_\nu,$$

or a domain

$$|\zeta| < \rho, \mathrm{Re}(\zeta) < c_\nu.$$

In this mapping the point s remains inside D_ρ. We observe that the number c_ν is independent of ρ.

Now let $O_s^{(\nu)}$ be a strip domain and let the point s remain in D_ρ. Since $O_s^{(\nu)}$ is divided into two domains by an arc I_ν on which $\mathrm{Im}(\zeta) =$ constant, we see that the portion of $O_s^{(\nu)}$ inside J_ρ and cut along I_ν maps into two domains in the ζ-plane, and each domain is of one of the following forms:

$$a_\nu < \mathrm{Re}(\zeta) < b_\nu, \mathrm{Im}(\zeta) > d_\nu, |\zeta| < \rho,$$

or

$$a_\nu < \mathrm{Re}(\zeta) < b_\nu, \mathrm{Im}(\zeta) < d_\nu, |\zeta| < \rho.$$

The numbers a_ν, b_ν, and d_ν are independent of ρ. The two domains in the ζ-plane have the same width, and we denote them by $H_\nu^{(1)}$, $H_\nu^{(2)}$ (see Figure 1).

If we remove the cross-cut I_ν from $O_s^{(\nu)}$, then the portion of $O_s^{(\nu)}$ which lies inside J_ρ is mapped into a single domain R_ν. The domain R_ν, illustrated in Figure 2, is obtained by fitting together the domains $H_\nu^{(1)}$, $H_\nu^{(2)}$ along the portions corresponding to I_ν. This domain is bounded by two vertical lines and by two circular arcs, each of radius ρ but not necessarily concentric. By suitable choice of the constant of integration, the domain R_ν will be that portion of the

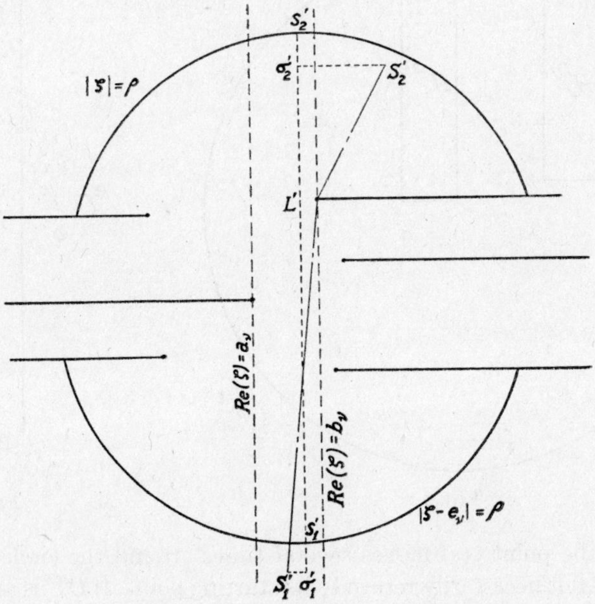

FIG. 3

Map of a strip domain and its extension. The strip domain maps into the region between the lines $\mathrm{Re}(\zeta) = a_\nu$ and $\mathrm{Re}(\zeta) = b_\nu$. The extended domain maps into the entire plane cut by the horizontal lines shown solid.

strip $a_\nu < \mathrm{Re}(\zeta) < b_\nu$ which lies under the arc $|\zeta| = \rho$ and above the arc $|\zeta - e_\nu| = \rho$. The number e_ν and the width of the strip are independent of ρ.

We shall denote an arc $\mathrm{Re}(\zeta) =$ constant, $-\infty < \mathrm{Im}(\zeta) < \infty$, by the letter L. An arc L which is not part of Γ_s must lie entirely in some strip or end domain $O_s^{(\nu)}$. Let L cut J_ρ at the two points s_1 and s_2, and we suppose that $\mathrm{Im}\,\zeta(s_1) < \mathrm{Im}\,\zeta(s_2)$ when $\mathrm{Im}(\zeta)$ is computed by moving along L from s_1 to s_2. Let S_1 and S_2 respectively be the points into which s_1 and s_2 are mapped by the function (7.2.6). Then the points S_1, S_2, s_1, s_2 lie in the extended domain $\Omega_s^{(\nu)}$ of $O_s^{(\nu)}$. Let an arc $\mathrm{Im}(\zeta) =$ constant be drawn in $\Omega_s^{(\nu)}$ from the point S_1 and let it meet the arc L at a point σ_1. Likewise let an arc $\mathrm{Im}(\zeta) =$ constant drawn in $\Omega_s^{(\nu)}$ from S_2 meet L at a point σ_2. The images in the ζ-plane of the points s_1, S_1, σ_1, s_2, S_2, σ_2 will be denoted by s_1', S_1', σ_1', s_2', S_2', σ_2'. The construction in the ζ-plane is illustrated in Figure 3 for a strip domain and in Figure 4 for an end domain. An end domain $O_s^{(\nu)}$ has an asymptotic angle at infinity equal to $\pi/(n-1)$ while its extension $\Omega_s^{(\nu)}$ has an asymptotic angle at infinity equal to

$2\pi/(n-1)$ since $\boldsymbol{\Omega}_s^{(\nu)}$ has an additional angular opening $\pi/2(n-1)$ on either side of $\boldsymbol{O}_s^{(\nu)}$. A strip domain $\boldsymbol{O}_s^{(\nu)}$ has two angles at infinity, each asymptotically equal to zero. Its extended domain has two angles at infinity, each asymptotic to $\pi/(n-1)$ with $\pi/2(n-1)$ on either side of the corresponding zero angle of the strip domain. Since J_ρ lies in $|s| > \sigma_0$, it is clear that for large σ_0 the extended domains must include the points S_1 and S_2 whether $\boldsymbol{O}_s^{(\nu)}$ is a strip or an end domain.

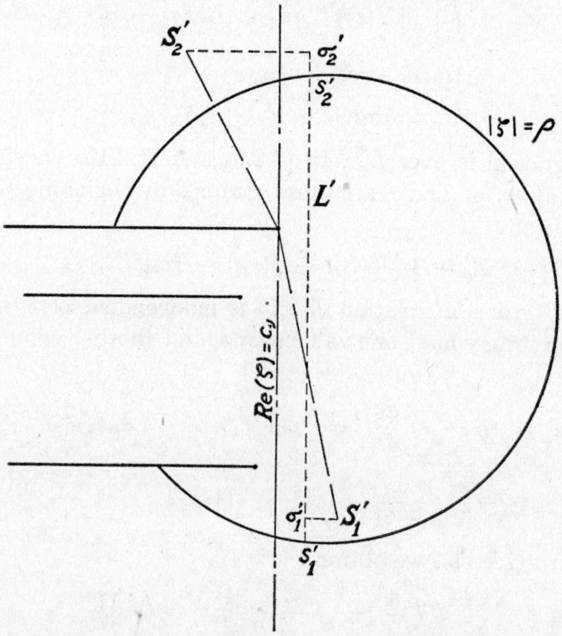

FIG. 4

Map of an end domain and its extension. The end domain maps into the half-plane $\text{Re}(\zeta) > c_\nu$. The extended domain maps into the entire plane cut along the horizontal lines drawn solid.

The polygonal line lying in the map of the extended domain and joining S_1' and S_2' is the image of the geodesic joining S_1 and S_2. The length of this line is the geodesic distance from S_1 to S_2 and we have

$$(7.2.8)\qquad \int_{S_1}^{S_2} |(N(s))^{1/2}\,ds| \geqq \int_{\sigma_1}^{\sigma_2} |(N(s))^{1/2}\,d(s)|$$
$$= \text{Im}\{\zeta(\sigma_2) - \zeta(\sigma_1)\} = \text{Im}\{\sigma_2' - \sigma_1'\}.$$

The integration from σ_1 to σ_2 is along the line L on which s_1 and s_2 lie. This arc L is plainly the geodesic from σ_1 to σ_2. The integration from S_1 to S_2 is over the geodesic, but then the inequality (7.2.8) is a fortiori true for any admissible path joining S_1 and S_2.

Let L_ρ be the portion of L joining s_1 and s_2; L_ρ lies inside $\boldsymbol{J}_\rho$ except for its end points. Let $\mathfrak{L}_\rho$ be the length of the map of L_ρ in the ζ-plane. Then

$$\mathfrak{L}_\rho = \int_{s_1}^{s_2} |(N(s))^{1/2}\,ds| = \operatorname{Im}(s_2' - s_1') \tag{7.2.9}$$

where the integration is over L_ρ. The arc L_ρ maps into an arc L_ρ^* in the S-plane under the mapping (7.2.6) and L_ρ^* joins S_1 and S_2. If $\mathfrak{L}_\rho^*$ is the length of the image of L_ρ^* in the ζ-plane, we have, using (7.2.8),

$$\begin{aligned}\mathfrak{L}_\rho^* &= \int_{S_1}^{S_2} |(N(s))^{1/2}\,ds| \geqq \operatorname{Im}(\sigma_2' - \sigma_1') \\ &= \operatorname{Im}(s_2' - s_1') + \operatorname{Im}(\sigma_2' - s_2' - \sigma_1' + s_1') \\ &= \mathfrak{L}_\rho + \operatorname{Im}(S_2' - s_2' - S_1' + s_1')\end{aligned} \tag{7.2.10}$$

where the integration is over L_ρ^*. In (7.2.9) and (7.2.10) we suppose that the values of $\operatorname{Im}(\zeta)$ at s_2, σ_1, and σ_2 are those obtained by beginning at s_1 and traversing the arc L. Then

$$\mathfrak{L}_\rho^* \geqq \mathfrak{L}_\rho + \operatorname{Im}[\zeta(S_2) - \zeta(s_2)] - \operatorname{Im}[\zeta(S_1) - \zeta(s_1)]. \tag{7.2.11}$$

The constant term c in relation (7.2.7) is independent of ρ, but in case $O_s^{(\nu)}$ is a strip domain it may have one value near s_1 and another value near s_2. In any case we have

$$\begin{aligned}\zeta(S) - \zeta(s) &= (S - s)\frac{d\zeta}{ds}\{1 + o(1)\} = \pm\frac{\alpha}{2}(A_{n-1})^{1/2}s^{-n+1}\{1 + o(1)\} \\ &= -\frac{\alpha}{2}\frac{A_{n-1}}{n-1}\frac{1}{\zeta(s)}\{1 + o(1)\}.\end{aligned}$$

Substituting into (7.2.11), we obtain

$$\begin{aligned}\mathfrak{L}_\rho^* &\geqq \mathfrak{L}_\rho + \operatorname{Im}\left\{-\frac{\alpha A_{n-1}}{2n-2}\frac{1}{s_2'} + \frac{\alpha A_{n-1}}{2n-2}\frac{1}{s_1'} + o\left(\frac{1}{\rho}\right)\right\} \\ &= \mathfrak{L}_\rho + \operatorname{Im}\left\{\frac{\alpha A_{n-1}}{2n-2}\frac{s_2' - s_1'}{s_1's_2'}\right\} + o\left(\frac{1}{\rho}\right).\end{aligned}$$

Now $s_2' - s_1' = i\mathfrak{L}_\rho$ and the points s_1' and s_2' each lie on circular arcs of radius ρ which have fixed centers as ρ becomes large. Also s_2' differs from the complex conjugate of s_1' by a constant which depends on the domain $O_s^{(\nu)}$ but is independent of ρ. Hence we have

$$\mathfrak{L}_\rho^* \geqq \mathfrak{L}_\rho + \frac{\gamma\mathfrak{L}_\rho}{\rho^2} + o\left(\frac{1}{\rho}\right), \tag{7.2.12}$$

where

$$\gamma = \operatorname{Re}\left(\frac{\alpha A_{n-1}}{2n-2}\right). \tag{7.2.13}$$

These relations are true whether $O_s^{(\nu)}$ is a strip or an end domain. The term $o(1/\rho)$ in (7.2.12) holds uniformly as ρ becomes large for all arcs L not belonging to $\boldsymbol{\Gamma}_s$ that cut J_ρ.

Let $\epsilon > 0$ be given. Then for sufficiently large ρ we have

$$\left[\mathfrak{L}_\rho\left(1+\frac{\gamma}{\rho^2}\right)-\frac{\epsilon}{\rho}\right]^2 \leqq (\mathfrak{L}_\rho^*)^2 = \left\{\int_{S_1}^{S_2} |\,(N(S))^{1/2}\,dS\,|\right\}^2$$

$$= \left\{\int_{s_1}^{s_2} \left|\,(N\,(S(s)))^{1/2}\,\frac{dS}{ds}\,ds\,\right|\right\}^2$$

where the last integration is over L_ρ. Since $d\zeta = (N(s))^{1/2}\,ds$, this becomes

$$\left[\mathfrak{L}_\rho\left(1+\frac{\gamma}{\rho^2}\right)-\frac{\epsilon}{\rho}\right]^2 \leqq \left\{\int\left|\left(\frac{N(S)}{N(s)}\right)^{1/2}\frac{dS}{ds}\,d\zeta\,\right|\right\}^2 \tag{7.2.14}$$

where $s = s(\zeta)$ and the integration is along the image of L_ρ in the ζ-plane. Then the inequality of Schwarz gives

$$\mathfrak{L}_\rho^2\left(1+\frac{\gamma}{\rho^2}\right)^2-\frac{2\epsilon}{\rho}\,\mathfrak{L}_\rho\left(1+\frac{\gamma}{\rho^2}\right) \leqq \left\{\int |\,d\zeta\,|\right\}\cdot\left\{\int\left|\,\frac{N(S)}{N(s)}\left(\frac{dS}{ds}\right)^2\,d\zeta\,\right|\right\}.$$

The first integral on the right is equal to $\mathfrak{L}_\rho$; dividing by $\mathfrak{L}_\rho$ and writing $\zeta = u + iv$, we obtain

$$\mathfrak{L}_\rho\left(1+\frac{\gamma}{\rho^2}\right)^2-\frac{2\epsilon}{\rho}\left(1+\frac{\gamma}{\rho^2}\right) \leqq \int\left|\,\frac{N(S)}{N(s)}\left(\frac{dS}{ds}\right)^2\,dv\,\right|. \tag{7.2.15}$$

Suppose that $O_s^{(\nu)}$ is the strip domain $a_\nu < \operatorname{Re}(\zeta) < b_\nu$. Each arc L in $O_s^{(\nu)}$ cuts J_ρ in two points s_1 and s_2. These points depend upon L and inequality (7.2.15) is valid for each line L, the integration being over the image of L in the ζ-plane. Thus (7.2.15) holds for $a_\nu < u < b_\nu$. Integrating both sides with respect to u, we obtain

$$\int_{a_\nu}^{b_\nu}\left\{\mathfrak{L}_\rho\left(1+\frac{\gamma}{\rho^2}\right)^2-\frac{2\epsilon}{\rho}\left(1+\frac{\gamma}{\rho^2}\right)\right\}du \leqq \iint\left|\,\frac{N(S)}{N(s)}\left(\frac{dS}{ds}\right)^2\,du\,dv\,\right|. \tag{7.2.16}$$

The integration on the right is over a domain R_ν in the ζ-plane which is the map of the portion of $O_s^{(\nu)}$ inside J_ρ when the cross-cut I_ν is removed. We may transform the integral on the right into one in the s-plane, obtaining

$$\int_{a_\nu}^{b_\nu}\left\{\mathfrak{L}_\rho\left(1+\frac{\gamma}{\rho^2}\right)^2-\frac{2\epsilon}{\rho}\left(1+\frac{\gamma}{\rho^2}\right)\right\}du \leqq \iint\left|\,N(S)\left(\frac{dS}{ds}\right)^2\,da\,\right|.$$

Here da is the element of area in the s-plane and the integration on the right is over the portion of $O_s^{(\nu)}$ lying inside J_ρ. The function $S(s)$ is schlicht in $O_s^{(\nu)}$; so we have

$$\int_{a_\nu}^{b_\nu}\left\{\mathfrak{L}_\rho\left(1+\frac{\gamma}{\rho^2}\right)^2-\frac{2\epsilon}{\rho}\left(1+\frac{\gamma}{\rho^2}\right)\right\}du \leqq \iint |\,N(S)\,|\,dA \tag{7.2.17}$$

where the integration on the right is over the image domain in the S-plane of the portion of $O_s^{(\nu)}$ inside J_ρ and dA is the element of area in the S-plane. Now $\mathfrak{L}_\rho$ is the length of the map in the ζ-plane of the portion of L between s_1 and s_2 ; so $\mathfrak{L}_\rho$ is the length of a vertical chord of the domain R_ν . Thus $\int \mathfrak{L}_\rho \, du$ is the area of the domain R_ν , the image in the ζ-plane of the portion of $O_s^{(\nu)}$ inside J_ρ . Since

$$\int du < \frac{3\rho}{2}, \qquad \left(1 + \frac{\gamma}{\rho^2}\right) < \frac{4}{3}$$

for large ρ, we have

$$(7.2.18) \qquad \left(1 + \frac{\gamma}{\rho^2}\right)^2 \mathfrak{A}_\nu - 4\epsilon \leqq \iint | N(S) | \, dA$$

where $\mathfrak{A}_\nu$ is the area of the map in the ζ-plane of the part of $O_s^{(\nu)}$ that lies inside J_ρ . The inequality (7.2.18) may be written

$$(7.2.19) \qquad \left(1 + \frac{\gamma}{\rho^2}\right)^2 \iint | N(s) | \, da - 4\epsilon \leqq \iint | N(S) | \, dA$$

where the integration on the left is over the portion of $O_s^{(\nu)}$ inside J_ρ and the integration on the right is over the map of this domain in the S-plane.

If $O_s^{(\nu)}$ is an end domain, we obtain an inequality that differs from (7.2.17) in having one of the limits of integration on the left a function of ρ. We then obtain inequality (7.2.19), where the integration on the left is over that part of $O_s^{(\nu)}$ which lies inside J_ρ and the integration on the right is over the image of this domain in the S-plane.

If there are k end and strip domains in all, we add the k inequalities (7.2.19) and obtain

$$\left(1 + \frac{\gamma}{\rho^2}\right)^2 \iint | N(s) | \, da \leqq \iint | N(S) | \, dA + 4k\epsilon.$$

The integration on the left is over the portion of each strip and end domain lying inside J_ρ . It is then numerically equal to the integral over the interior of J_ρ . Let (7.2.6) map J_ρ into the closed Jordan curve J_ρ^* in the S-plane. The integral on the right of the above inequality is over a portion of the interior of J_ρ^*; so the inequality is a fortiori true if the integration on the right is extended over the whole of the interior of J_ρ^* . Replacing S by s on the right, we obtain

$$(7.2.20) \qquad \frac{2\gamma}{\rho^2} \iint_G | N(s) | \, da \leqq 4k\epsilon + \iint_{G^*} | N(s) | \, da - \iint_G | N(s) | \, da$$

where G is the interior of J_ρ and G^* is the interior of J_ρ^* . Each integral represents an area in the ζ-plane.

Let r be so large that the circle $| s | < r$ contains all zeros of $N(s)$ in its interior and the arc λ on which the points t_1 and t_2 lie cuts $| s | = r$ in precisely one point. In the domain $| s | > r$ cut along λ from a point of $| s | = r$ to $s = \infty$, the function $\zeta(s)$ is single-valued and we write

$$(7.2.21)\qquad \zeta(s) = \frac{(A_{n-1})^{1/2}}{n-1}\, s^{n-1}\left\{1 + \frac{\tau_1}{s^2} + \cdots\right\} + (a + ib)\log s + c.$$

The domain $|s| > r$ is cut by precisely $2n - 2$ arcs $\text{Im}(\zeta) = 0$, each of which extends from some point of $|s| = r$ to $s = \infty$ and intersects the circle $|s| = r$ at precisely one point. If ρ is large enough, these $2n - 2$ arcs together with the arc λ divide the ring-shaped region between $|s| = r$ and J_ρ into $2n - 1$ subdomains. Of these there are $2n - 3$, each of which is bounded by two arcs $\text{Im}(\zeta) = 0$, a portion of $|s| = r$, and a part of J_ρ. Each of these $2n - 3$ domains is mapped by (7.2.21) into a domain bounded by two arcs $\text{Im}(\zeta) = 0$, a semi-circular arc of $|\zeta| = \rho$, and an arc corresponding to $|s| = r$. The two remaining domains have λ as a bounding arc and each maps into a domain in the ζ-plane bounded by an arc $\text{Im}(\zeta) = 0$, an arc $\text{Re}(\zeta) = \text{constant}$, an arc $|\zeta| = \rho$, and an arc corresponding to a piece of $|s| = r$. Let these two domains abutting λ be called Δ_1 and Δ_2.

We subtract the circle $|s| \leqq r$ from the regions of integration G, G^* of the two integrals on the right of (7.2.20). If ρ is sufficiently large, each of the resulting domains will be ring-shaped. Let G_1 and G_1^* denote these ring-shaped domains, each of which is supposed cut along λ.

In estimating the value of the integral over G_1^* we cut the exterior of $|s| = r$ along the arc λ^* into which λ is mapped by the function (7.2.6). We have shown that

$$(7.2.22)\qquad \zeta(S) = \zeta(s) - \frac{\alpha}{2}\frac{A_{n-1}}{n-1}\cdot\frac{1}{\zeta(s)} + o\left(\frac{1}{\zeta(s)}\right).$$

As $\zeta(s)$ moves over the circle $|\zeta| = \rho$, the point

$$\zeta(s) - \frac{\alpha}{2}\frac{A_{n-1}}{n-1}\cdot\frac{1}{\zeta(s)}$$

moves over an ellipse whose semi-axes are of lengths

$$\rho + \left|\frac{\alpha A_{n-1}}{2n-2}\right|\frac{1}{\rho}, \qquad \rho - \left|\frac{\alpha A_{n-1}}{2n-2}\right|\frac{1}{\rho}.$$

The area of this ellipse is equal to

$$\pi\left\{\rho^2 - \left|\frac{\alpha A_{n-1}}{2n-2}\right|^2\frac{1}{\rho^2}\right\}.$$

As s moves over λ, the point S moves over λ^* and the point $\zeta(S)$ is given by (7.2.22). This change in λ adds an area to one of the domains Δ_1, Δ_2 and subtracts an area from the other. The areas added and subtracted because of the motion from the image of λ to the image of λ^* differ by an amount $o(1)$. It readily follows that the difference of the two integrals on the right side of (7.2.20) is $o(1)$ as ρ becomes large, and the value of the integral on the left side of (7.2.20) is $(n - 1)\pi\rho^2 + O(\rho)$ as ρ becomes large. Thus we obtain from (7.2.20)

$$2\pi\,(n - 1)\,\gamma \leqq 4k\epsilon + o(1). \tag{7.2.23}$$

By choosing ϵ small and ρ large it is seen that

$$\operatorname{Re}(\alpha A_{n-1}) \leqq 0. \tag{7.2.24}$$

To consider the case of equality we use the known result that if there is an $h > 0$ such that

$$a \leqq c < c + h \leqq d < d + h \leqq b$$

and a measurable function $f(x)$ defined in the interval (a, b) such that

$$f(x) \leqq m,\ c \leqq x < c + h,$$
$$f(x) \geqq M,\ d \leqq x < d + h,$$

where $M > m$, then

$$\left(\int_a^b f(x)\,dx\right)^2 \leqq (b - a)\int_a^b f(x)^2\,dx - \frac{h}{4}(b - a)(M - m)^2. \tag{7.2.25}$$

To prove inequality (7.2.25) we note that

$$(b - a)\int_a^b f(x)^2\,dx - \left(\int_a^b f(x)\,dx\right)^2 = \frac{1}{2}\int_a^b\int_a^b \{f(x) - f(y)\}^2\,dx\,dy$$

$$\geqq \frac{1}{2}\int_{y=a}^b\int_{t=0}^h \{[f(c + t) - f(y)]^2 + [f(d + t) - f(y)]^2\}\,dt\,dy$$

$$\geqq \frac{h}{4}(b - a)(M - m)^2$$

since

$$[f(c + t) - z]^2 + [f(d + t) - z]^2 \geqq [f(d + t) - f(c + t)]^2/2 \geqq (M - m)^2/2$$

for all real z.

If

$$\left|\left(\frac{N(S(s))}{N(s)}\right)^{1/2}\frac{dS}{ds}\right|$$

is not equal to a constant on some curve L_ρ, then it will be greater than some constant M in one neighborhood and less than another constant m in some other neighborhood where $M > m$. Then from (7.2.14) we shall have

$$\mathfrak{L}_\rho^2\left(1 + \frac{\gamma}{\rho^2}\right)^2 - \frac{2\epsilon}{\rho}\,\mathfrak{L}_\rho\left(1 + \frac{\gamma}{\rho^2}\right) \leqq \mathfrak{L}_\rho \int \left|\frac{N(S)}{N(s)}\left(\frac{dS}{ds}\right)^2 d\zeta\right| - \frac{h}{4}\,\mathfrak{L}_\rho(M - m)^2 \tag{7.2.26}$$

for some number $h > 0$. Dividing by $\mathfrak{L}_\rho$ and observing that (7.2.26) holds in some interval of u while (7.2.15) holds for all u in the range, we see that there is a number $\delta > 0$ such that

$$\int_{a_\nu}^{b_\nu} \left\{ \mathfrak{L}_\rho \left(1 + \frac{\gamma}{\rho^2}\right)^2 - \frac{2\epsilon}{\rho}\left(1 + \frac{\gamma}{\rho^2}\right)\right\} du \leqq \iint | N(S) \,|\, dA - \frac{h\delta}{4}(M - m)^2.$$

In place of (7.2.20) we then obtain

$$\frac{2\gamma}{\rho^2}\iint_G | N(s) \,|\, da \leqq 4k\epsilon + \iint_{G^*} | N(s) \,|\, da - \iint_G | N(s) \,|\, da - \frac{h\delta}{4}(M - m)^2.$$

Thus finally we obtain the inequality

$$2\pi\,(n - 1)\,\gamma \leqq 4k\epsilon + o(1) - \frac{h\delta}{4}(M - m)^2.$$

Choosing ϵ small and then ρ so large that the term $o(1)$ is small, we see that

$$2\pi\,(n - 1)\,\gamma \leqq -\frac{h\delta}{4}(M - m)^2.$$

Thus $\gamma < 0$ unless

$$\left| \frac{N(S(s))}{N(s)} \left(\frac{dS}{ds}\right)^2 \right|$$

is a constant on each arc L. But this implies that $S = s$ identically and so $\psi(w) = w$.

7.3. We show now that every $\mathfrak{D}_n$-function belongs to the boundary of V_n.

THEOREM IV. *There is a one–one correspondence between points on the boundary of V_n and $\mathfrak{D}_n$-functions. Every $\mathfrak{D}_n$-function belongs to some boundary point of V_n and to any given boundary point of V_n there belongs one and only one $\mathfrak{D}_n$-function. There are no other boundary functions of Class $\mathfrak{S}$.*

At the end of Chapter I we proved that if a point is interior to V_n, then it belongs to more than one function. From Theorem IV we see that if a point belongs to more than one function, then it is an interior point. The interior points of V_n are thus characterized by the property of belonging to more than one function of class $\mathfrak{S}$.

Theorem IV states that to each point of the boundary there belongs precisely one $\mathfrak{D}_n$-function. Before taking up the proof of Theorem IV we remark that a more elementary argument shows that to each point of the everywhere dense set E of the boundary of V_n defined in Lemma VIII there belongs exactly one $\mathfrak{D}_n$-function. A point of the boundary of V_n belongs to E if there is a sphere having only this point in common with V_n. Given any point of E, we can construct a real function $F(a_2, \overline{a_2}, \cdots, a_n, \overline{a_n})$ which is continuous together with its first derivatives in an open set containing V_n and which has a maximum at the given point. It follows at once that one and only one function of class $\mathfrak{S}$ can belong to any point of $\boldsymbol{E}$. For if there were two, say $f_1(z)$ and $f_2(z)$, they would both maximize the same function F. Since both functions belong to the same point and therefore have the same coefficients $a_2, a_3, \cdots, a_n$, the vectors $\mathfrak{A} = (A_1, A_2, \cdots, A_{n-1})$ and $\mathfrak{B} = (B_1, B_2, \cdots, B_{n-1})$ defined by the formulas of Lemmas VII and VIII would be identical. Then $w_1 = f_1(z)$ and $w_2 = f_2(z)$ would satisfy the same $\mathfrak{D}_n$-equation

$$\left(\frac{z}{w}\frac{dw}{dz}\right)^2 P(w) = Q(z). \tag{7.3.1}$$

But it may be verified that there is one and only one function $w = f(z)$ having the given numbers $a_1 = 1, a_2, a_3, \cdots, a_n$ as its first n coefficients and satisfying a differential equation of the form (7.3.1). Thus $f_1(z) = f_2(z)$.

PROOF OF THEOREM IV: Let

$$f(z) = z + a_2 z^2 + a_3 z^3 + \cdots$$

be a $\mathfrak{D}_n$-function satisfying the equation

$$\left(\frac{z}{w}\frac{dw}{dz}\right)^2 P(w) = Q(z)$$

where

$$P(w) = \sum_{\nu=1}^{n-1} \frac{A_\nu}{w^\nu}, \qquad Q(z) = \sum_{\nu=-(n-1)}^{n-1} \frac{B_\nu}{z^\nu}.$$

By Theorem III, $f(z)$ is schlicht in $|z| < 1$ and $w = f(z)$ maps $|z| < 1$ onto the entire w-plane minus a subcontinuum containing $w = \infty$ of $\boldsymbol{\Gamma}_w$. Here $\boldsymbol{\Gamma}_w$ is the set of all possible arcs

$$\operatorname{Re} \int (P(w))^{1/2} \frac{dw}{w} = \text{constant}$$

issuing from $w = \infty$ and each finite zero of $P(w)$. Let

$$z = f^{-1}(w) = \sum_{\nu=1}^{\infty} c_\nu w^\nu$$

be the function inverse to f. Then

$$w = \sum_{\nu=1}^{\infty} a_\nu z^\nu = \sum_{\nu=1}^{\infty}\sum_{k=\nu}^{\infty} a_\nu c_k^{(\nu)} w^k = \sum_{k=1}^{\infty}\left(\sum_{\nu=1}^{k} a_\nu c_k^{(\nu)}\right) w^k,$$

and so

$$\sum_{\nu=1}^{k} a_\nu c_k^{(\nu)} = 0, \qquad k > 1. \tag{7.3.2}$$

Now let

$$g(z) = z + b_2 z^2 + b_3 z^3 + \cdots$$

be any function which is regular and schlicht in $|z| < 1$ with

$$b_\nu = a_\nu, \qquad \nu = 2, 3, \cdots, n-1.$$

Then

$$g\{f^{-1}(w)\} = \sum_{\nu=1}^{\infty} \sum_{k=\nu}^{\infty} b_\nu c_k^{(\nu)} w^k = \sum_{k=1}^{\infty} \left(\sum_{\nu=1}^{k} b_\nu c_k^{(\nu)} \right) w^k$$

$$= \sum_{k=1}^{n-1} \left(\sum_{\nu=1}^{k} b_\nu c_k^{(\nu)} \right) w^k + \sum_{\nu=1}^{n} b_\nu c_n^{(\nu)} w^n + \sum_{k=n+1}^{\infty} \left(\sum_{\nu=1}^{k} b_\nu c_k^{(\nu)} \right) w^k.$$

Since $b_\nu = a_\nu$ for $\nu = 1, 2, \cdots, n-1$ ($a_1 = b_1 = 1$), we obtain by (7.3.2)

$$g\{f^{-1}(w)\} = w + (b_n - a_n) w^n + \sum_{k=n+1}^{\infty} d_k w^k.$$

The function $g\{f^{-1}(w)\}$ is regular and schlicht in the domain which is the image of $|z| < 1$ by $w = f(z)$, and so by Lemma XXIII

(7.3.3) $$\operatorname{Re}\{(b_n - a_n) A_{n-1}\} \leqq 0$$

with equality if and only if $g\{f^{-1}(w)\} = w$. It then follows that $f(z)$ does not belong to an interior point of V_n. Thus the point belonging to $f(z)$ is a boundary point of V_n.

Now consider any boundary point of V_n. By Lemma VIII there is at least one $\mathfrak{D}_n$-function belonging to this boundary point. If there were two, say

$$g(z) = \sum_{\nu=1}^{\infty} a_\nu z^\nu, \qquad f(z) = \sum_{\nu=1}^{\infty} b_\nu z^\nu,$$

where $a_\nu = b_\nu$, $\nu = 1, 2, \cdots, n$, then according to the above argument $g\{f^{-1}(w)\} = w$, $f(z) = a(z)$. Thus there is only one $\mathfrak{D}_n$-function belonging to the given point.

Finally, the above argument shows that every 2-dimensional cross-section of V_n obtained by holding $a_2, a_3, \cdots, a_{n-1}$ fixed and letting a_n vary is convex. This follows from (7.3.3).

CHAPTER VIII

RELATIONS BETWEEN $P(w)$ AND $Q(z)$

8.1. We have shown that there is a one–one correspondence between boundary points of V_n and $\mathfrak{D}_n$-functions. However, there is no one–one correspondence between $\mathfrak{D}_n$-functions and $\mathfrak{D}_n$-equations. We shall see that there are special functions which satisfy more than one equation and special equations which have as solutions a family of $\mathfrak{D}_n$-functions depending on a parameter.

In this chapter we investigate some of the properties of differential equations of the form

$$\left(\frac{z}{w}\frac{dw}{dz}\right)^2 P(w) = Q(z) \tag{8.1.1}$$

where

$$P(w) = P_n(w) = \sum_{\nu=1}^{n-1} \frac{A_\nu}{w^\nu}, \qquad Q(z) = Q_n(z) = \sum_{\nu=-(n-1)}^{n-1} \frac{B_\nu}{z^\nu}. \tag{8.1.2}$$

If (8.1.1) is a $\mathfrak{D}_n$-equation, in which case $Q(z) \geqq 0$ on $|z| = 1$ with at least one zero there, then the coefficient B_0 is uniquely determined by $B_1, B_2, \cdots, B_{n-1}$. Thus a $\mathfrak{D}_n$-equation is determined by the pair of vectors $\mathfrak{A} = (A_1, A_2, \cdots, A_{n-1})$, $\mathfrak{B} = (B_1, B_2, \cdots, B_{n-1})$.

A question that now arises is the following: Given $\mathfrak{A}$, does there exist a $\mathfrak{B}$ such that (8.1.1) is a $\mathfrak{D}_n$-equation which has a $\mathfrak{D}_n$-function $w = f(z) = z + \cdots$ as solution? Conversely, given $\mathfrak{B}$, does there exist an $\mathfrak{A}$ such that (8.1.1) has these properties? If the answer to either of these questions is affirmative, we can also ask under what circumstances the vector to be determined is unique.

8.2. Before considering these questions we ask if the existence of a regular normalized solution of (8.1.1) in a neighborhood of $z = 0$ imposes any restrictions on $\mathfrak{A}$ and $\mathfrak{B}$. The answer to this question is contained in the following lemma:

LEMMA XXIV. *Suppose that $n \geqq 2$ and let $P(w)$ and $Q(z)$ be defined by* (8.1.2) *where $A_{n-1} = B_{n-1} \neq 0$. Consider solutions of the differential equation*

$$\left(\frac{z}{w}\frac{dw}{dz}\right)^2 P(w) = Q(z)$$

which are regular in a neighborhood of the origin and of the form

$$w = z + a_2 z^2 + \cdots .$$

Case 1. *If n is odd, $n = 2m + 1$, there is a solution if and only if the constant term p_0 in the expansion*

$$(P(w))^{1/2} = (A_{n-1})^{1/2} w^{-m} + p_{-m+1} w^{-m+1} + \cdots + p_0 + \cdots$$

and the constant term q_0 in the expansion

$$(Q(z))^{1/2} = (B_{n-1})^{1/2} z^{-m} + q_{-m+1} z^{-m+1} + \cdots + q_0 + \cdots$$

are equal under the assumption that $(A_{n-1})^{1/2} = (B_{n-1})^{1/2}$. If these coefficients are equal, there are infinitely many solutions. All solutions have the same coefficients $a_2, a_3, \cdots, a_m$ but a_{m+1} is arbitrary. The remaining coefficients are then determined by the parameter a_{m+1}.

Case 2. *If n is even, there is one and only one solution.*

PROOF: We consider first the case when n is odd, $n = 2m + 1$. We have

$$\int (P(w))^{1/2} \frac{dw}{w} = \int (Q(z))^{1/2} \frac{dz}{z}$$

and so

$$\alpha_{-m} w^{-m} + \alpha_{-m+1} w^{-m+1} + \cdots + \alpha \log w + \cdots$$
$$= \beta_{-m} z^{-m} + \beta_{-m+1} z^{-m+1} + \cdots + \beta \log z + c + \cdots$$

where $\alpha_{-m} = \beta_{-m}$. In order to have a regular solution we must have $\alpha = \beta$. Then

$$w^{-m}\{1 + wF(w) + \lambda w^m\} + \alpha' \log \frac{w}{z} = z^{-m}\{1 + zG(z)\}$$

where

$$\lambda = -c/\alpha_{-m}$$

is a constant of integration, and the functions $F(w)$ and $G(z)$ are uniquely determined by $\mathfrak{A}$ and $\mathfrak{B}$. Thus

$$1 + wF(w) + \lambda w^m + \alpha' w^m \log \frac{w}{z} = \left(\frac{w}{z}\right)^m \{1 + zG(z)\}$$

or

$$\left(\frac{w}{z}\right)^m = \frac{1 + wF(w) + \lambda w^m + \alpha' w^m \log (w/z)}{1 + zG(z)}. \tag{8.2.1}$$

Let $F(w)$ be regular in $|w| < r_1$ and $G(z)$ in $|z| < r_1$, and let r_1 be so small that $|zG(z)| < 1/2$ in $|z| < r_1$. After substituting $w = z(1 + \zeta)$ in (8.2.1) and subtracting 1 from each side we obtain an equation of the form

$$(1 + \zeta)^m - 1 = zp(z, \zeta). \tag{8.2.2}$$

If $|z| \leqq r_1/2$ and $|\zeta| \leqq 1/2$, then $|w| \leqq 3r_1/4$; so the function $p(z, \zeta)$ is regular and bounded by some number M in the region $|z| \leqq r_1/2$, $|\zeta| \leqq 1/2$. We observe that the number M and the function $p(z, \zeta)$ depend on λ. If r_2 is the minimum of $r_1/2$ and $1/(2M)$, then $zp(z, \zeta)$ is bounded by 1/2 in $|z| \leqq r_2$, $|\zeta| \leqq 1/2$. Thus if we add 1 to each side of (8.2.2) and take the mth root, we have

$$1 + \zeta = 1 + \frac{1}{m} zp(z, \zeta) + \frac{1}{2}\left(\frac{1}{m}\right)\left(\frac{1}{m} - 1\right) z^2 p(z, \zeta)^2 \cdots$$

or

$$\zeta = zq(z, \zeta). \tag{8.2.3}$$

The function $q(z, \zeta)$ is regular in the two variables z and ζ in the region $|z| \leqq r_2$, $|\zeta| \leqq 1/2$, and is bounded by some constant in this region.

To construct a sequence of functions whose limit is a solution of (8.2.3), suppose that

$$|q| + \left|\frac{\partial q}{\partial \zeta}\right| < N$$

in the region $|z| \leqq r_3$, $|\zeta| \leqq 1/4$. Choose r_3 such that $Nr_3 < 1/4$. Let $\zeta_0(z) \equiv 0$. If $\zeta_{m-1}(z)$ has been defined for some $m \geqq 1$ and is regular and bounded by 1/4 in $|z| < r_3$, let

$$\zeta_m(z) = zq(z, \zeta_{m-1}(z)).$$

Then $\zeta_m(z)$ is regular and bounded by 1/4 in the circle $|z| < r_3$ and in this domain

$$|\zeta_{m+1}(z) - \zeta_m(z)| = |z\{q(z, \zeta_m) - q(z, \zeta_{m-1})\}| < N|z| \cdot |\zeta_m - \zeta_{m-1}|.$$

Since $N|z| < 1/4$, it follows that the functions $\zeta_m(z)$ tend uniformly to a limit $\zeta(z)$ in the circle $|z| < r_3$. This limit is an analytic function of z in this circle and satisfies (8.2.3) there. Thus $w = z(1 + \zeta)$ is a solution of (8.2.1) and is therefore a solution of the differential equation.

To investigate uniqueness in the case $n = 2m + 1$ we return to the relation (8.2.1), namely

$$\{1 + zG(z)\}w^m = z^m\{1 + wF(w) + \lambda w^m + \alpha' w^m \log(w/z)\}. \tag{8.2.4}$$

Let

$$1 + zG(z) = \sum_{\nu=0}^{\infty} b_\nu z^\nu, \qquad 1 + wF(w) = \sum_{\nu=0}^{\infty} c_\nu w^\nu$$

and

$$w = \sum_{\nu=1}^{\infty} a_\nu z^\nu, \qquad w^k = \sum_{\nu=k}^{\infty} a_\nu^{(k)} z^\nu$$

where $b_0 = c_0 = 1$ and $a_1 = 1$. Then

$$\log \frac{w}{z} = \log\left(1 + \sum_{\nu=1}^{\infty} a_{\nu+1} z^\nu\right) = \sum_{\nu=1}^{\infty} e_\nu z^\nu.$$

Substituting into (8.2.4), we obtain

$$\left(1+\sum_{\nu=1}^{\infty} b_\nu z^\nu\right)\left(\sum_{\nu=m}^{\infty} a_\nu^{(m)} z^\nu\right) = z^m + z^m \sum_{k=1}^{\infty}\sum_{\nu=k}^{\infty} c_k a_\nu^{(k)} z^\nu$$
$$+ \lambda z^m \sum_{\nu=m}^{\infty} a_\nu^{(m)} z^\nu + \alpha' z^m \left(\sum_{\nu=m}^{\infty} a_\nu^{(m)} z^\nu\right)\left(\sum_{\nu=1}^{\infty} e_\nu z^\nu\right).$$

The coefficient of z^p on each side of this equation is zero if $p < m$ and is 1 if $p = m$. Letting p be greater than m, we obtain

$$\sum_{\nu=m}^{p} a_\nu^{(m)} b_{p-\nu} = \sum_{k=1}^{p-m} c_k a_{p-m}^{(k)} + \lambda a_{p-m}^{(m)} + \alpha' \sum_{\nu=m}^{p-m-1} a_\nu^{(m)} e_{p-\nu-m}.$$

The term $a_{p-m}^{(m)}$ is zero for $p < 2m$ and is 1 for $p = 2m$. The last summation is zero for $p < 2m + 1$. Now $a_\nu^{(k)}$ is zero for $\nu < k$, is equal to 1 for $\nu = k$, and is a polynomial in $a_2, a_3, \cdots, a_{\nu+1-k}$ for $\nu > k$ where the coefficient of $a_{\nu+1-k}$ is equal to k. Thus the recursion formula expresses a_{p+1-m} in terms of $1, a_2, a_3, \cdots, a_{p-m}, \lambda$ for $p > m$. If $p < 2m$, the formula does not contain λ; so $a_2, a_3, \cdots, a_m$ are uniquely determined. The integration constant λ determines a_{m+1} and is itself determined by a_{m+1}. The higher coefficients are uniquely determined by a_{m+1}. Here we note that e_j is a polynomial in $a_2, a_3, \cdots, a_{j+1}$.

Now consider the case in which n is even, $n = 2m$. Separating variables and integrating each side of the differential equation, we obtain

$$w^{-m+1/2}\{1 + \alpha_1 w + \alpha_2 w^2 + \cdots\} = z^{-m+1/2}\{1 + \beta_1 z + \beta_2 z^2 + \cdots\} + c$$

where c is the constant of integration. Then

$$(8.2.5) \qquad 1 + \alpha_1 w + \cdots = (w/z)^{m-1/2}\{1 + \beta_1 z + \cdots\} + c w^{m-1/2}.$$

If $w = z + a_2 z^2 + \cdots$ is regular in a neighborhood of $z = 0$, then so is the function $(w/z)^{m-1/2}$. It follows from (8.2.5) that $cw^{m-1/2}$ must be regular and so $c = 0$. Then

$$\left(\frac{w}{z}\right)^{m-1/2} = \frac{1 + wF(w)}{1 + zG(z)}.$$

This equation has one and only one solution of the form $w = z + a_2 z^2 + \cdots$ which is regular in a neighborhood of the origin, as can be seen by squaring and referring to the previous case. This completes the proof of Lemma XXIV.

Let n be odd, $n = 2m + 1$. Then

$$(P(w))^{1/2} = \frac{(A_{2m})^{1/2}}{w^m}\left\{1 + \frac{A_{2m-1}}{A_{2m}} w + \cdots + \frac{A_m}{A_{2m}} w^m + \cdots\right\}^{1/2},$$

$$(Q(z))^{1/2} = \frac{(B_{2m})^{1/2}}{z^m}\left\{1 + \frac{B_{2m-1}}{B_{2m}} z + \cdots + \frac{B_m}{B_{2m}} z^m + \cdots\right\}^{1/2}.$$

Under the assumption that $(A_{2m})^{1/2} = (B_{2m})^{1/2}$, the condition that the constant term in these expansions be equal may be written as an equation

$$\Pi_m(A_m, A_{m+1}, \cdots, A_{2m}) = \Pi_m(B_m, B_{m+1}, \cdots, B_{2m}) \tag{8.2.6}$$

where $\Pi_m(x_m, x_{m+1}, \cdots, x_{2m})$ is a well-determined polynomial in the $m + 1$ quantities $x_m, x_{m+1}, \cdots, x_{2m}$ and it is clear that the term x_m occurs linearly. In the simplest case $n = 3$, $m = 1$, we obtain the relation $A_1 = B_1$.

At the beginning of **8.1** we stated that there are special $\mathfrak{D}_n$-equations which have a one-parameter family of $\mathfrak{D}_n$-functions as solutions. If $A_{n-1} = B_{n-1} \neq 0$, we see by Lemma XXIV that n must be an odd number. An example is provided by the equation

$$\left(\frac{z}{w}\frac{dw}{dz}\right)^2 \frac{1}{w^{2m}} = 2 + \frac{1}{z^{2m}} + z^{2m}, \qquad m \geqq 1, \tag{8.2.7}$$

which is satisfied by the functions

$$w = \frac{z}{(1 - ie^{i\theta}z^m)^{1/m}(1 - ie^{-i\theta}z^m)^{1/m}}, \qquad \theta \text{ real}. \tag{8.2.8}$$

We observe that $n = 2m + 1$ is an odd number, as required.

8.3. Let $\mathfrak{A} = (A_1, A_2, \cdots, A_{n-1})$ and $\mathfrak{B} = (B_1, B_2, \cdots, B_{n-1})$ denote any pair of vectors satisfying $A_{n-1} = B_{n-1} \neq 0$. If n is odd, we suppose in addition that the condition (8.2.6) is satisfied. We denote a solution of the corresponding $\mathfrak{D}_n$-equation by

$$w = f(z) = z + a_2z^2 + a_3z^3 + \cdots. \tag{8.3.1}$$

In Chapter II we obtained a set of numbers $F_2, F_3, \cdots, F_n$ which were related to the A_ν, B_ν and coefficients a_ν by the formulas

$$A_\nu = \sum_{k=\nu+1}^{n} a_k^{(\nu+1)} F_k, \qquad \nu = 1, 2, \cdots, n - 1, \tag{8.3.2}$$

$$B_\nu = \sum_{k=1}^{n-\nu} k a_k F_{k+\nu}, \qquad \nu = 1, 2, \cdots, n - 1, \tag{8.3.3}$$

$$B_0 = \sum_{k=2}^{n} (k - 1) a_k F_k. \tag{8.3.4}$$

We assume that $n \geqq 3$ and that $A_{n-1} = B_{n-1} = F_n \neq 0$. We investigate the relations which exist between these three equations and the four vectors

$$\begin{aligned} &\mathfrak{A} = (A_1, \cdots, A_{n-1}), \qquad \mathfrak{B} = (B_1, \cdots, B_{n-1}), \\ &\mathfrak{F} = (F_2, \cdots, F_n), \qquad \mathfrak{a} = (a_2, a_3, \cdots, a_{n-1}). \end{aligned} \tag{8.3.5}$$

We observe first that we can solve the system of equations (8.3.2) either for $\mathfrak{F}$ in terms of $\mathfrak{A}$ and $\mathfrak{a}$ or for $\mathfrak{a}$ in terms of $\mathfrak{A}$ and $\mathfrak{F}$. The system is linear in the F_ν with non-vanishing determinant but it is not linear in the a_ν. However, for $\nu > k$, $a_\nu^{(k)}$ is a polynomial in $a_2, a_3, \cdots, a_{\nu+1-k}$ which is linear in $a_{\nu+1-k}$ with the coefficient of $a_{\nu+1-k}$ equal to k, and it is then readily seen that the system (8.3.2) may be solved for $a_2, \cdots, a_{n-1}$ in terms of $\mathfrak{A}$ and $\mathfrak{F}$. Similarly the system (8.3.5) may be solved either for $\mathfrak{F}$ in terms of $\mathfrak{B}$ and $\mathfrak{a}$ or for $\mathfrak{a}$ in terms of $\mathfrak{B}$ and $\mathfrak{F}$; it is linear in both sets of variables and has a non-vanishing determinant with respect to each.

Suppose that (8.3.1) satisfies a $\mathfrak{D}_n$-equation in some neighborhood of the origin. Then it can be shown that if either one of the two equations (8.3.2), (8.3.3) is valid, so is the other. From Lemma XXIV we observe that a_n is determined from the $\mathfrak{D}_n$-equation if $\mathfrak{A}$, $\mathfrak{B}$, and $\mathfrak{a}$ are given. Also, when n is even, $\mathfrak{a}$ is determined by $\mathfrak{A}$, $\mathfrak{B}$. By considering the various cases we see that given any two of the four vectors (8.3.5) plus the equations (8.3.2), (8.3.3), then the other two vectors are determined with the one important exception that $\mathfrak{A}$ and $\mathfrak{B}$ do not determine $\mathfrak{a}$ and $\mathfrak{F}$ when n is odd. This is a purely algebraic remark and does not imply that the corresponding quantities define a function of class $\mathfrak{S}$.

The significance of the vector $\mathfrak{F}$ is intuitively clear if we assume that the boundary of V_n is differentiable in the neighborhood of the point $(a_2, a_3, \cdots, a_n)$. For let $w = f(z) = z + a_2z^2 + a_3z^3 + \cdots$ belong to this point and let it satisfy a $\mathfrak{D}_n$-equation given by the vectors $\mathfrak{A}$, $\mathfrak{B}$. If $\mathfrak{F}$ is computed from either (8.3.2) or (8.3.3), then it satisfies the other also, for in this case $\mathfrak{F}$ is unique, and

$$\operatorname{Re}\{F_2\,\delta a_2 + F_3\,\delta a_3 + \cdots + F_n\,\delta a_n\} = 0$$

where δa_k denotes the first variation of a_k as defined in Chapter II. Thus it would seem that $(F_2, F_3, \cdots, F_n)$, or more strictly $(\bar{F}_2, \bar{F}_3, \cdots, \bar{F}_n)$, defines the normal to the boundary of V_n at the point $(a_2, a_3, \cdots, a_n)$.

8.4. Given a non-null vector $\mathfrak{A}$, let $\mathbf{\Gamma}_w$ be the system of all arcs

$$\operatorname{Re}\int (P(w))^{1/2}\frac{dw}{w} = \text{constant}$$

issuing from any finite zero of $P(w)$ and from $w = \infty$. Suppose that there is a $\mathfrak{D}_n$-function $f(z)$ and a vector $\mathfrak{B}$ such that $w = f(z)$ satisfies the equation

$$\left(\frac{z}{w}\frac{dw}{dz}\right)^2 P(w) = Q(z) \tag{8.4.1}$$

where $Q(z) \geqq 0$ on $|z| = 1$ with at least one zero there. Then $w = f(z)$ maps $|z| < 1$ onto the complement of a subcontinuum $\mathbf{C}_w$ of $\mathbf{\Gamma}_w$. The subcontinuum $\mathbf{C}_w$ contains $w = \infty$ and is of mapping radius 1. We now prove:

LEMMA XXV. *Let a non-null vector $\mathfrak{A}$ be given and let $\mathbf{C}_w$ be a subcontinuum of $\mathbf{\Gamma}_w$ containing $w = \infty$ and of mapping radius* 1. *Then there exists a vector $\mathfrak{B}$ and a function $w = f(z) = z + a_2z^2 + a_3z^3 + \cdots$ mapping $|z| < 1$ onto the complement of $\mathbf{C}_w$ such that $w = f(z)$ satisfies the $\mathfrak{D}_n$-equation* (8.4.1).

We remark that there is a single arc of $\mathbf{\Gamma}_w$ issuing from $w = \infty$ if and only if $A_1 \neq 0$. Beginning at $w = \infty$, trace out a portion γ_ρ of this arc which has a mapping radius ρ. It is readily seen that $\boldsymbol{C}_w$ will be uniquely determined if and only if $A_1 \neq 0$ and γ_ρ contains no finite zero of $P(w)$ for $\rho > 1$.

PROOF: We may suppose without loss of generality that $A_n \neq 0$. Given $\mathfrak{A}$, $A_n \neq 0$, let

$$\zeta = \int (P(w))^{1/2} \frac{dw}{w}.$$

Now $\boldsymbol{C}_w$ is a finite set of analytic arcs and so the function $w = f(z)$ is regular on $|z| = 1$ except perhaps at a finite number of points where, however, the limit of $f(z)$ exists (though perhaps infinite) when the approach is from within $|z| < 1$. If z_0 is a point of $|z| = 1$, then there is a neighborhood N of z_0 such that $\zeta(f(z))$ is regular and bounded in the part of N which lies in $|z| < 1$. Since $\zeta(f(z))$ has constant real part on $|z| = 1$, it is regular in a complete neighborhood of z_0. Thus $\zeta(f(z))$ is regular everywhere on $|z| = 1$ and $z(d\zeta/dz)$ is regular and real on $|z| = 1$. Since

$$z \frac{d\zeta}{dz} = z \frac{d\zeta}{dw} \frac{dw}{dz} = \frac{z}{w} \frac{dw}{dz} (P(w))^{1/2}$$

where $w = f(z)$, we see that

$$\left(\frac{z}{w} \frac{dw}{dz}\right)^2 P(w)$$

is regular and non-negative on $|z| = 1$.

There are numbers $B_1, B_2, \cdots, B_{n-1}$ such that the function

$$\left(\frac{z}{w} \frac{dw}{dz}\right)^2 P(w) - \sum_{\nu=1}^{n-1} \left(\frac{B_\nu}{z^\nu} + \bar{B}_\nu z^\nu\right)$$

is regular in a neighborhood of the origin. It is then regular in $|z| \leqq 1$ and real on $|z| = 1$. Then it is regular in the closed z-plane and so must be a constant and the constant is real. Thus

$$\left(\frac{z}{w} \frac{dw}{dz}\right)^2 P(w) = B_0 + \sum_{\nu=1}^{n-1} \left(\frac{B_\nu}{z^\nu} + \bar{B}_\nu z^\nu\right).$$

Since the left side of this equation is non-negative on $|z| = 1$, the right side is also; so in particular $B_0 > 0$. Finally $\boldsymbol{C}_w$ has at least one tip of a slit and at the corresponding point on $|z| = 1$ we have $f'(z) = 0$. Hence the left side has at least one zero on $|z| = 1$; so the right side also has at least one zero on $|z| = 1$.

8.5. We now turn our attention to the right side of a $\mathfrak{D}_n$-equation and ask to what extent it can be chosen arbitrarily and to what extent it determines the left side.

If $w = z + a_2 z^2 + a_3 z^3 + \cdots$ is regular in $|z| < 1$ and satisfies the $\mathfrak{D}_n$-equation

$$\left(\frac{z}{w}\frac{dw}{dz}\right)^2 P(w) = Q(z) \tag{8.5.1}$$

where

$$P(w) = \sum_{\nu=1}^{n-1} \frac{A_\nu}{w^\nu}, \qquad Q(z) = \sum_{\nu=-(n-1)}^{n-1} \frac{B_\nu}{z^\nu}, \tag{8.5.2}$$

then it maps $|z| < 1$ onto the complement of a subcontinuum C_w of Γ_w which contains $w = \infty$ and has mapping radius 1. As the point z moves once around $|z| = 1$, say in the counterclockwise direction, the point $w = f(z)$ moves over C_w covering each arc of C_w twice, once in each sense. Thus each arc of C_w corresponds to two arcs of $|z| = 1$. If C_w consists of m arcs and their end points, then the unit circle $|z| = 1$ consists of $2m$ open arcs

$$I_1', I_2', \cdots, I_m', \qquad I_1'', I_2'', \cdots, I_m''$$

and their end points, the arcs being identified in pairs in the sense that I_ν' and I_ν'' map into the two edges of the same arc of C_w. If p' is one end point of I_ν' and p'' is one end point of I_ν'', the points p' and p'' being so chosen that one is the right end point of its interval and the other the left end point of its interval as seen from the origin $z = 0$, the point z' of I_ν' and the point z'' of I_ν'' that map into the same point in the w-plane satisfy the equation

$$\int_{p'}^{z'} (Q(t))^{1/2} \frac{dt}{t} = \int_{p''}^{z''} (Q(t))^{1/2} \frac{dt}{t}. \tag{8.5.3}$$

As z' moves over I_ν', the point z'' moves over I_ν'', and one of the arcs is described in the clockwise direction, the other in the counterclockwise direction. Since dt/t is pure imaginary and $(Q(t))^{1/2}$ is real on $|t| = 1$, this means that the positive square root of Q is taken on one side of (8.5.3) and the negative square root of Q on the other.

If two arcs I_ν' and I_ν'' which are identified have a common end point at z_0, let $w_0 = f(z_0)$. If w_0 is a finite point, then it is the tip of a slit in the w-plane and $f'(z_0) = 0$; so $Q(z_0) = 0$. If $w_0 = \infty$, then there can be only one arc of C_w which extends to infinity; so the function $1/f(z)$ has a double zero at z_0 while its derivative has a simple zero there. It therefore follows that if $A_1 \neq 0$, then $Q(z_0) \neq 0$ while if $A_1 = 0$, then $Q(z_0) = 0$. Thus with one possible exception, the common end point of a pair of abutting arcs which are identified is a zero of $Q(z)$.

It is also clear that no pair of identified arcs separates another pair. That is, if $j \neq \nu$, then there is an arc of $|z| = 1$ that contains I_j' and I_j'' but does not contain either I_ν' or I_ν''.

Thus if $w = f(z)$ is a function of class $\mathfrak{S}$ which satisfies a $\mathfrak{D}_n$-equation with right side $Q(z)$, then the unit circumference $|z| = 1$ is the sum of $2m$ open arcs

$$I_1', I_2', \cdots, I_m', \qquad I_1'', I_2'', \cdots, I_m''$$

and their end points. The arcs are identified in pairs and satisfy the following three conditions:

(i) The point z' of I'_ν and the point z'' of I''_ν are identified by relation (8.5.3) where z' describes I'_ν, say clockwise, as z'' describes I''_ν counterclockwise.

(ii) No pair of identified arcs separates another identified pair.

(iii) With one possible exception, the common end point of a pair of abutting arcs which are identified is a zero of $Q(z)$.

It is to be shown that if $Q(z)$ is given and is non-negative on $|z| = 1$ with at least one zero there and if the unit circle $|z| = 1$ is dissected into $2m$ arcs and their end points in such a way as to satisfy conditions (i)–(iii) above, then there is a function $w = f(z)$ of class $\mathfrak{S}$ which maps identified points into the same point in the w-plane. We say briefly that $f(z)$ realizes the given identification. This function satisfies a $\mathfrak{D}_n$-equation whose right side is the given function $Q(z)$.

It is clear that if the function $Q(z)$ is given, then there is at least one dissection of $|z| = 1$ into arcs and their end points such that the arcs are identified in pairs according to (i), (ii), (iii).

The dissection of $|z| = 1$ into arcs and their end points can always be made in such a way that all zeros of $Q(z)$ that lie on $|z| = 1$ are among the end points of the arcs. For if $Q(z)$ vanishes at a point of one of the arcs, then that arc and the one with which it is identified are the sum of several subarcs and their end points. The subarcs may then be identified to give the same identification of points as before. Therefore, whenever convenient we shall without loss of generality make the additional hypothesis:

(iv) The zeros of $Q(z)$ that lie on $|z| = 1$ are among the end points of the arcs.

The unit circle $|z| \leqq 1$ with its circumference divided into $2m$ arcs is a topological polygon. The arcs on $|z| = 1$ are its sides and the end points of arcs are its vertices. Two points which are identified with each other may be said to be equivalent. There are the following classes of equivalent points: an interior point of $|z| < 1$ which is equivalent only to itself; an interior point of a side which is equivalent to exactly one other point (also an interior point of a side); a vertex which can be equivalent to many other vertices, to one other, or only to itself (in the last case the vertex is a common end point of two identified arcs). The set which arises from the polygon by identification of equivalent points is a closed surface or a closed 2-dimensional manifold. The "points" of the surface are the "classes of equivalent points" of the polygon.

If we introduce the metric on $|z| = 1$ which is defined by

$$|d\zeta|^2 = \left|\frac{Q(z)}{z^2}\right| |dz|^2 = Q(z)\, d\theta^2, \tag{8.5.4}$$

then we see that in the identification of a pair of arcs I'_ν, I''_ν the metric (8.5.4) is preserved. That is to say, any subarc of I'_ν of length $\mathfrak{L}'$ in the metric (8.5.4) is mapped into a subarc of I''_ν of length $\mathfrak{L}'' = \mathfrak{L}'$ in this metric. Conversely, if the identification of arcs I'_ν, I''_ν is such that the metric (8.5.4) is preserved, then the identification is also given by formula (8.5.3).

Condition (i) above states that as z' traverses I'_ν clockwise, the identified point z'' traverses I''_ν counterclockwise. This condition is equivalent to saying that the closed surface resulting from the identification is orientable. Condition (ii) is plainly equivalent to the assertion that the closed surface is of genus zero or, in other words, that it is topologically equal to the 2-sphere.

Thus the conditions (i) to (iii) may be stated in the alternative form:

(i)′ The circle $|z| \leqq 1$ with $2m$ arcs on $|z| = 1$ identified in pairs is equivalent to the sphere.

(ii)′ In the identification of any pair of arcs the metric (8.5.4) is preserved.

(iii)′ If two identified arcs abut, the common end point is a zero of $Q(z)$ with at most one exception.

By introducing suitable local uniformizing variables or local uniformizers the circle $|z| \leqq 1$ with points identified in the above manner becomes a closed Riemann surface of genus zero. Corresponding to the above three classes of equivalent points, we introduce local uniformizers as follows, and we shall suppose that (iv) is true.

If z is an interior point of $|z| < 1$, z is itself a local uniformizer.

If z'_0 is an interior point of an arc I'_ν on $|z| = 1$, let z''_0 be the identified point of the arc I''_ν. Then z'_0 and z''_0 are connected by the mapping (8.5.3), and a complete neighborhood of z'_0 is mapped into a complete neighborhood of z''_0. Since z' moves clockwise on I'_ν as z'' moves counterclockwise on I''_ν, we see that the half-neighborhood at z'_0 which is interior to $|z| < 1$ is mapped into the exterior half-neighborhood at z''_0 while the half-neighborhood at z'_0 exterior to the unit circle is mapped into the interior half-neighborhood at z''_0. Let

$$Z'(z') = \int_{p'}^{z'} (Q(t))^{1/2} \frac{dt}{t}, \qquad Z''(z'') = \int_{p''}^{z''} (Q(t))^{1/2} \frac{dt}{t}. \tag{8.5.5}$$

Denoting the half-neighborhood at z'_0 which is interior to the unit circle by N' and the half-neighborhood at z''_0 interior to $|z| < 1$ by N'', we define the local uniformizer τ at z'_0 (or z''_0) by the formula

$$\tau = \begin{cases} Z'(z') - Z'(z'_0), & z' \subset N', \\ Z''(z'') - Z''(z''_0), & z'' \subset N''. \end{cases} \tag{8.5.6}$$

The uniformizer τ maps $N' + N''$ in a one–one manner into a complete neighborhood of $\tau = 0$, identified points going into the same point.

Finally, we have to define a local uniformizer at a point of the surface corresponding to a class of equivalent vertices which form a cycle in the terminology of Poincaré. If p is a vertex, it is the left end point of an arc J of the set $I'_1, \cdots, I'_m, I''_1, \cdots, I''_m$. Here left and right are as seen from the origin. The arc J is identified with another arc K of the set, and under this identification p corresponds to the right end point q of K. We say that p precedes q or that q follows p. If I'_ν and I''_ν are identified arcs, it thus follows that the left end point of either arc precedes the right end point of the other. Each vertex is clearly followed by a unique vertex and is preceded by a unique vertex.

Beginning at any vertex p_1, let it be followed by a vertex which we call p_2. Let p_3 be the vertex which follows p_2, and so on. There are only finitely many vertices; so ultimately one of them must be repeated, and the first vertex in the sequence to be repeated must be p_1 since each vertex has a unique predecessor. Thus there is a finite chain of vertices

$$p_1, p_2, \cdots, p_{k-1}$$

such that $p_1 = p_k$ and p_i precedes p_{i+1}. This chain forms a cycle or class of equivalent vertices. A cycle may contain only one vertex. The same cycle is clearly generated by every vertex belonging to it. We see that the set of all vertices on $|z| = 1$ is the sum of finitely many cycles.

Let the vertices $p_1, p_2, \cdots, p_{k-1}$, where $p_1 = p_k$, form a cycle, and let $Q(z)$ have a zero of precise order $2\sigma_\nu \geqq 0$ at p_ν. The vertex p_ν is the end point of two arcs; let the arc of which it is the left end point be I_ν' and let the arc of which it is the right end point be I_ν''. The identification which carries I_1' into I_2'' carries p_1 into p_2 and, generally, the identification which carries I_ν' into $I_{\nu+1}''$ carries p_ν into $p_{\nu+1}$. Finally as I_{k-1}' goes into $I_k'' = I_1''$, p_{k-1} goes into $p_k = p_1$. Let the mapping from I_ν' into $I_{\nu+1}''$ be defined by

$$\int_{p_\nu}^{z'} (Q(t))^{1/2} \frac{dt}{t} = \int_{p_{\nu+1}}^{z''} (Q(t))^{1/2} \frac{dt}{t}. \tag{8.5.7}$$

As z' traverses I_ν', z'' traverses $I_{\nu+1}''$. Write

$$Z_\nu(z) = \int_{p_\nu}^{z} (Q(t))^{1/2} \frac{dt}{t}, \qquad \nu = 1, 2, \cdots, k. \tag{8.5.8}$$

Let $Q^{1/2}$ in the integrand of Z_1 be negative on I_1'. Then $Q^{1/2}$ in the integrand of Z_2 is positive on I_2'', and so Z_2 is uniquely defined by continuation throughout a neighborhood of p_2. Continuing in this way, we define uniquely the k functions Z_ν, $\nu = 1, 2, \cdots, k$.

Let N_ν be the half-neighborhood at p_ν lying in $|z| < 1$. Since $Q(z)$ has a zero at p_ν of order $2\sigma_\nu \geqq 0$, there are $2\sigma_\nu + 2$ arcs $\text{Re}(Z_\nu) = \text{constant}$ emerging from p_ν, and they form $2\sigma_\nu + 2$ angular openings at p_ν, each of which maps into a half-neighborhood with angular opening π in the plane of Z_ν. Under the mapping (8.5.7) (with $\nu = 1$) the set N_1 maps into a set N_{12} at p_2 which covers $\sigma_1 + 1$ of the angular openings at p_2. The sets N_{12} and N_2 lie on either side of the arc I_2'' near p_2 and a point moves counterclockwise about p_2 in passing from N_{12} to N_2. Under the mapping (8.5.7) (with $\nu = 2$) the sets N_{12} and N_2 are carried into sets N_{13} and N_{23} at p_3 and these sets cover $\sigma_1 + 1$ and $\sigma_2 + 1$ of the angular openings at p_3 respectively. A point moves counterclockwise about p_3 in passing from N_{13} to N_{23} to N_3.

This process may be continued until the point returns to $p_1 = p_k$. The sets $N_1, N_2, \cdots, N_{k-1}$ map into sets $N_{1k}, N_{2k}, \cdots, N_{k-1,k}$ respectively and $N_{\nu k}$ covers $\sigma_\nu + 1$ of the angular openings at p_1 formed by the arcs $\text{Re}(Z_1) = \text{constant}$

emerging from p_1. These sets fit together in such a way that a point moves counterclockwise about p_1 in passing from N_{1k} to N_{2k} to $\cdots$ to $N_{k-1,k}$ to N_1.

Let

$$\sigma = \sum_{\nu=1}^{k-1} (\sigma_\nu + 1). \tag{8.5.9}$$

The mapping defined by

$$\tau^{\sigma/2} = Z_1(z)$$

then carries N_{1k}, N_{2k}, $\cdots$, $N_{k-1,k}$ into sets R_1, R_2, $\cdots$, R_{k-1} in the τ-plane and R_ν covers an angular opening with central angle equal to $2\pi(\sigma_\nu + 1)/\sigma$ at $\tau = 0$. The sum of these angles is 2π; so the domains R_1, R_2, $\cdots$, R_{k-1} together with the radial lines which lie on their boundary form a complete neighborhood of the point $\tau = 0$. It is more convenient, however, to define the local uniformizer τ by the equivalent formula

$$\tau = \{Z_\nu(z)\}^{2/\sigma}, \qquad z \subset N_\nu \tag{8.5.10}$$

The branches are here chosen in such a way that $N_1 + N_2 + \cdots + N_{k-1}$ maps in a one–one way into a complete neighborhood of $\tau = 0$.

We have thus defined a local uniformizer at each point of the closed surface. As in Chapter II, a function F defined in a neighborhood of a point of the surface will be said to be regular-analytic there if it has a development in terms of the local uniformizer τ which is of the form

$$F = b_0 + b_1\tau + b_2\tau^2 + \cdots.$$

With this definition the surface becomes a Riemann surface.

8.6. Lemma XXVI. *There exists a function $w = F$ on the Riemann surface which is regular at every point of it except for a simple pole of residue 1 at $z = 0$ and this function maps the surface onto the closed w-plane in a one–one manner. Any two functions F with these properties differ by a constant.*

Most of Lemma XXVI is a consequence of Lemma III, and we have only to add a few remarks to make the proof of Lemma XXVI complete.

From Lemma III we obtain a function F which is regular at points of the Riemann surface corresponding to a pair of identified points z' and z'' which are interior points of their arcs I'_ν, I''_ν. We show now that F is regular at every point of the Riemann surface. Let z_0 be any point on $|z| = 1$ and let τ be a local uniformizer at z_0. The half-neighborhood at z_0 interior to $|z| < 1$ and the corresponding half-neighborhoods at the points on $|z| = 1$ which are equivalent to z_0 appear in the τ-plane as sectors coming together at $\tau = 0$. Any radial line separating two adjacent sectors is the image of a pair of identified arcs on $|z| = 1$. When expressed as a function of τ, F is regular in the interior of each sector. Moreover, as τ approaches a point on one of these radial lines from either side, F takes the same value and we see that F in any sector is the analytic continuation of F in the adjacent sectors. Therefore F is regular and single-valued throughout the neighborhood of $\tau = 0$. Thus

$$F = \sum_{-\infty}^{\infty} b_\nu \tau^\nu.$$

Since, however, the area of the map by F of any ring $\epsilon < |\tau| < \rho_0$, $\epsilon > 0$, is bounded uniformly in ϵ, we see that all negative powers of τ have coefficients equal to zero and so

$$F = \sum_{\nu=0}^{\infty} b_\nu \tau^\nu$$

near $\tau = 0$. This shows that F is regular at each point of the closed Riemann surface. The remainder of the lemma follows from Lemma III. Since F is schlicht, it follows that $b_1 \neq 0$.

8.7. We recall that two points of $|z| < 1$ are equivalent if they are identified with each other or belong to the same cycle. Moreover, we say that any point of $|z| < 1$ is identified with itself. The points of $|z| \leqq 1$ are thus divided into classes of equivalent points. A class contains only finitely many points of $|z| \leqq 1$ and it corresponds to a single point of the closed Riemann surface resulting from the identification. When we speak of the class z, we mean the class of points of $|z| \leqq 1$ which are equivalent to z.

Given any point z_0 on $|z| = 1$, let m denote the number of points which belong to the class z_0. If z_0 is an interior point of an arc I'_ν, $m = 2$ while if z_0 is a vertex (end point of an arc I'_ν), then $m \geqq 1$. If $m = 1$, z_0 is the common end point of abutting arcs which are identified with one another. Let $z_0, z_1, \cdots, z_{m-1}$ be the m points of the class z_0 and write

$$\sigma = \sum_{\nu=0}^{m-1} (\sigma_\nu + 1)$$

where $2\sigma_\nu$, $\sigma_\nu \geqq 0$, is the precise order of the zero of $Q(z)$ at z_ν. When we wish to emphasize the dependence of σ on the class z_0, we write $\sigma = \sigma(z_0)$. We remark that, by postulate (iii) concerning the identification, $\sigma(z_0) = 1$ for at most one point on $|z| = 1$.

Let τ be the local uniformizer belonging to the class z_0. In terms of τ we have

$$F = \sum_{\nu=0}^{\infty} b_\nu \tau^\nu, \qquad b_1 \neq 0. \tag{8.7.1}$$

In the half-neighborhood N_0 at z_0 which is interior to $|z| < 1$ we may express τ as a function of z_0:

$$\tau = (z - z_0)^{2(\sigma_0+1)/\sigma} \sum_{\nu=0}^{\infty} c_\nu (z - z_0)^\nu, \qquad c_0 \neq 0. \tag{8.7.2}$$

Thus

$$t = \tau^{\sigma/2(\sigma_0+1)} = \sum_{\nu=1}^{\infty} d_\nu(z - z_0), \qquad d_1 \neq 0,$$

is a regular function of z near $z = z_0$ and we have

$$F = \sum_{\nu=0}^{\infty} b_\nu t^{(2(\sigma_0+1)/\sigma)\nu}, \qquad b_1 \neq 0.$$

This formula is valid by analytic continuation for all small t. Replacing t by its development in terms of $z - z_0$, we obtain

(8.7.3)
$$F = b_0 + \sum_{\nu=2(\sigma_0+1)}^{\infty} e_\nu(z - z_0)^{\nu/\sigma}, \qquad e_{2(\sigma_0+1)} \neq 0.$$

When $\sigma = 1$ or 2, we see from (8.7.2) that τ is a regular function of $z - z_0$ and then F is a regular function of z at z_0. In particular $\sigma = 2$ at interior points of arcs I'_ν, I''_ν; so F is regular there.

We remark that $w = F$ maps $|z| < 1$ onto the w-plane minus a continuum which is composed of a finite set of arcs. Each arc is analytic everywhere including its end points. For the portion of any arc in the neighborhood of an end point is the image by (8.7.1) of a radial line in the plane of the local uniformizer τ. A point where two or more arcs intersect will be called a knot.

If $z_0, z_1, \cdots, z_{m-1}$ are the m points of a class which map into a knot, then near z_k we have

(8.7.4)
$$F = b_0 + \sum_{\nu=2(\sigma_k+1)}^{\infty} e_\nu(z - z_k)^{\nu/\sigma}, \qquad e_{2(\sigma_k+1)} \neq 0.$$

The point z_k corresponds to an angular opening between two adjacent branches or arcs at the knot and the angle between these two branches is equal to $2\pi(\sigma_k + 1)/\sigma$. The angle will equal 2π only if $\sigma_k + 1 = \sigma$ and then $m = 1$, $k = 0$, and $F'(z_0) = 0$.

8.8. Let $Q(z)$ always denote a function of the form

$$Q(z) = \sum_{\nu=-(n-1)}^{n-1} \frac{B_\nu}{z^\nu}$$

which is non-negative on $|z| = 1$ with at least one zero there. We assume that not all coefficients B_ν are zero.

Of the classes of equivalent points on $|z| = 1$ we shall distinguish one and denote it by $\boldsymbol{\Sigma}$. If there is a point on $|z| = 1$ for which $\sigma(z) = 1$, then this point is a cycle containing one point and will be taken to be $\boldsymbol{\Sigma}$. On the other hand, if $\sigma(z) \geqq 2$ on $|z| = 1$, then any class of equivalent points can be chosen as $\boldsymbol{\Sigma}$. Thus we suppose that $\boldsymbol{\Sigma}$ satisfies the following condition:

(v) $\boldsymbol{\Sigma}$ is any class of identified points on $|z| = 1$ subject only to the restriction that if $\sigma(z) = 1$ at some point of $|z| = 1$, then $\boldsymbol{\Sigma}$ is that point.

By making a further subdivision of arcs we shall always suppose that the points of $\boldsymbol{\Sigma}$ are vertices. We define a canonical identification of arcs on $|z| = 1$ to be one satisfying conditions (i), (ii), and (iii) in which every cycle of equivalent points different from $\boldsymbol{\Sigma}$ and containing precisely two points corresponds to interior points of identified arcs. This canonical subdivision satisfies condition (v) but not necessarily (iv). It is clear that every identification of points can be converted into a canonical one without loss of generality.

Two functions Q will be considered equal if their ratio is identically equal to a positive constant. Each Q, a canonical identification of arcs on $|z| = 1$, and $\boldsymbol{\Sigma}$ such that (i), (ii), (iii), and (v) are satisfied define a point in a space $\boldsymbol{S}_n$.

Given a point of $\boldsymbol{S}_n$, we say that a function of class $\mathcal{S}$

$$w = f(z) = z + a_2 z^2 + \cdots$$

belongs to it if it is regular at every point of the Riemann surface $\mathcal{R}$ except for a simple pole at the point of $\mathcal{R}$ corresponding to $\boldsymbol{\Sigma}$.

THEOREM V. *To each point of* $\boldsymbol{S}_n$ *there belongs a unique* $\mathfrak{D}_n$*-function* $w = f(z)$. *It satisfies the differential equation*

$$\left(\frac{z}{w}\frac{dw}{dz}\right)^2 P(w) = Q(z) \tag{8.8.1}$$

where Q *is given by the point of* $\boldsymbol{S}_n$ *and*

$$P(w) = \frac{A_1}{w} + \frac{A_2}{w^2} + \cdots + \frac{A_{n-1}}{w^{n-1}}.$$

To each $\mathfrak{D}_n$*-function there corresponds a point of* $\boldsymbol{S}_n$ *and if* $f(z)$ *is not algebraic, it belongs to a unique point of* $\boldsymbol{S}_n$.

PROOF: Given a point of $\boldsymbol{S}_n$, let F be the function of Lemma XXVI. If w_0 is any value taken by F on $|z| = 1$, the function

$$f(z) = \frac{1}{F(z) - w_0}$$

will belong to class $\mathcal{S}$. If z_0 is a point on $|z| = 1$ where $F(z_0) \neq w_0$, then from (8.7.3) we obtain locally

$$f(z) = f(z_0) + \sum_{\nu=2(\sigma_0+1)}^{\infty} \alpha_\nu (z - z_0)^{\nu/\sigma}, \qquad \alpha_{2(\sigma_0+1)} \neq 0. \tag{8.8.2}$$

On the other hand if $F(z_0) = w_0$, then near z_0

$$f(z) = \sum_{\nu=-2(\sigma_0+1)}^{\infty} \alpha_\nu (z - z_0)^{\nu/\sigma}, \qquad \alpha_{-2(\sigma_0+1)} \neq 0. \tag{8.8.3}$$

Let z' be an interior point of an arc I' and let z'' be the corresponding interior point of I''. Then

$$\int_{p'}^{z'} (Q(t))^{1/2} \frac{dt}{t} = \int_{p''}^{z''} (Q(t))^{1/2} \frac{dt}{t}.$$

If Z is the common value of these integrals, then locally we may express z' as a function Z, $z' = z'(Z)$ and similarly $z'' = z''(Z)$. Now

$$f(z'(Z)) = f(z''(Z)). \tag{8.8.4}$$

Differentiating both sides with respect to Z, we obtain

$$\frac{df(z')}{dz'}\frac{dz'}{dZ} = \frac{df(z'')}{dz''}\frac{dz''}{dZ},$$

that is,

$$z'\frac{df(z')}{dz'}\frac{1}{(Q(z'))^{1/2}} = z''\frac{df(z'')}{dz''}\frac{1}{(Q(z''))^{1/2}}. \tag{8.8.5}$$

Writing $w = f(z)$, we see that the expression

$$P(w) = Q(z) \Big/ \left(\frac{z}{w}\frac{dw}{dz}\right)^2$$

takes equal values on opposite edges of the slits in the plane of $w = f$ and is therefore a single-valued function of w. At a finite point of a slit or at a knot we see from formula (8.8.2) that $P(w)$ behaves like

$$(z - z_0)^{2(\sigma_0+1)[1-2/\sigma]} \quad \text{or} \quad (w - w_0)^{\sigma-2}$$

where $\sigma = \sigma(z_0)$. If $\sigma(z_0) \geqq 2$, P remains finite while if $\sigma(z_0) > 2$, P vanishes. If $\sigma(z_0) = 1$, P becomes infinite. At $w = \infty$, P behaves like

$$(z - z_0)^{2(\sigma_0+1)} \quad \text{or} \quad 1/w^{\sigma}.$$

We may suppose without loss of generality that the coefficient B_{n-1} of Q is not zero. Then P has a pole of order $n - 1$ at $w = 0$.

It is clear that the function P will have the form

$$P(w) = \frac{A_1}{w} + \frac{A_2}{w^2} + \cdots + \frac{A_{n-1}}{w^{n-1}} \tag{8.8.6}$$

if and only if $\sigma \geqq 2$ at every point of $|z| = 1$ where f is finite. (This is the basis of the definition (v) of $\mathbf{\Sigma}$.) If this is the case, let z_0 be a point on $|z| = 1$ where $f = \infty$. Then

$$A_1 = A_2 = \cdots = A_{\mu-1} = 0$$

where $\mu = \sigma(z_0)$, $f(z_0) = \infty$. This shows that there is at least one $\mathfrak{D}_n$-function belonging to the point of $\mathbf{S}_n$. Lemma XXVI shows that it is unique.

Suppose now that the same $\mathfrak{D}_n$-function $w = f(z)$ belongs to two different points of $\mathbf{S}_n$. Then the identifications of points on $|z| = 1$ as well as $\mathbf{\Sigma}$ are equal and so the functions Q_1, Q_2 at the two points are different; that is, Q_1/Q_2 is not equal to a positive constant. Then $w = f$ must satisfy two $\mathfrak{D}_n$-equations

$$\left(\frac{z}{w}\frac{dw}{dz}\right)^2 P_1(w) = Q_1(z), \qquad \left(\frac{z}{w}\frac{dw}{dz}\right)^2 P_2(w) = Q_2(z).$$

We see (dividing one equation by the other) that $w = F$ satisfies an algebraic equation

$$\frac{P_1(w)}{P_2(w)} = \frac{Q_1(z)}{Q_2(z)}.$$

That is, $w = f$ is an algebraic function. This completes the proof of Theorem V.

We add a remark about the relation between functions belonging to distinct points of $\boldsymbol{S}_n$ which differ only in the choice of $\boldsymbol{\Sigma}$. That is, we assume that Q is the same and that the identification subject to the postulates (i), (ii), and (iii) is the same at the points. Then the function $\sigma(z)$ is the same at the points of $\boldsymbol{S}_n$. It is clear from the proof of Theorem V or from **8.5** that if $\sigma(z) = 1$ at some point of $|z| = 1$, then that point must be mapped into infinity by the $\mathfrak{D}_n$-function. Thus consider the case where $\sigma(z) \geqq 2$ on $|z| = 1$. If for some choice of $\boldsymbol{\Sigma}$, $w = f(z)$ is a $\mathfrak{D}_n$-function belonging to the corresponding point of $\boldsymbol{S}_n$ and satisfying a $\mathfrak{D}_n$-equation

$$\left(\frac{z}{w}\frac{dw}{dz}\right)^2 P(w) = Q(z)$$

where

$$P(w) = \sum_{\nu=1}^{n-1} \frac{A_\nu}{w^\nu},$$

let c be a point lying on the image of $|z| = 1$ by $w = f(z)$. Then the function

$$w_1 = \frac{cf(z)}{c - f(z)} = z + \cdots$$

is also a $\mathfrak{D}_n$-function belonging to a new point of $\boldsymbol{S}_n$ which differs from the original point only in the choice of $\boldsymbol{\Sigma}$. As c traverses the whole of the image of $|z| = 1$ by $w = f(z)$, $\boldsymbol{\Sigma}$ sweeps over all possible cycles of points on $|z| = 1$ belonging to the given identification. It may be verified that the function w_1 satisfies the $\mathfrak{D}_n$-equation

$$\left(\frac{z}{w_1}\frac{dw_1}{dz}\right)^2 P_1(w_1) = Q(z)$$

where

$$P_1(w_1) = \sum_{\nu=1}^{n-1} \frac{A_\nu^{(1)}}{w_1^\nu}$$

and

$$A_\nu^{(1)} = \sum_{k=\nu}^{n-1} A_k \binom{k-2}{\nu-2} c^{\nu-k}.$$

8.9. The total realization of the identification as expressed by Theorem V may also be realized step-wise. Instead of considering that all points on $|z| = 1$ are identified, we may suppose that only some of the arcs are identified. For the sake of simplicity, let there be two arcs I', I'' of $|z| = 1$ which abut at a zero z_1 of $Q(z)$ on $|z| = 1$ and are identified by the relation

$$\int_{z_1}^{z'} (Q(t))^{1/2} \frac{dt}{t} = \int_{z_1}^{z''} (Q(t))^{1/2} \frac{dt}{t} \tag{8.9.1}$$

where z' is a point of I' and z'' is a point of I''. Here the square root of Q is positive in one integral, negative in the other. The arc of $|z| = 1$ complementary to $I' + I''$ is an unidentified arc I which we suppose does not degenerate to a single point. We now allow the possibility that $Q(z)$ vanishes at interior points of I' and I''. At a pair of identified points z_0' and z_0'' belonging to I' and I'' respectively, we introduce a local uniformizer τ as above. If $Q(z)$ does not vanish at z_0' or at z_0'', then the uniformizer is defined by (8.5.6). If $Q(z)$ vanishes at either or both of the points, let it have order $2\sigma'$, $\sigma' \geqq 0$, at z_0' and order $2\sigma''$ at z_0''. Writing $\sigma = (\sigma' + 1) + (\sigma'' + 1)$, let

$$Z'(z') = \int_{z_0'}^{z'} (Q(t))^{1/2} \frac{dt}{t}, \qquad Z''(z'') = \int_{z_0''}^{z''} (Q(t))^{1/2} \frac{dt}{t},$$

and denote the half-neighborhood interior to $|z| = 1$ at z_0' by N', that at z_0'' by N''. We define

$$\tau = \begin{cases} [Z'(z')]^{2/\sigma}, & z' \subset N', \\ [Z'(z'')]^{2/\sigma}, & z'' \subset N''. \end{cases}$$

At the common end point z_1 of I' and I'', let N be a half-neighborhood interior to $|z| = 1$ and suppose that Q has a zero of order $2\sigma_1$, $\sigma_1 > 0$, at z_1. Let

$$\tau = [Z(z)]^{2/\sigma}, \qquad z \subset N,$$

where

$$Z(z) = \int_{z_1}^{z} (Q(t))^{1/2} \frac{dt}{t}$$

and $\sigma = \sigma_1 + 1$. By introducing these local uniformizers the circle $|z| < 1$ plus the part of $|z| = 1$ complementary to the closed arc I becomes a Riemann surface.

There is a function F which is regular at each point of the Riemann surface except for a simple pole of residue 1 at $z = 0$ and which maps the Riemann surface onto the plane minus a single slit parallel to the real axis. This is almost a corollary of Lemma III, and we have only to investigate the behavior of F as a function of the local uniformizer corresponding to points where Q vanishes on the arc of $|z| = 1$ complementary to the closed interval I. From such an investigation it readily follows that F is regular at each point of the Riemann surface.

LEMMA XXVII. *There is a function*

$$v(z) = u\{z + b_2 z^2 + b_3 z^3 + \cdots\}, \qquad u > 0, \tag{8.9.2}$$

which is regular for $|z| < 1$ *and maps* $|z| < 1$ *onto* $|v| < 1$ *minus a piecewise analytic slit whose two edges correspond to* I' *and* I'' *respectively, identified points being brought together. The function* $v(z)$ *satisfies a differential equation of the form*

$$\left(\frac{z}{v}\frac{\partial v}{\partial z}\right)^2 R(v) = Q(z) \tag{8.9.3}$$

where

$$R(v) = \sum_{\nu=-(n-1)}^{n-1} \frac{\beta_\nu}{v^\nu}, \qquad \beta_{-\nu} = \bar{\beta}_\nu. \tag{8.9.4}$$

Here $R(v)$ *is non-negative on* $|v| = 1$ *and has at least one zero there.*

PROOF: Let z_2' and z_2'' be the non-abutting end points of I' and I''. Let

$$Z'(z') = \int_{z_2'}^{z'} (Q(t))^{1/2} \frac{dt}{t}, \qquad Z''(z'') = \int_{z_2''}^{z''} (Q(t))^{1/2} \frac{dt}{t},$$

and let N' and N'' be the half-neighborhoods at z_2' and z_2'' which are interior to $|z| = 1$. If $2\sigma'$, $\sigma' \geqq 0$, is the order of the zero of Q at z_2', and $2\sigma''$ that of Q at z_2'', let $\sigma = (\sigma' + 1) + (\sigma'' + 1)$ and define

$$\tau = \begin{cases} [Z'(z')]^{1/\sigma}, & z' \subset N', \\ [Z''(z'')]^{1/\sigma}, & z'' \subset N''. \end{cases} \tag{8.9.5}$$

Now map the exterior of the slit in the plane of F onto the interior of the unit circle $|v| < 1$ in such a way that the point at infinity goes into the origin with direction preserved. We then obtain a function

$$v(z) = u\{z + b_2 z^2 + b_3 z^3 + \cdots\}, \qquad u > 0,$$

which is regular for $|z| < 1$ and maps $|z| < 1$ onto $|v| < 1$ minus a piecewise analytic slit in such a way that identified points of I' and I'' are brought together.

In the plane of the variable τ defined by (8.9.5) the two pieces of the unidentified arc I near z_2' and z_2'' appear as segments of the same straight line meeting at $\tau = 0$. The two half-neighborhoods N' and N'' appear as adjacent sectors lying on one side of this straight line. The function v considered as a function of τ is then schlicht and regular on one side of the straight line and of modulus unity on the straight line itself. By reflection it follows that v is regular and schlicht in a complete neighborhood of $\tau = 0$. We have

$$\tau = \begin{cases} (z - z_2')^{(\sigma'+1)/\sigma} \sum_{\nu=0}^{\infty} c_\nu'(z - z_2')^\nu , & z \subset N', \\ (z - z_2'')^{(\sigma''+1)/\sigma} \sum_{\nu=0}^{\infty} c_\nu''(z - z_2'')^\nu, & z \subset N'', \end{cases}$$

where $c_0' \neq 0$, $c_0'' \neq 0$. Since $\partial v/\partial\tau$ is regular and non-vanishing at $\tau = 0$, we have

$$\frac{\partial v}{\partial z} \sim \begin{cases} (z - z_2')^{(\sigma'+1)/\sigma-1} & \text{near} \quad z_2' , \\ (z - z_2'')^{(\sigma''+1)/\sigma-1} & \text{near} \quad z_2'' . \end{cases} \tag{8.9.6}$$

The symbol $\sim$ as here used indicates behavior in so far as order of magnitude is concerned. In a similar fashion but simpler we find that near a point z_0' of I'

$$\frac{\partial v}{\partial z} \sim (z - z_0')^{2(\sigma'+1)/\sigma-1} \tag{8.9.6$'$}$$

where $2\sigma'$ is the order of the zero of Q at z_0', $2\sigma''$ that at z_0'', and $\sigma = (\sigma' + 1) + (\sigma'' + 1)$. Near z_1 we have

$$\frac{\partial v}{\partial z} \sim (z - z_1). \tag{8.9.6$''$}$$

By analytic continuation across I the function $v(z)$ is schlicht and single-valued in the whole z-plane minus the closure of $I' + I''$, and it maps the complement of the closure of $I' + I''$ onto the complement of a piecewise analytic slit γ which is symmetrical about the unit circumference $|z| = 1$ and intersects the unit circumference at a point $v_2 = v(z_2') = v(z_2'')$. Then the expression

$$R(v) = Q(z) \Big/ \left(\frac{z}{v} \frac{\partial v}{\partial z}\right)^2 \tag{8.9.7}$$

considered as a function of v is single-valued across γ since the identification is defined by Q. At the point z_1 where I' and I'' abut, we see that $R(v)$ is finite since the zero of $Q(z)$ at z_1 cancels the double zero of $(\partial v/\partial z)^2$. At a point z_0' of I' we have

$$R(v) \sim (z - z_0')^{2(\sigma'+1)(1-2/\sigma)} \sim (v - v_0')^{\sigma-2}$$

where $v_0' = v(z_0')$. Since $\sigma \geqq 2$, we see that $R(v)$ is bounded at points corresponding to points of I' and I''. Finally, at the point z_2' we have

$$R(v) \sim (z - z_2')^{2(\sigma'+1)(1-1/\sigma)} \sim (v - v_2)^{2(\sigma-1)}.$$

Since $\sigma \geqq 2$, this implies that $R(v)$ vanishes at v_2. It follows that $R(v)$ is a rational function which is non-negative on $|v| = 1$ with at least one zero there. By considering the behavior near $v = 0$, $z = 0$, we see that $R(v)$ is of the form (8.9.4).

By using Lemma XXVII we now show how to realize the identification continuously. Given an identification satisfying (i), (ii), (iii), and a distinguished cycle Σ satisfying condition (v), let $I_1', I_1'', I_2', I_2'', \cdots, I_k', I_k''$ be the pairs of identified arcs on $|z| = 1$. By the definition of Σ it is clear that there is always one pair of abutting arcs with a common end point $e^{-i\alpha}$ not in Σ. Without loss of generality let this pair of arcs be denoted by I_1' and I_1''. It then follows that $Q(z)$ must be zero at $z = e^{-i\alpha}$. Let I' and I'' be identified subarcs of I_1' and I_1'' abutting at $e^{-i\alpha}$. If $k > 1$, we may take I' and I'' equal to I_1' and I_1'' respectively. According to Lemma XXVII, there is a function

$$v(z, u) = u\{z + b_2 z^2 + b_3 z^3 + \cdots\}, \tag{8.9.8}$$

where u is some number between 0 and 1 which identifies I' and I'' and satisfies an equation

$$\left(\frac{z}{v}\frac{\partial v}{\partial z}\right)^2 Q(v, u) = Q(z). \tag{8.9.9}$$

Here we have written $Q(v, u) = R(v)$. The function

$$Q(v, u) = \sum_{\nu=-(n-1)}^{n-1} \frac{B_\nu(u)}{v^\nu}, \qquad B_{-\nu}(u) = \overline{B_\nu(u)},$$

is non-negative on $|v| = 1$ and vanishes at the point $e^{-i\alpha(u)}$ on $|v| = 1$ which is the image of the two disjoint end points of I' and I''. The function $v(z, u)$ maps $|z| < 1$ onto $|v| < 1$ minus a piecewise analytic arc whose two edges correspond to I' and I'' respectively, identified points being brought together. The function transports the part of $|z| = 1$ complementary to $I' + I''$ into the entire circumference $|v| = 1$ and it carries any pair of identified points on this complementary set into a new pair of identified points, arcs into arcs and vertices into vertices. Since

$$\left|\frac{Q(v, u)}{v^2}\right| |dv|^2 = \left|\frac{Q(z)}{z^2}\right| |dz|^2, \tag{8.9.10}$$

the new identification on $|v| = 1$ is established in the metric defined by $Q(v, u)$. The new identification satisfies postulates (i), (ii), and (iii), and the cycle Σ is mapped by v into a new distinguished cycle whose points are vertices in the new identification since the points of the cycle Σ are vertices in the original identification. The new distinguished cycle also satisfies condition (v). The function $v(z, u) = u\{z + \cdots\}$, $u > 0$, is uniquely determined by the identified arcs I', I''.

Given proper subarcs I', I'' of I_1', I_1'', let the corresponding function $v(z, u)$ map the arcs $I_1' - I'$, $I_1'' - I''$, $I_2', I_2'', \cdots, I_k', I_k''$ onto the identified arcs

J_1', J_1'', J_2', J_2'', $\cdots$, J_k', J_k'' respectively. The common end point $v = e^{-i\alpha(u)}$ of J_1', J_1'' is a zero of $Q(v, u)$. If J', J'' are identified subarcs of J_1', J_1'' and abut at $e^{-i\alpha(u)}$, then there is a function

$$h = \gamma\{v + \cdots\}, \qquad 0 < \gamma < 1,$$

which identifies J' and J''. The composite mapping from z to v and from v to h is a function $v(z, u\gamma)$ of the form (8.9.8) with leading coefficient $u\gamma$ and it identifies subarcs of I_1' and I_1'' which include I' and I''. It follows that u decreases as I' and I'' increase. It is clear that as J' and J'' shrink to their common end point, the function h tends to the identity mapping and γ tends toward 1. Thus u decreases continuously as I' and I'' increase. We are therefore justified in expressing the dependence of v on u as in (8.9.8).

If $k > 1$, the arcs I' and I'' may be allowed to increase continuously until they coincide with I_1' and I_1''. This defines a function $v(z, u)$ in the interval $u_1 \leqq u \leqq 1$, $u_1 > 0$. The function $v_1 = v(z, u_1)$ maps I_1' and I_1'' into the opposite edges of a slit interior to the unit circle and it maps I_2', I_2'', $\cdots$, I_k', I_k'' into a new set of identified arcs J_2', J_2'', $\cdots$, J_k', J_k'' lying on the unit circumference in the plane of v_1. The identification in the plane of v_1 which is established in the metric defined by $Q(v_1, u_1)$ satisfies postulates (i), (ii), and (iii); and the cycle $\boldsymbol{\Sigma}$ is mapped by v_1 into a new cycle whose points are among the end points of the arcs J_2', J_2'', $\cdots$, J_k', J_k''. However, if the non-abutting end points of I_1' and I_1'' are points of the cycle $\boldsymbol{\Sigma}$, then the new cycle corresponding to $\boldsymbol{\Sigma}$ contains one point less. If $k > 2$, the process may be repeated, giving rise to a family of functions which are regular in $|v_1| < 1$ and so regular in $|z| < 1$. The functions (8.9.8) are thus defined in an interval $u_2 \leqq u \leqq 1$, $0 < u_2 < u_1 < 1$, and satisfy (8.9.9). Proceeding in this way, we realize more and more of the original identification, and finally after $k - 1$ steps we arrive at an identification in the metric defined by $Q(v_{k-1}, u_{k-1})$ which consists of only two identified arcs K_1', K_1'' with end points at a and b, $|a| = |b| = 1$. One of the two points, say a, is not an image of $\boldsymbol{\Sigma}$ whereas the other, b, is the complete image of $\boldsymbol{\Sigma}$. Then a is a zero of $Q(v_{k-1}, u_{k-1})$. Now let K' and K'' be identified subarcs of K_1' and K_1'' abutting at a, and let

$$v = \frac{u}{u_{k-1}}\{v_{k-1} + \cdots\} = u\{z + \cdots\}, \qquad 0 < u < 1,$$

realize the identification of K' and K''. Increase the lengths of the identified arcs K' and K'' continuously and let them tend toward K_1' and K_1'' respectively. Then u decreases monotonically and we shall show that its limit is zero.

A subsequence of the functions u/v tends toward a limit F. By introducing a local uniformizer τ at the point b, it is seen that F is regular in a neighborhood of $\tau = 0$, therefore continuous at $\tau = 0$, and so continuous at $v_{k-1} = b$. The function F is then regular at each point of the closed Riemann surface in the z-plane. By Lemma XXVI the function $w = F$ maps the closed Riemann surface onto the closed w-plane in a one–one way. In $|z| < 1$ the function $|u/v|$ is bounded below by u and it follows that u tends to zero as K' and K'' tend to K_1' and K_1''. Then $v(z)/u$ tends to a function

$$w = f = z + a_2 z^2 + a_3 z^3 + \cdots$$

which maps the closed Riemann surface onto the closed w-plane. It satisfies the differential equation

$$\left(\frac{z}{w}\frac{dw}{dz}\right)^2 P(w) = Q(z) \tag{8.9.11}$$

where $Q(z)$ is the given function Q and

$$P(w) = \sum_{\nu=1}^{n-1} \frac{A_\nu}{w^\nu}.$$

This provides a more constructive proof of the existence of the function $f(z)$ described in Theorem V.

We have shown that a subsequence of the functions $u/v(z)$ tends toward a limit F, where F is regular on the closed Riemann surface except for a simple pole of residue 1 at $z = 0$. We now show that $u/v(z)$ tends toward F on every subsequence as u tends to zero. Under the mapping from the v_{k-1}-plane to the plane of $w = F$, a small arc containing the point b and complementary to $K' + K''$ maps into a small slit λ in the w-plane and λ shrinks to a point (the origin) as u tends to zero. Considered as a function of w, the function

$$\frac{u}{v(z, u)} = w + b_0 + \frac{b_1}{w} + \cdots, \qquad v(z) = u\,\{z + \cdots\},$$

is regular outside the slit $\lambda = \lambda(u)$ except for a pole at $w = \infty$. It is also schlicht in the complement of λ and maps the complement of λ onto the exterior of the circle of radius u with center at the origin. Then it follows that $u/v(z, u)$ tends to $w + c$, where c is a constant. But since the origin maps into the origin, we see that $c = 0$. Then $v(z, u)/u$ tends to $f(z)$ as u tends to zero.

For small u the mapping from the plane of u/v to the plane of the limit function F is schlicht in the exterior of the circle of radius u with center at the origin and maps the exterior of this circle into the exterior of the slit λ. The mapping is regular in the exterior of the circle of radius u except for a pole at infinity. The slit λ is regular even at its end points because it is the image by F of a straight-line segment in the plane of the local uniformizer τ at a point of the closed Riemann surface in the z-plane. It is an arc of the locus on which

$$\operatorname{Re} \int \left(\sum_{\nu=1}^{n-1} A_\nu w^\nu\right)^{1/2} \frac{dw}{w} = \text{constant}.$$

Magnifying each plane by the factor $1/u$ we obtain a mapping of the exterior of the unit circle in the plane of $1/v$ onto the exterior of a slit $\Lambda(u)$ in the plane of F/u, infinity going into infinity with unit magnification and preservation of direction there. The slit $\Lambda(u)$ is a magnification of $\lambda(u)$ and, since $\lambda(u)$ is analytic, it follows that $\Lambda(u)$ approaches a straight-line segment Λ_0 of length 4 with one end point at the origin. The points $\bar{a}$ and $\bar{b}$ which are the end points of the identified arcs in the plane of $1/v$ map into the end points of the slit $\Lambda(u)$ with $\bar{b}$ mapping into the end point at the origin. If Λ_0 makes angle τ with the positive real axis, then $\bar{a}$ approaches $e^{i\tau}$, $\bar{b}$ approaches $-e^{i\tau}$ as u tends to zero. Thus as u tends to zero,

$$\text{(8.9.12)} \qquad \lim a = e^{-i\tau}, \qquad \lim b = -e^{-i\tau}.$$

If the identification is canonical, the order of the step-wise realization of the identification is uniquely determined if and only if the total number of identified arcs is precisely two. The realization to which we refer is one that tends to a $\mathfrak{D}_n$-function with the cycle $\mathbf{\Sigma}$ mapping into the point at infinity.

CHAPTER IX

LÖWNER CURVES[1]

9.1. Given any Q-function

$$Q(z) = \sum_{\nu=-(n-1)}^{n-1} \frac{B_\nu}{z^\nu}, \qquad B_{-\nu} = \bar{B}_\nu, \tag{9.1.1}$$

and corresponding identification of points on $|z| = 1$ satisfying conditions (i), (ii), (iii), let $\boldsymbol{\Sigma}$ be a cycle of points on $|z| = 1$ satisfying (v). We suppose that there are $2k$ arcs identified in pairs, $k \geqq 1$, and that the points of $\boldsymbol{\Sigma}$ are among the end points of the arcs. The continuous realization of the identification described in **8.9** gives rise to a family of functions

$$v(z, u) = u\,\{z + b_2(u)z^2 + b_3(u)z^3 + \cdots\}, \qquad 0 < u \leqq 1, \tag{9.1.2}$$

which realizes more and more of the given identification as u tends to zero. The function $v(z, u)$ maps $|z| < 1$ onto $|v| < 1$ cut along finitely many analytic arcs. It satisfies the differential equation

$$\left(\frac{z}{v}\frac{\partial v}{\partial z}\right)^2 Q(v, u) = Q(z) \tag{9.1.3}$$

where

$$Q(v, u) = \sum_{\nu=-(n-1)}^{n-1} \frac{B_\nu(u)}{v^\nu}, \qquad B_{-\nu}(u) = \overline{B_\nu(u)}, \tag{9.1.4}$$

and $Q(v, u)$ is non-negative on $|v| = 1$ with a zero at $v = e^{-i\alpha(u)}$. If

$$0 < u' < u'' \leqq 1$$

and $v' = v(z, u')$, $v'' = v(z, u'')$, then v' is a regular function of v'' in $|v''| < 1$. As u tends to zero, the function $v(z, u)/u$ tends toward $f(z)$ where $f(z)$ is the $\mathfrak{D}_n$-function belonging to the given point of S_n. The function $w = f(z)$ satisfies a differential equation

$$\left(\frac{z}{w}\frac{dw}{dz}\right)^2 P(w) = Q(z) \tag{9.1.5}$$

where $Q(z)$ is the given Q function and

$$P(w) = \sum_{\nu=1}^{n-1} \frac{A_\nu}{w^\nu}. \tag{9.1.6}$$

According to (8.9.12), $e^{-i\alpha(u)}$ tends to a limit as u tends to zero. There are k intervals $u_{\nu-1} < u < u_\nu$, $\nu = 1, 2, \cdots, k$, where $u_0 = 0 < u_1 < u_2 < \cdots < u_k = 1$,

[1] A remark of historical character is to be found at the end of the present chapter.

such that over any one of these intervals the function $v(z, u)$ identifies more and more of the same pair of identified arcs in the original identification on $|z| = 1$ as u decreases. We shall show that $v(z, u)$ is continuous in $0 \leqq u \leqq 1$ and has a continuous derivative in each of the subintervals. At the points u_ν the one-sided derivatives exist but the left and right derivatives may not be the same.

THEOREM VI. *Let $w = f(z)$ be a $\mathfrak{D}_n$-function belonging to a given point of $\mathbf{S}_n$ and satisfying*

$$\left(\frac{z}{w}\frac{dw}{dz}\right)^2 P(w) = Q(z). \tag{9.1.7}$$

Then there is a function

$$v(z, u) = u\{z + b_2(u)z^2 + b_3(u)z^3 + \cdots\}, \qquad 0 < u \leqq 1,$$

which is regular and schlicht for $|z| < 1$, maps $|z| < 1$ onto $|v| < 1$ minus a finite set of analytic arcs, and satisfies

$$\left(\frac{z}{v}\frac{\partial v}{\partial z}\right)^2 Q(v, u) = Q(z) \tag{9.1.8}$$

where

$$Q(v, u) = \sum_{\nu=-(n-1)}^{n-1} \frac{B_\nu(u)}{v^\nu}, \qquad B_{-\nu}(u) = \overline{B_\nu(u)},$$

is non-negative on $|v| = 1$ with at least one zero there. Moreover, $v(z, u)$ is continuous with respect to u and, except at finitely many points, is differentiable and satisfies

$$u\frac{\partial v}{\partial u} = v\frac{1 + e^{i\alpha}v}{1 - e^{i\alpha}v} \tag{9.1.9}$$

where $\alpha = \alpha(u)$ is a piecewise continuous function of u and $Q(e^{-i\alpha(u)}, u) = 0$. The terminal conditions are

$$v(z, 1) = z, \qquad \lim_{u\to 0}\frac{v(z, u)}{u} = f(z),$$

$$Q(z, 1) = Q(z), \qquad \lim_{u\to 0} Q(uw, u) = P(w).$$

We remark that (9.1.9) is the differential equation derived by Löwner for functions mapping $|z| < 1$ onto $|v| < 1$ minus any Jordan arc not passing through the origin. However, the equation (9.1.9) for the special class of functions under consideration is almost an automatic consequence of the continuous realization of the identification.

PROOF: Let

$$h(z, u', u'') = v\{v^{-1}(z, u''), u'\} = \frac{u'}{u''}\{z + \cdots\} \tag{9.1.10}$$

where $0 < u' < u'' \leqq 1$ and where $v^{-1}(z, u'')$ denotes the function inverse to $v(z, u'')$. Since the function h is single-valued across the slit or slits in the plane of $v(z, u'')$, it is regular in $|z| < 1$ and maps $|z| < 1$ onto the interior of the unit circle minus one or more slits in the plane of $v(z, u')$ which correspond to the part of the unit circumference identified by v' but not by v''. The function $v(z, u)$ satisfies an equation of the form (9.1.8) according to Lemma XXVII; so $h(z, u', u'')$ satisfies the equation

$$\left(\frac{z}{h}\frac{\partial h}{\partial z}\right)^2 Q(h, u') = Q(z, u''). \tag{9.1.11}$$

If either u' or u'' is fixed and the other is sufficiently near, then h maps $|z| < 1$ onto $|h| < 1$ minus a single slit extending into $|h| < 1$ from a point $h_0 = \exp(-i\alpha(u'))$. Two abutting arcs J' and J'', identified using $Q(z, u'')$, are mapped by h into the opposite edges of the slit λ. The arc J of $|z| = 1$ complementary to J' plus J'' maps into the whole circumference $|h| = 1$. By reflection on J, h is defined and is schlicht over the entire z-plane cut along J' plus J''. The function h maps J' plus J'' into a slit γ composed of λ plus the image of λ on $|h| = 1$. As $u'' - u'$ tends to zero with either u' or u'' fixed, h tends to z uniformly in every circle $|z| \leqq r < 1$ since the function $\log h(z)/z$ tends to zero at the origin and has real part equal to or less than zero on $|z| = 1$. It follows that the coefficients $b_\nu(u)$ of $v(z, u)$ and the coefficients $B_\nu(u)$ of $Q(z, u)$ depend continuously on u in $0 < u \leqq 1$. The zeros of $Q(z, u)$ then depend continuously on u, $0 < u \leqq 1$. According to the definition of $e^{-i\alpha(u)}$ given in **8.9**, we know that $e^{-i\alpha(u)}$ can jump from one zero of $Q(z, u)$ to another only at the points u_ν. Then since $e^{-i\alpha(u)}$ is a zero of $Q(z, u)$ and since $\alpha(u)$ has a limit as u approaches zero, it follows that $\alpha(u)$ is a piecewise continuous function of u whose only possible discontinuities are at the points u_ν, $\nu = 1, 2, \cdots, k - 1$.

We shall first show that for u'' fixed and u' tending toward u'' from below, h has a left derivative. Let $u' = (1 - \delta)u''$, $\delta > 0$. The slit γ lies on the set of arcs $\mathbf{\Gamma}_h$ defined by $\operatorname{Re}(\zeta) =$ constant where

$$\zeta = \int (Q(h, u'' - u''\delta))^{1/2} \frac{dh}{h},$$

and γ intersects the circumference at $h_0 = \exp(-i\alpha(u'))$. As δ tends toward zero, the point h_0 tends toward a limit χ_0 since Q varies continuously, and the slit γ shrinks to the point χ_0. If u'' is not one of the points u_ν, then $\chi_0 = \exp(-i\alpha(u''))$. Near $z = \infty$ we have

$$h(z) = \frac{z}{1 - \delta} + b_0 + \frac{b_1}{z} + \cdots$$

where $h(z) = h(z, u'' - u''\delta)$. Let z be any point of the plane not on the closure of J' plus J'' and let C_1 be a contour enclosing J' plus J'' but not containing z in its interior. Let C_2 be a large contour enclosing both z and J' plus J''. As C_2 becomes large, the integral

$$\frac{1}{2\pi i}\int_{C_2} h(t)\left\{\frac{1}{t}-\frac{1}{t-z}\right\}dt = -\frac{z}{2\pi i}\int \frac{h(t)}{t(t-z)}\,dt$$

tends to $-z/(1-\delta)$. But

$$\frac{1}{2\pi i}\int_{C_2} h(t)\left\{\frac{1}{t}-\frac{1}{t-w}\right\}dt = -h(z)+\frac{1}{2\pi i}\int_{C_1} h(t)\left\{\frac{1}{t}-\frac{1}{t-z}\right\}dt;$$

so

$$h(z) = \frac{z}{1-\delta}-\frac{z}{2\pi i}\int_{C_1}\frac{h(t)}{t(t-z)}\,dt.$$

Let t_1 and t_2 be two points on the opposite edges of the arc J' plus J'' in the z-plane, t_1 on the edge in $|z|<1$, t_2 on the edge in $|z|>1$. Then

$$h(t_2)-h(t_1) = |h(t_2)-h(t_1)|\,(\chi_0(1)+o(1)).$$

Hence

$$h(z) = \frac{z}{1-\delta}-\frac{z}{2\pi}\int_{\theta_0''}^{\theta_0'}\{h(t_2)-h(t_1)\}\,\frac{d\theta}{e^{i\theta}-z}$$

where $e^{i\theta_0'}$, $e^{i\theta_0''}$ are the non-abutting end points of J' and J'', and the integration is along J' plus J''. We have

$$\frac{h(z)-z}{\delta} = \frac{z}{1-\delta}-\frac{z}{2\pi}\int_{\theta_0''}^{\theta_0'}\frac{h(t_2)-h(t_1)}{\delta}\,\frac{d\theta}{e^{i\theta}-z},$$

$$\frac{h(z)-z}{\delta z} = \frac{1}{1-\delta}-\frac{e^{-i\alpha(u'')}}{e^{-i\alpha(u'')}-z}\cdot\frac{1}{2\pi}\int_{\theta_0''}^{\theta_0'}\frac{|h(t_2)-h(t_1)|}{\delta}\,K_\delta(\theta,z)\,d\theta,$$

where

$$K_\delta(\theta,z) = \frac{h(t_2)-h(t_1)}{|h(t_2)-h(t_1)|}\,\frac{e^{-i\alpha(u'')}-z}{e^{i\theta}-z}\,\bar{\chi}_0.$$

As δ tends to zero, $K_\delta(\theta, z)$ clearly tends uniformly to 1 in any domain which excludes a neighborhood of χ_0. Taking $z = 0$, we obtain, since $h(z) = (1-\delta)z+\cdots$,

$$-1 = \frac{1}{1-\delta}-\frac{1}{2\pi}\int_{\theta_0''}^{\theta_0'}\frac{|h(t_2)-h(t_1)|}{\delta}\,K_\delta(\theta,0)\,d\theta$$

and so

$$\frac{1}{2\pi}\int_{\theta_0''}^{\theta_0'}\frac{|h(t_2)-h(t_1)|}{\delta}\,K_\delta(\theta,z)\,d\theta \to 2$$

as δ tends to zero. Thus finally

$$\frac{h(z)-z}{\delta z}\to 1-\frac{2\chi_0}{\chi_0-z} = -\frac{1+\bar{\chi}_0 z}{1-\bar{\chi}_0 z}.$$

In other words, for u'' fixed and u' tending to u'' from below we have, if u'' is not one of the points u_ν,

$$u'' \frac{h(z, u', u'') - z}{u'' - u'} \to -z \frac{1 + e^{i\alpha(u'')}z}{1 - e^{i\alpha(u'')}z}. \tag{9.1.12}$$

If u'' is one of the points u_ν, then the limit exists and (9.1.12) is true if we interpret $\alpha(u'')$ as $\alpha(u'' - 0)$. By a similar argument we may show that

$$u' \frac{h(z, u', u'') - z}{u'' - u'} \to - z \frac{1 + e^{i\alpha(u')}z}{1 - e^{i\alpha(u')}z} \tag{9.1.12$'$}$$

as u'' tends to u', u' fixed. Thus the one-sided derivatives exist at all points of $0 < u \leqq 1$.

By (9.1.10) we have

$$h(v(z, u''), u', u'') = v(z, u').$$

Replacing z in (9.1.12) and (9.1.12)$'$ by $v(z, u'')$, we obtain

$$u \frac{v(z, u') - v(z, u'')}{u'' - u'} \to -v(z, u) \frac{1 + e^{i\alpha(u)}v(z, u)}{1 - e^{i\alpha(u)}v(z, u)}.$$

Thus

$$u \frac{\partial v(z, u)}{\partial u} = v(z, u) \frac{1 + e^{i\alpha(u)}v(z, u)}{1 - e^{i\alpha(u)}v(z, u)}. \tag{9.1.13}$$

Dividing each side of this equation by u, we see that $\partial v/\partial u$ approaches $f(z)$ as u approaches zero. Since the limit exists as u approaches zero, it follows that the one-sided derivative exists at $u = 0$.

9.2. For future reference we obtain Löwner's integral representation for the coefficients $b_2(u)$, $b_3(u)$, and $b_4(u)$ of the function $v(z, u)$. Equating like powers of z in equation (9.1.9), we obtain

$$\frac{db_k}{du} = 2 \sum_{\nu=1}^{k-1} u^{\nu-1} e^{i\nu\alpha} b_k^{(\nu+1)}, \qquad k = 2, 3, \cdots,$$

where $b_k^{(\nu+1)}$ is the coefficient of z^k in the expression

$$\{z + b_2 z^2 + b_3 z^3 + \cdots\}^{\nu+1} = \sum_{k=\nu+1}^{\infty} b_k^{(\nu+1)} z^k.$$

In particular,

$$\frac{db_2}{du} = 2e^{i\alpha},$$

$$\frac{db_3}{du} = 2ue^{2i\alpha} + 4b_2 e^{i\alpha},$$

$$\frac{db_4}{du} = 2u^2 e^{3i\alpha} + 6b_2 ue^{2i\alpha} + 2(b_2^2 + 2b_3)e^{i\alpha}.$$

By integration we obtain

$$
(9.2.1)\qquad
\begin{aligned}
b_2(u) &= -2\int_u^1 e^{i\alpha(v)}\,dv,\\
b_3(u) &= -2\int_u^1 ve^{2i\alpha(v)}\,dv + 4\left(\int_u^1 e^{i\alpha(v)}\,dv\right)^2,\\
b_4(u) &= -2\int_u^1 v^2e^{3i\alpha(v)}\,dv + 12\int_u^1\!\!\int_u^1 v_1e^{2i\alpha(v_1)}e^{i\alpha(v_2)}\,dv_1\,dv_2\\
&\quad -4\int_{v_2=u}^1\int_{v_1=v_2}^1 v_1e^{2i\alpha(v_1)}e^{i\alpha(v_2)}\,dv_1\,dv_2 - 8\left(\int_u^1 e^{i\alpha(v)}\,dv\right)^3,
\end{aligned}
$$

where $b_2(0) = a_2$, $b_3(0) = a_3$, $b_4(0) = a_4$. The formula for $b_4(0)$ will be used in Chapter XIV.

Taking logarithms of both sides of (9.1.8) and differentiating with respect to u, we have

$$u\frac{\partial}{\partial u}\log\left\{\left(\frac{z}{v}\frac{\partial v}{\partial z}\right)^2 Q(v,u)\right\} = 0,$$

and so using (9.1.9),

$$(9.2.2)\qquad u\frac{\partial}{\partial u}\{\log Q(v,u)\} = -2u\frac{\partial\log(v'/v)}{\partial u} = -\frac{4e^{i\alpha}v}{(1-e^{i\alpha}v)^2}$$

where $v' = \partial v/\partial z$. By (9.1.9) we have also

$$u\frac{\partial}{\partial u}\{Q(v,u)\} = v\frac{\partial Q}{\partial v}\frac{1+e^{i\alpha}v}{1-e^{i\alpha}v} + u\frac{\partial Q}{\partial u}.$$

Hence, substituting into (9.2.2) and replacing v by z, we have

$$(9.2.3)\qquad u\frac{\partial Q(z,u)}{\partial u} = -\frac{4e^{i\alpha}z}{(1-e^{i\alpha}z)^2}Q(z,u) - z\frac{\partial Q(z,u)}{\partial z}\frac{1+e^{i\alpha}z}{1-e^{i\alpha}z}.$$

Here

$$
(9.2.4)\qquad
\begin{aligned}
&Q(z,1) = Q(z) = \sum_{\nu=-n+1}^{n-1}\frac{B_\nu}{z^\nu}, \qquad B_{-\nu} = \bar{B}_\nu,\\
&\lim_{u\to 0} Q(uw,u) = P(w) = \sum_{\nu=1}^{n-1}\frac{A_\nu}{w^\nu},
\end{aligned}
$$

where $Q(z)$ and $P(w)$ are rational functions of the $\mathfrak{D}_n$-equation (9.1.7).

Equating coefficients of like powers of z in (9.2.3), we obtain the differential equations satisfied by the coefficients $B_k(u)$, namely

$$(9.2.5)\qquad u\frac{dB_k(u)}{du} = kB_k(u) + 2\sum_{\nu=1}^{n-k-1}(k-\nu)B_{k+\nu}(u)e^{i\nu\alpha(u)}.$$

This formula is valid for $-n + 1 \leqq k \leqq n - 1$ with the standard notation that $\sum_\nu^\mu$ is zero in case $\mu < \nu$. The case $k = n - 1$ gives

$$u \frac{dB_{n-1}(u)}{du} = (n - 1)B_{n-1}(u).$$

Assume without loss of generality that $B_{n-1} \neq 0$ in the given function $Q(z)$. Then

$$B_{n-1}(u) = B_{n-1}(1)\, u^{n-1} = B_{n-1}u^{n-1}. \tag{9.2.6}$$

Since $Q(z, u)$ has a zero of precise order 2 at $e^{-i\alpha(u)}$ except for finitely many values of u, we have

$$\sum_{\nu=-n+1}^{n-1} \nu B_\nu(u)e^{i\nu\alpha(u)} = 0. \tag{9.2.7}$$

It is sometimes convenient to write

$$C_k(u) = B_k(u)u^{-k}. \tag{9.2.8}$$

The differential equation for $C_k(u)$ is

$$\frac{dC_k(u)}{du} = 2 \sum_{\nu=1}^{n-k-1} (k - \nu)C_{k+\nu}(u)u^{\nu-1}e^{i\nu\alpha(u)}, \tag{9.2.9}$$

which is valid for $-n + 1 \leqq k \leqq n - 1$. In case $k = n - 1$ we have

$$C_{n-1}(u) = B_{n-1}(1) = B_{n-1}. \tag{9.2.10}$$

If

$$V(z, u) = v(z, u)/u = z + b_2(u)z^2 + \cdots,$$

then $V(z, u)$ belongs to class $\mathfrak{S}$ and satisfies the terminal conditions

$$V(z, 1) = z, \qquad V(z, 0) = f(z), \tag{9.2.11}$$

where $f(z)$ is the $\mathfrak{D}_n$-function belonging to the given point of $\boldsymbol{S}_n$. The function $V(z, u)$ satisfies the differential equation

$$\left(\frac{z}{V}\frac{\partial V}{\partial z}\right)^2 \sum_{\nu=-n+1}^{n-1} \frac{C_\nu(u)}{V^\nu} = Q(z). \tag{9.2.12}$$

The terminal conditions on the $C_\nu(u)$ are

$$\begin{aligned} &C_\nu(1) = B_\nu, && \nu = -n + 1, -n + 2, \cdots, n - 1, \\ &C_\nu(0) = A_\nu, && \nu = 1, 2, \cdots, n - 1, \\ &C_\nu(0) = 0, && \nu \leqq 0. \end{aligned} \tag{9.2.13}$$

It may be verified from (9.2.8) that $C_{-\nu}(u) = \bar{C}_\nu(u)u^{2\nu}$. Except at finitely many values of u, $Q(z, u)$ has a zero of precise order 2 at the point $e^{-i\alpha}$, $\alpha = \alpha(u)$; so

$$\sum_{\nu=-n+1}^{n-1} \nu u^\nu C_\nu(u) e^{i\nu\alpha(u)} = 0. \tag{9.2.14}$$

Each of the sets of equations (9.2.5), (9.2.7) and (9.2.9), (9.2.14) constitutes a closed system. They link the coefficients of the function $Q(z)$ of (9.1.7) with the coefficients of the function $P(w)$ of (9.1.7).

Except for finitely many u, $Q(z, u)$ has a zero of precise order 2 at $e^{-i\alpha(u)}$. Let

$$z_1, z_2, \cdots, z_{n-2}, 1/\bar{z}_1, 1/\bar{z}_2, \cdots, 1/\bar{z}_{n-2},$$

be the remaining zeros of $Q(z, u)$, multiple zeros being written multiply. The function

$$|z_\nu| \frac{(z - z_\nu)(z - 1/\bar{z}_\nu)}{-zz_\nu}$$

is non-negative on $|z| = 1$; so

$$Q(z, u) = |B_{n-1}(u)| \frac{(z - e^{-i\alpha})^2}{-ze^{-i\alpha}} \prod_{\nu=1}^{n-2} |z_\nu| \frac{(z - z_\nu)(z - 1/\bar{z}_\nu)}{-zz_\nu}. \tag{9.2.15}$$

Then

$$|B_{n-1}(u)| e^{i\alpha}(-1)^{n-1} \prod_{\nu=1}^{n-2} \frac{|z_\nu|}{z_\nu} = \bar{B}_{n-1}(u),$$

that is,

$$\alpha(u) = \sum_{\nu=1}^{n-2} \arg(z_\nu) + (n-1)\pi - \arg B_{n-1}(u). \tag{9.2.16}$$

Thus the chosen zero $e^{-i\alpha}$ is determined by $B_{n-1}(u)$ and the remaining zeros z_ν, $\nu = 1, 2, \cdots, n-2$.

To obtain a differential equation for $z_\nu(u)$ we obtain

$$Q(z_\nu(u), u) = 0.$$

Differentiating this relation with respect to u, we obtain

$$u \frac{\partial Q}{\partial z_\nu} \frac{\partial z_\nu}{\partial u} + u \frac{\partial Q}{\partial u} = 0$$

and so

$$u \frac{\partial z_\nu}{\partial u} = -u \frac{\partial Q}{\partial u} \Big/ \frac{\partial Q}{\partial z_\nu} = z_\nu \frac{1 + e^{i\alpha} z_\nu}{1 - e^{i\alpha} z_\nu}$$

by (9.2.3) since $Q(z_\nu, u) = 0$. Thus

$$u\,\frac{\partial \log z_\nu}{\partial u} = \frac{1 + e^{i\alpha} z_\nu}{1 - e^{i\alpha} z_\nu}, \qquad \nu = 1, 2, \cdots, n - 2. \tag{9.2.17}$$

Writing $z_\nu = \rho_\nu e^{i\varphi_\nu}$, we obtain, differentiating both sides of (9.2.16) with respect to u,

$$u\,\frac{d\alpha}{du} = \operatorname{Im} \sum_{\nu=1}^{n-2} \frac{1 + e^{i\alpha} z_\nu}{1 - e^{i\alpha} z_\nu} = 2 \sum_{\nu=1}^{n-2} \rho_\nu \frac{\sin(\alpha + \varphi_\nu)}{1 - 2\rho_\nu \cos(\alpha + \varphi_\nu) + \rho_\nu^2}. \tag{9.2.18}$$

If $|z_\nu| < 1$, we see from (9.2.17) that

$$u\,\frac{d \log |z_\nu|}{du} = \operatorname{Re}\left\{\frac{1 + e^{i\alpha} z_\nu}{1 - e^{i\alpha} z_\nu}\right\} > 0;$$

thus $|z_\nu|$ is a strictly monotonic function of u which decreases as u decreases. Indeed, if $|z_\nu| < 1$, the inequality

$$\frac{u}{|z_\nu|}\,\frac{d\,|z_\nu|}{du} \geqq \frac{1 - |z_\nu|}{1 + |z_\nu|}$$

shows that

$$z_\nu(u) = o(u) \tag{9.2.19}$$

as u tends to zero. If $|z_\nu| = 1$, $z_\nu = e^{i\varphi_\nu}$, then from (9.2.17) we have

$$u\,\frac{d\varphi_\nu}{du} = \cot\left(\frac{\alpha + \varphi_\nu}{2}\right).$$

The angle $\varphi_\nu = \varphi_\nu(u)$ will decrease or increase according as $-\alpha < \varphi_\nu < -\alpha + \pi$ or $-\alpha - \pi < \varphi_\nu < -\alpha$. In other words, the roots of modulus unity move toward the point $e^{-i\alpha}$ as u decreases. However, the point $e^{-i\alpha(u)}$ may move away from $z_\nu(u)$ so fast that the distance between the points is increasing.

The equations (9.2.17) and (9.2.18) constitute a closed non-linear system of differential equations of the first order. There will be singularities wherever a root $z_\nu = e^{i\varphi_\nu}$ on $|z| = 1$ meets the point $e^{-i\alpha}$, but this can occur for only finitely many u. In Chapter XIV we shall indicate how this system of equations in the case $n = 4$ can be used to obtain information concerning V_4.

We remark that it is sometimes convenient to introduce a new variable ζ_ν in place of z_ν, namely

$$z_\nu = u\zeta_\nu.$$

Then the differential equation for ζ_ν is

$$\frac{d\zeta_\nu}{du} = \frac{2e^{i\alpha}\zeta_\nu^2}{1 - ue^{i\alpha}\zeta_\nu}, \qquad \nu = 1, 2, \cdots, n - 2. \tag{9.2.20}$$

From (9.2.16) it follows that

$$\alpha(u) = \sum_{\nu=1}^{n-2} \arg(\zeta_\nu) + (n - 1)\pi - \arg B_{n-1}(u). \tag{9.2.21}$$

For small u the function $Q(z, u)$ has a zero of precise order 2 at $e^{-i\alpha}$ and a zero of precise order $2j$, $j \geqq 0$, at a point $e^{-i\beta}$ which is the image of the distinguished cycle $\boldsymbol{\Sigma}$ by $v(z, u)$. The remaining zeros of $Q(z, u)$ lie in $|z| < 1$ or in $|z| > 1$. Thus for small u the function

$$Q(uz, u) = \sum_{\nu=-(n-1)}^{n-1} \frac{C_\nu(u)}{z^\nu}$$

has precisely $n-2-j$ zeros in $|z| < 1/u$. Since $\zeta_\nu(u)$ is a zero of $Q(uz, u)$, it follows from (9.2.19) that $n-2-j$ of the $\zeta_\nu(u)$ tend to the set of $n-2-j$ finite zeros of $P(w)$ as u tends toward zero. There are $j \geqq 0$ of the $\zeta_\nu(u)$ that tend toward infinity as u tends toward zero. It follows that $A_{j+1} \neq 0$ but $A_\nu = 0$ for $\nu < j + 1$.

We shall make a more complete investigation of the equations developed in this paragraph in Chapter XI.

9.3. The continuous realization of the identification described in the preceding paragraphs gives rise to a one-parameter family of functions

$$v(z, u) = u\{z + b_2(u)z^2 + b_3(u)z^3 + \cdots\} \tag{9.3.1}$$

which defines a curve in the space of schlicht functions extending from an arbitrary given $\mathfrak{D}_n$-function $f(z)$ to the function z. The point $(b_2(u), b_3(u), \cdots, b_n(u))$ defines a curve in V_n extending from the point $(a_2, a_3, \cdots, a_n)$ belonging to $f(z)$ to the origin. It is also possible to define a curve lying on the boundary of V_n and extending from any given $\mathfrak{D}_n$-function $f(z)$ to the Koebe function

$$f_0(z) = \frac{z}{(1 - e^{i\theta}z)^2}\,. \tag{9.3.2}$$

Given a $\mathfrak{D}_n$-function $f(z)$ and a corresponding family of functions (9.3.1) such that

$$\lim_{u \to 0} \frac{v(z, u)}{u} = f(z), \tag{9.3.3}$$

let

$$w(v, u) = f(z(v)) = \frac{1}{u}\{v + a_2(u)v^2 + a_3(u)v^3 + \cdots\} \tag{9.3.4}$$

where $z(v) = z(v, u)$ is the function inverse to $v(z, u)$. Since two points on opposite edges of the slit in the plane of $v(z, u)$ map by $z(v, u)$ into two points on $|z| = 1$ which are identified by $f(z)$, we see that $w(v, u)$ is single-valued across the slits in the plane of $v(z, u)$. It follows that $w(v, u)$ is regular and schlicht in $|v| < 1$ and maps $|v| < 1$ into the w-plane minus a subcontinuum $\boldsymbol{C}(u)$ of the image $\boldsymbol{C}$ of $|z| = 1$ by $f(z)$. The family of subcontinua $\boldsymbol{C}(u)$, $0 < u \leqq 1$, satisfies the following two conditions:

(a) $\boldsymbol{C}(u)$ contains $w = \infty$ and has (internal) mapping radius $1/u$.

(b) $\boldsymbol{C}(u') \subset \boldsymbol{C}(u'')$ for $0 < u' < u'' \leqq 1$.

According to the scheme described at the end of Chapter VIII in which the continuous realization of the identification leads finally to a single pair of identified arcs, the subcontinuum $C(u)$ consists of a single analytic arc for small u. By (9.3.4) we have

$$w(v(z, u), u) = f(z). \tag{9.3.5}$$

It may be shown from this relation and (9.1.9) that $\partial w/\partial u$ exists, except perhaps at finitely many u, and that

$$u \frac{\partial w(v, u)}{\partial u} = -v \frac{\partial w}{\partial v} \frac{1 + e^{i\alpha} v}{1 - e^{i\alpha} v}.$$

Hence, replacing v by z, we have

$$u \frac{\partial w(z, u)}{\partial u} = -z \frac{\partial w}{\partial z} \frac{1 + e^{i\alpha} z}{1 - e^{i\alpha} z} \tag{9.3.6}$$

where $\alpha = \alpha(u)$. Differentiating (9.3.5) with respect to z, we obtain

$$\frac{\partial w}{\partial z} = \frac{\partial w}{\partial v} \frac{\partial v}{\partial z} = \frac{df}{dz},$$

and so by (9.1.7),

$$\left(\frac{z}{w} \frac{\partial w}{\partial v} \frac{\partial v}{\partial z}\right)^2 P(w) = Q(z).$$

But by (9.1.8)

$$\left(\frac{z}{v} \frac{\partial v}{\partial z}\right)^2 Q(v, u) = Q(z)$$

and so

$$\left(\frac{v}{w} \frac{\partial w}{\partial v}\right)^2 P(w) = Q(v, u).$$

Since $f = f(z) = w(v, u)$ by (9.3.5), we obtain (replacing v by z)

$$\left(\frac{z}{w} \frac{\partial w}{\partial z}\right)^2 P(w) = Q(z, u), \qquad w = w(z, u). \tag{9.3.7}$$

Let

$$f(z, u) = uw(z, u) = z + a_2(u)z^2 + a_3(u)z^3 + \cdots. \tag{9.3.8}$$

The function $f(z, u)$ satisfies the equations

$$\left(\frac{z}{f} \frac{\partial f}{\partial z}\right)^2 P\left(\frac{f}{u}\right) = Q(z, u) \tag{9.3.9}$$

and

$$u \frac{\partial f}{\partial u} = f - z \frac{\partial f}{\partial z} \frac{1 + e^{i\alpha} z}{1 - e^{i\alpha} z}, \tag{9.3.10}$$

where $f(z, 1) = f(z)$ is the given $\mathfrak{D}_n$-function. It follows from (9.3.10) that the coefficients $a_k(u)$ satisfy the differential equations

(9.3.11) $$\frac{d}{du}[u^{k-1}a_k(u)] = -2u^{k-2}\sum_{\nu=1}^{k-1} \nu a_\nu(u)e^{i(k-\nu)\alpha(u)}.$$

We now find the limit of $f(z, u)$ as u tends to zero. The function $w = f(z, u)$ maps $|z| < 1$ onto the complement of an arc $C^*(u)$ containing $w = \infty$ on which

$$\mathrm{Re}\int\left(P\left(\frac{w}{u}\right)\right)^{1/2}\frac{dw}{w} = \text{constant},$$

and all finite zeros of $P(w/u)$ will lie in $|w| < 1/4$ if u is small. Suppose that

$$P(w) = \sum_{\nu=1}^{n-1}\frac{A_\nu}{w^\nu}$$

where $A_k \neq 0$ but $A_\nu = 0$ for $\nu < k$, $1 \leqq k \leqq n - 1$. Then as u approaches zero, the arc $C^*(u)$ approaches a straight line on which

$$\mathrm{Re}\ ((A_k)^{1/2}w^{-k/2}) = 0.$$

If $\arg(A_k) = \lambda$, the straight-line segment is the ray

$$w = re^{i\varphi}$$

where $1/4 \leqq r < \infty$ and

$$\varphi = \frac{\lambda + (2\nu + 1)\pi}{k}.$$

Hence

(9.3.12) $$\lim_{u\to 0} f(z, u) = \frac{z}{(1 + e^{-i\varphi}z)^2}.$$

The point $e^{-i\alpha(u)}$ of (9.3.10) maps into the tip of the slit in the plane of $f(z, u)$; so it follows from (9.3.12) that

$$\alpha(0) = -\varphi, \qquad a_k(0) = (-1)^{k-1}ke^{-i(k-1)\alpha(0)}.$$

For the sake of completeness we remark that

(9.3.13) $$w\{h(z, u', u''), u'\} = w(z, u''),$$

where h is defined by (9.1.10) and w by (9.3.5), $0 < u' < u'' \leqq 1$. Differentiating this equation with respect to u', u'' fixed, and writing $w = w(h, u')$, $\alpha' = \alpha(u')$, we have

$$u'\frac{\partial w}{\partial h}\frac{\partial h}{\partial u'} + u'\frac{\partial w}{\partial u'} = 0.$$

Hence

$$u'\frac{\partial w}{\partial h}\frac{\partial h}{\partial u'} = h\frac{\partial w}{\partial h}\frac{1 + e^{i\alpha'}h}{1 - e^{i\alpha'}h}$$

by (9.3.6) or

(9.3.14)
$$u' \frac{\partial h}{\partial u'} = h \frac{1 + e^{i\alpha'} h}{1 - e^{i\alpha'} h}.$$

Differentiating (9.3.13) with respect to u'', u' fixed, writing $\alpha'' = \alpha(u'')$, and using (9.3.6), we obtain

(9.3.14)′
$$u'' \frac{\partial h}{\partial u''} = -z \frac{\partial h}{\partial z} \frac{1 + e^{i\alpha''} z}{1 - e^{i\alpha''} z}.$$

The idea of combining Löwner's method with variational methods seems to have been discovered independently and approximately simultaneously by M. Schiffer and the authors although from somewhat different points of view. The point of view adopted by Schiffer [24d] when interpreted in terms of the coefficient region V_n leads to curves which lie on the boundary of V_n as described in **9.3.** A program of calculations begun in January, 1947, under a project sponsored by the Office of Naval Research has been based on the method of Theorem VI. These calculations will be described in detail in Chapter XIV.

Chapter X

LINEAR FORMS AND THE SUPPORTING SURFACE

10.1. Let

$$F = 2 \operatorname{Re} \{\lambda_2 b_2 + \lambda_3 b_3 + \cdots + \lambda_n b_n\} \tag{10.1.1}$$

where $\lambda_2, \lambda_3, \cdots, \lambda_n$ are given constants not all zero and let the maximum of F in V_n occur at the point $(a_2, a_3, \cdots, a_n)$. If the maximum value is M, then V_n lies entirely on one side of the $(2n - 3)$-dimensional hyperplane $F = M$; that is to say, $F = M$ is a supporting hyperplane. The subset K_n of V_n in which a supporting hyperplane can touch V_n is called the supporting surface (more properly, the supporting set; we use the word surface rather loosely). The set V_n is not convex if $n > 2$. For the points $(2, 3, 4, \cdots, n)$ and $(-2, 3, -4, \cdots, \pm n)$ belong to V_n since they correspond to the functions $z/(1 - z)^2$ and $z/(1 + z)^2$. But their average $(0, 3, 0, \cdots)$ does not belong to V_n since the Löwner formulas (9.2.1) show that on the boundary of V_n (and hence for all points of V_n)

$$| b_3 - b_2^2 | \leqq 1.$$

It follows that K_n is a proper subset of the boundary of V_n for $n > 2$. We shall prove that the set K_n is connected and that any function $w = f(z)$ belonging to a point of K_n maps $| z | < 1$ onto the w-plane minus a single analytic slit extending to infinity which contains no finite critical points and therefore is unforked. Part of this will follow from Lemma XXVIII below, which is an extension of an important lemma of Schiffer [24b]. Schiffer considers a function

$$w = f(z) = z + a_2 z^2 + \cdots$$

of class $\mathfrak{S}$ which imparts to $| a_n |$ its maximum possible value. Such a function maps $| z | < 1$ onto the complement of a set of arcs in the w-plane which satisfy the equation

$$\left(\frac{1}{w}\frac{dw}{dt}\right)^2 \frac{1}{a_n} S_n\left(\frac{1}{w}\right) + 1 = 0$$

where t is real and

$$S_n\left(\frac{1}{w}\right) = \sum_{\nu=2}^{n} \frac{a_n^{(\nu)}}{w^{\nu-1}} \tag{10.1.2}$$

(a result also due to Schiffer; see [24a]). It was shown by Schiffer that if the function (10.1.2) has a zero at a finite point of one of the slits in the w-plane, then all the slits lie on a single straight line through the origin, $w = 0$. His method remains unchanged for domains of the variable z other than the unit circle, including multiply-connected domains. The authors have given in [22c] a formulation in terms of linear forms which is based on the same method, and Golusin [6d] has recently given a similar formulation.

The methods which will be developed in later sections of the present chapter will allow us to strengthen the conclusions of Lemma XXVIII. For the method of Schiffer shows that if there is a finite critical point, then the map of $|z| = 1$ by $w = f(z)$ lies on two rays forming an angle π at the origin. Our subsequent investigations will show that the map of $|z| = 1$ in fact lies on a single radial line.

10.2 LEMMA XXVIII. *Let*

$$F = 2 \operatorname{Re} \{\lambda_2 b_2 + \lambda_3 b_3 + \cdots + \lambda_n b_n\}$$

where $\lambda_2, \lambda_3, \cdots, \lambda_n$ *are given constants, and let the maximum of* F *in* V_n *occur at the point* $(a_2, a_3, \cdots, a_n)$. *Then the function* $w = f(z)$ *belonging to this point satisfies a* $\mathfrak{D}_n$*-equation*

$$\left(\frac{z}{w}\frac{dw}{dz}\right)^2 P(w) = Q(z)$$

and $w = f(z)$ *maps* $|z| < 1$ *onto the complement of a set of arcs* $\mathbf{C}$ *in the* w*-plane. If* $P(w) = 0$ *at a finite point of* $\mathbf{C}$, *then* $\mathbf{C}$ *lies on a straight line through* $w = 0$.

Thus $\mathbf{C}$ can have no finite critical points, for $P(w) = 0$ at each finite critical point of $\mathbf{C}$.

PROOF: Writing

$$F = \lambda_2 b_2 + \bar{\lambda}_2 \bar{b}_2 + \cdots + \lambda_n b_n + \bar{\lambda}_n \bar{b}_n,$$

we see that F satisfies the conditions of Lemma VII. In the notation of that lemma we have

$$F_k = \lambda_k,$$

and so

$$P(w) = \sum_{\nu=1}^{n-1} \frac{A_\nu}{w^\nu} \tag{10.2.1}$$

where

$$A_\nu = \sum_{k=\nu+1}^{n} a_k^{(\nu+1)} \lambda_k . \tag{10.2.2}$$

The set of arcs $\mathbf{C}$ is a subcontinuum of $\mathbf{\Gamma}_w$, $\mathbf{\Gamma}_w$ being the system of loci $\operatorname{Re}(\zeta) =$ constant issuing from $w = \infty$ and the finite zeros of $P(w)$ where

$$\zeta = \int (P(w))^{1/2} \frac{dw}{w} .$$

Let C contain a point w_0, $P(w_0) = 0$. The function

$$g(z) = \frac{f(z)}{1 - \dfrac{f(z)}{w_0}}$$

belongs to class $\mathfrak{S}$ and

$$g(z) = \sum_{\nu=1}^{\infty} \frac{f(z)^\nu}{w_0^{\nu-1}} = \sum_{\nu=1}^{\infty}\sum_{k=\nu}^{\infty} \frac{a_k^{(\nu)} z^k}{w_0^{\nu-1}} = \sum_{k=1}^{\infty} \alpha_k z^k$$

where

$$\alpha_k = \sum_{\nu=1}^{k} \frac{a_k^{(\nu)}}{w_0^{\nu-1}}.$$

We have

$$\sum_{k=2}^{n} \lambda_k \alpha_k = \sum_{k=2}^{n}\sum_{\nu=1}^{k} \lambda_k \frac{a_k^{(\nu)}}{w_0^{\nu-1}} = \sum_{k=2}^{n} \lambda_k a_k + \sum_{k=2}^{n}\sum_{\nu=2}^{k} \frac{\lambda_k a_k^{(\nu)}}{w_0^{\nu-1}}$$

$$= \sum_{k=2}^{n} \lambda_k a_k + \sum_{\nu=1}^{n-1}\sum_{k=\nu+1}^{n} \frac{\lambda_k a_k^{(\nu+1)}}{w_0^{\nu}} = \sum_{k=2}^{n} \lambda_k a_k + P(w_0) = \sum_{k=2}^{n} \lambda_k a_k$$

since $P(w_0) = 0$ by hypothesis. It therefore follows that the function F takes its maximum in V_n at $(\alpha_2, \alpha_3, \cdots, \alpha_n)$ as well as at the point $(a_2, a_3, \cdots, a_n)$. Hence $s = g(z)$ is a solution of the $\mathfrak{D}_n$-equation

$$\left(\frac{z}{s}\frac{ds}{dz}\right)^2 P^*(s) = Q^*(z)$$

where

(10.2.3) $$P^*(s) = \sum_{\nu=1}^{n-1} \frac{A_\nu^*}{s^\nu}, \qquad A_\nu^* = \sum_{k=\nu+1}^{n} a_k^{(\nu+1)} \lambda_k .$$

Thus $s = g(z)$ maps $|z| = 1$ onto a subcontinuum C^* of the system of arcs Γ_s^*, and Re (ζ^*) = constant on C^* where

(10.2.4) $$\zeta^* = \int (P^*(s))^{1/2} \frac{ds}{s}.$$

The set C^* in the s-plane is the image of C in the w-plane under the transformation

(10.2.5) $$s = \frac{w}{1 - w/w_0}.$$

Substituting from (10.2.5) into (10.2.4), we obtain

$$\zeta^* = \int \left\{\sum_{\nu=0}^{n-1} \frac{A_\nu'}{w^\nu}\right\}^{1/2} \frac{dw}{w(1 - w/w_0)}$$

where

(10.2.6) $$\sum_{\nu=0}^{n-1} \frac{A_\nu'}{w^\nu} = \sum_{\nu=1}^{n-1} \frac{A_\nu^*}{w^\nu}\left(1 - \frac{w}{w_0}\right)^\nu.$$

On the locus $\boldsymbol{C}$ we therefore have Re (ζ) = constant and Re (ζ^*) = constant. From (10.2.6) we have

$$A'_\nu = \frac{1}{2\pi i}\int \sum_{\mu=1}^{n-1} A^*_\mu \left(1 - \frac{w}{w_0}\right)^\mu w^{\nu-\mu-1}\, dw$$

where the integration is counterclockwise around a curve, say a circle, enclosing $w = 0$. Substituting for the A^*_μ from (10.2.3), we obtain

$$\text{(10.2.7)}\qquad \begin{aligned} A'_\nu &= \frac{1}{2\pi i}\int \sum_{\mu=1}^{n-1}\sum_{k=\mu+1}^{n} \alpha_k^{(\mu+1)}\lambda_k \left(1 - \frac{w}{w_0}\right)^\mu w^{\nu-\mu-1}\, dw \\ &= \frac{1}{2\pi i}\int \sum_{k=2}^{n}\sum_{\mu=2}^{k} \alpha_k^{(\mu)}\lambda_k \left(1 - \frac{w}{w_0}\right)^{\mu-1} w^{\nu-\mu}\, dw. \end{aligned}$$

For sufficiently small u and z

$$\sum_{\mu=1}^{\infty} u^\mu g(z)^\mu = \sum_{\mu=1}^{\infty}\sum_{k=\mu}^{\infty} u^\mu \alpha_k^{(\mu)} z^k = \sum_{k=1}^{\infty}\sum_{\mu=1}^{k} \alpha_k^{(\mu)} u^\mu z^k.$$

But

$$\begin{aligned} \sum_{\mu=1}^{\infty} u^\mu g(z)^\mu &= \frac{ug(z)}{1 - ug(z)} = \frac{uf(z)}{1 - f(z)(u + 1/w_0)} \\ &= u\sum_{m=1}^{\infty}\left(u + \frac{1}{w_0}\right)^{m-1} f(z)^m = u\sum_{m=1}^{\infty}\sum_{k=m}^{\infty} a_k^{(m)}\left(u + \frac{1}{w_0}\right)^{m-1} z^k \\ &= \sum_{k=1}^{\infty}\sum_{m=1}^{k} a_k^{(m)} u \left(u + \frac{1}{w_0}\right)^{m-1} z^k; \end{aligned}$$

so

$$\sum_{\mu=1}^{k} \alpha_k^{(\mu)} u^\mu = \sum_{m=1}^{k} a_k^{(m)} u \left(u + \frac{1}{w_0}\right)^{m-1},$$

or

$$\sum_{\mu=2}^{k} \alpha_k^{(\mu)} u^{\mu-1} = \sum_{m=2}^{k} a_k^{(m)} \left(u + \frac{1}{w_0}\right)^{m-1} + a_k - \alpha_k .$$

This equation, proved for small u, holds for all u. If we write

$$u = \left(1 - \frac{w}{w_0}\right)\Big/ w = \frac{1}{w} - \frac{1}{w_0}$$

and substitute in (10.2.7), we obtain after some simplification

$$\begin{aligned} A'_\nu &= \frac{1}{2\pi i}\int w^{\nu-1}\sum_{k=2}^{n}\lambda_k(a_k - \alpha_k)\, dw + \frac{1}{2\pi i}\int \sum_{m=2}^{n}\sum_{k=m}^{n}\lambda_k a_k^{(m)} w^{\nu-m}\, dw \\ &= \sum_{k=\nu+1}^{n} \lambda_k a_k^{(\nu+1)} = A_\nu , \end{aligned}$$

and $A'_0 = 0$ since $\sum \lambda_k(a_k - \alpha_k) = 0$.

On any arc of $\boldsymbol{C}$, ζ and ζ^* each have constant real part; so $d\zeta$ and $d\zeta^*$ are pure imaginary. Since

$$d\zeta = (P(w))^{1/2}\frac{dw}{w}, \qquad d\zeta^* = (P(w))^{1/2}\frac{dw}{w(1 - w/w_0)},$$

the division of one equation by the other shows that $1 - (w/w_0)$ is real on each arc of $\boldsymbol{C}$. Thus $\boldsymbol{C}$ lies on the straight line

$$\operatorname{Im}(w/w_0) = 0.$$

10.3. The preceding lemma is based on a variation in the large. We now obtain additional properties of the supporting surface $\boldsymbol{K}_n$ based on variations in the large of a different type. We begin by proving that the supporting surface is connected. The supporting surface is clearly a closed set.

LEMMA XXIX. *The supporting surface* $\boldsymbol{K}_n$ *is connected and contains the curve* $(2e^{i\theta}, 3e^{2i\theta}, \cdots, ne^{i(n-1)\theta})$, *$\theta$ real.*

PROOF: It clearly contains each point of the curve $(2e^{i\theta}, 3e^{2i\theta}, \cdots, ne^{i(n-1)\theta})$ as may be seen by choosing $\lambda_2 = e^{-i\theta}$, $\lambda_3 = \lambda_4 = \cdots = \lambda_n = 0$ in (10.1.1).

Let $(a_2, a_3, \cdots, a_n)$ be any point of $\boldsymbol{K}_n$, and let

$$w = f(z) = z + a_2z^2 + a_3z^3 + \cdots + a_nz^n + \cdots \tag{10.3.1}$$

be the function belonging to this point. The image of $|z| < 1$ by $w = f(z)$ will be denoted by R. There are constants $\lambda_2, \lambda_3, \cdots, \lambda_n$ not all zero such that

$$2\operatorname{Re}\left(\sum_{\nu=2}^{n}\lambda_\nu b_\nu\right) \leqq M \tag{10.3.2}$$

for all points $(b_2, b_3, \cdots, b_n)$ of $\boldsymbol{V}_n$ with equality if $b_2 = a_2, \cdots, b_n = a_n$.

We begin with the following observation. Let

$$\mu_\nu = \sum_{k=\nu}^{n} a_k^{(\nu)}\lambda_k, \qquad \nu = 2, 3, \cdots, n. \tag{10.3.3}$$

If

$$\sum_{\nu=1}^{\infty}\beta_\nu w^\nu, \qquad \beta_1 = 1,$$

is regular and schlicht in R, then

$$\operatorname{Re}\left(\sum_{\nu=2}^{n}\mu_\nu\beta_\nu\right) \leqq 0 \tag{10.3.4}$$

with equality for $\beta_2 = \beta_3 = \cdots = \beta_n = 0$. For the function

$$\sum_{\nu=1}^{\infty}\beta_\nu w^\nu = \sum_{\nu=1}^{\infty}\sum_{k=\nu}^{\infty}\beta_\nu a_k^{(\nu)} z^k = \sum_{k=1}^{\infty} b_k z^k$$

where

$$b_k = \sum_{\nu=1}^{k} \beta_\nu a_k^{(\nu)}$$

belongs to class $\mathfrak{S}$. Hence from (10.3.2)

$$M \geqq 2 \operatorname{Re} \sum_{k=2}^{n} \lambda_k b_k = 2 \operatorname{Re} \sum_{k=2}^{n} \sum_{\nu=1}^{k} \lambda_k \beta_\nu a_k^{(\nu)}$$

$$= 2 \operatorname{Re} \sum_{\nu=2}^{n} \sum_{k=\nu}^{n} \lambda_k \beta_\nu a_k^{(\nu)} + 2 \operatorname{Re} \sum_{k=2}^{n} \lambda_k a_k = 2 \operatorname{Re} \sum_{\nu=2}^{n} \mu_\nu \beta_\nu + M.$$

If the boundary of R is $\boldsymbol{C}$, let $\boldsymbol{C}_1$ be a subcontinuum of $\boldsymbol{C}$ containing the point at infinity and of mapping radius $1/\rho$, $0 < \rho < 1$. Let R_1 be the complement of $\boldsymbol{C}_1$, $R_1 \supset R$. If $\zeta = \rho w$, then R_1 is mapped into a domain R^* in the ζ-plane of mapping radius unity. If

$$\sum_{\nu=1}^{\infty} \tau_\nu \zeta^\nu, \qquad \tau_1 = 1,$$

is regular and schlicht in R^*, then

(10.3.5)
$$\operatorname{Re} \left(\sum_{\nu=2}^{n} \mu_\nu \rho^{\nu-1} \tau_\nu \right) \leqq 0.$$

For the function

$$\sum_{\nu=1}^{\infty} \tau_\nu \rho^{\nu-1} w^\nu$$

is regular and schlicht in R_1 and hence in R. The statement then follows from (10.3.4).

Let

(10.3.6)
$$z = \sum_{\nu=1}^{\infty} \delta_\nu \zeta^\nu, \qquad \delta_1 = 1,$$

map R^* onto $|z| < 1$. If

$$q(z) = \sum_{\nu=1}^{\infty} c_\nu z^\nu$$

is any function of class $\mathfrak{S}$, then

$$q(z) = \sum_{\nu=1}^{\infty} \sum_{k=\nu}^{\infty} c_\nu \delta_k^{(\nu)} \zeta^k = \sum_{k=1}^{\infty} \gamma_k \zeta^k$$

where

$$\gamma_k = \sum_{\nu=1}^{k} c_\nu \delta_k^{(\nu)}.$$

From (10.3.5) we have

$$0 \geqq \operatorname{Re}\left(\sum_{k=2}^{n} \mu_k \rho^{k-1} \gamma_k\right) = \operatorname{Re}\left(\sum_{k=2}^{n} \sum_{\nu=1}^{k} \mu_k c_\nu \delta_k^{(\nu)} \rho^{k-1}\right)$$

$$= \operatorname{Re}\left(\sum_{\nu=2}^{n} \sum_{k=\nu}^{n} \mu_k c_\nu \delta_k^{(\nu)} \rho^{k-1}\right) + \operatorname{Re}\left(\sum_{k=2}^{n} \mu_k \rho^{k-1} \delta_k\right)$$

$$= \operatorname{Re}\left(\sum_{\nu=2}^{n} \lambda_\nu^* c_\nu\right) - \frac{1}{2} N$$

where

$$\lambda_\nu^* = \sum_{k=\nu}^{n} \mu_k \delta_k^{(\nu)} \rho^{k-1}, \qquad N = -2 \operatorname{Re}\left(\sum_{k=2}^{n} \mu_k \rho^{k-1} \delta_k\right).$$

Thus

$$2 \operatorname{Re}\left(\sum_{\nu=2}^{n} \lambda_\nu^* c_\nu\right) \leqq N \tag{10.3.7}$$

for all functions

$$q(z) = \sum_{\nu=1}^{\infty} c_\nu z^\nu$$

of class $\mathfrak{S}$. Equality occurs in (10.3.7) if $q(z) = \zeta$ maps $|z| < 1$ onto R^*, for in this case $\gamma_2 = \gamma_3 = \cdots = 0$. This means that the point $(a_2^*, a_3^*, \cdots, a_n^*)$ belongs to $\boldsymbol{K}_n$ where

$$f^*(z) = z + a_2^* z^2 + \cdots + a_n^* z^n + \cdots$$

maps $|z| < 1$ onto R^*. Since the subcontinuum can eventually be chosen as a single arc on the boundary of R and since the map of this arc in the ζ-plane approaches a straight line as ρ becomes small, it follows that $f^*(z)$ approaches a function of the form

$$\frac{z}{(1 - e^{i\theta} z)^2} = z + 2e^{i\theta} z^2 + 3e^{2i\theta} z^3 + \cdots + ne^{i(n-1)\theta} z^n + \cdots.$$

Thus every point of $\boldsymbol{K}_n$ is connected in $\boldsymbol{K}_n$ to some point of the curve $(2e^{i\theta}, 3e^{2i\theta}, \cdots, ne^{i(n-1)\theta})$, θ real.

We prove next that any function $w = f(z)$ belonging to a point of the supporting surface $\boldsymbol{K}_n$ maps $|z| < 1$ onto the plane minus a single analytic slit. This result strengthens and complements Lemma XXVIII.

Lemma XXX. *Any function $w = f(z)$ belonging to a point of the supporting surface $\boldsymbol{K}_n$ maps $|z| < 1$ onto the plane minus a single analytic slit.*

Proof: If there is a function $w = f(z)$ belonging to a point of the supporting surface $\boldsymbol{K}_n$ which maps $|z| < 1$ onto the plane minus more than one analytic slit extending to $w = \infty$, let $\boldsymbol{C}(\rho)$ be a subcontinuum of the boundary in the w-plane of mapping radius $1/\rho$ and let $\boldsymbol{C}(\rho)$ contain segments of precisely two arcs. If $\zeta = \rho w$, let $\boldsymbol{C}^*(\rho)$ be the image of $\boldsymbol{C}(\rho)$ in the plane of ζ. By choosing $\boldsymbol{C}(\rho)$ properly we see that $\boldsymbol{C}^*(\rho)$ will tend toward a pair of radial segments forming an angle greater than zero at $\zeta = \infty$ and whose distances from the origin are arbitrary subject to the condition that the mapping radius is unity. In particular, one of the radial segments may be chosen to lie in any given neighborhood of $\zeta = \infty$. Since $\boldsymbol{K}_n$ is a closed set, the mapping function $f^*(z)$ corresponding to the limit $\boldsymbol{C}_0^*$ of $\boldsymbol{C}^*$ belongs to a point of the supporting surface. The function $f^*(z)$ is of the form

$$(10.3.8)\quad f^*(z) = \frac{z}{(1 - e^{-i\alpha} z)^{2\mu}(1 - e^{-i\beta} z)^{2\gamma}} = z + a_2^* z^2 + \cdots + a_n^* z^n + \cdots$$

where

$$(10.3.9)\qquad \mu > 0, \qquad \gamma > 0, \qquad \mu + \gamma = 1.$$

The two angles formed by the radial segments of $\boldsymbol{C}_0^*$ intersecting at infinity are equal to $2\mu\pi$ and $2\gamma\pi$, and γ and μ are determined at least up to a finite number of choices by the original boundary in the plane of $w = f(z)$. Likewise the value of the linear form $\alpha\mu + \beta\gamma$ is determined by $\boldsymbol{C}_0^*$. With μ and γ held fixed and with this linear relation satisfied, we have one degree of freedom and so either α or β is completely at our disposal.

We shall show that the point $(a_2^*, a_3^*, \cdots, a_n^*)$ lies interior to the convex hull of the curve $(2e^{i\theta}, 3e^{2i\theta}, \cdots, ne^{i(n-1)\theta})$, θ real, if one of the radial segments is chosen to lie in a sufficiently small neighborhood of infinity. It is clearly sufficient to consider any rotation of the function (10.3.8), namely

$$g^*(z) = e^{-i\varphi} f^*(ze^{i\varphi}) = \frac{z}{(1 - e^{i(\varphi-\alpha)} z)^{2\mu}(1 - e^{i(\varphi-\beta)} z)^{2\gamma}}$$
$$= z + a_2^* e^{i\varphi} z^2 + \cdots + a_n^* e^{i(n-1)\varphi} z^n + \cdots.$$

We thus may suppose that $\varphi - \alpha = -\epsilon\gamma$, $\varphi - \beta = \epsilon\mu$, and we write

$$(10.3.10)\qquad g^*(z) = \frac{z}{(1-z)^2}\left(\frac{1 - e^{-i\epsilon\gamma} z}{1 - z}\right)^{-2\mu}\left(\frac{1 - e^{i\epsilon\mu} z}{1 - z}\right)^{-2\gamma}$$

since $\mu + \gamma = 1$. As ϵ approaches zero,

$$\left(\frac{1 - e^{-i\epsilon\gamma} z}{1 - z}\right)^{-2\mu} = 1 - 2\mu\left(i\epsilon\gamma + \frac{\epsilon^2\gamma^2}{2}\right)\frac{z}{1 - z} - \mu(2\mu + 1)\epsilon^2\gamma^2\frac{z^2}{(1 - z)^2} + O(\epsilon^3)$$

and similarly for the last factor. Then

$$(10.3.11)\qquad g^*(z) = \frac{z}{(1 - z)^2} - \epsilon^2\gamma\mu\frac{z^2}{(1 - z)^4} + O(\epsilon^3),$$

and so

(10.3.12) $$e^{i(k-1)\varphi} a_k^* = k\left\{1 - \frac{1}{6}\epsilon^2\gamma\mu(k^2-1)\right\} + O(\epsilon^3)$$

for $k = 2, 3, \cdots, n$.

We show that for small ϵ the coefficients a_k^*, $k = 2, 3, \cdots, n$, can be written in the form

(10.3.13) $$e^{i(k-1)\varphi} a_k^* = k\int_{-\pi}^{\pi} e^{i(k-1)\theta}\, d\chi^*(\theta)$$

where $\chi^*(\theta)$ is a strictly increasing function of θ satisfying

(10.3.14) $$\int_{-\pi}^{\pi} d\chi^*(\theta) = 1.$$

Let $\chi_1(\theta)$ be the step-function which is zero for $\theta < 0$ and equal to 1 for $\theta > 0$. Let $\chi_2(\theta)$ be the function which is equal to $-1/2$ for $\theta < -\delta$, equal to $\theta/2\delta$ for $-\delta \leqq \theta \leqq \delta$, and equal to $1/2$ for $\theta > \delta$. Finally let $\chi_3(\theta)$ be the function which vanishes for $\theta = 0$ and whose derivative $d\chi_3/d\theta$ is equal to

$$\frac{1}{2\pi}\operatorname{Re}\left(\frac{1 + re^{i\theta}}{1 - re^{i\theta}}\right), \qquad r = 1 - \frac{\delta^2}{3}.$$

Here $\delta^2 = 3\gamma\mu\epsilon^2$. If

$$\chi(\theta) = \{\chi_1(\theta) + \chi_2(\theta) + \chi_3(\theta)\}/3,$$

then

(10.3.15) $$k\int_{-\pi}^{\pi} e^{i(k-1)\theta}\, d\chi(\theta) = e^{i(k-1)\varphi} a_k^* + O(\epsilon^3)$$

for $k = 2, 3, \cdots, n$ and

$$\int_{-\pi}^{\pi} d\chi(\theta) = 1.$$

The function has a derivative except at three points, and its derivative is bounded below by a positive number times ϵ^2. If the error term in (10.3.15) is $b_k\epsilon^3$, let

$$d\chi_4(\theta)/d\theta = -\frac{\epsilon^3}{2\pi}\sum_{k=2}^{n-1}\frac{1}{k}\,[b_k e^{-i(k-1)\theta} + \bar{b}_k e^{i(k-1)\theta}]$$

and define $\chi_4(0) = 0$. Then if

$$\chi^*(\theta) = \chi(\theta) + \chi_4(\theta),$$

it follows that (10.3.14) is satisfied, and, for small ϵ, $\chi^*(\theta)$ is a strictly increasing function of θ since its derivative is positive. Then relation (10.3.13) is satisfied for the strictly increasing function $\chi(\theta)$. Hence the point $(a_2^*, a_3^*, \cdots, a_n^*)$ is an interior point of the convex hull of the curve $(2e^{i\theta}, 3e^{2i\theta}, \cdots, ne^{i(n-1)\theta})$, θ real, and so is a fortiori an interior point of the convex hull of V_n.

Although Lemmas XXVIII and XXX restrict the functions which belong to points on the supporting surface K_n, they do not completely characterize these functions. The set of points belonging to functions characterized by Lemmas XXVIII and XXX is not closed whereas K_n is closed. Let a $\mathfrak{D}_n$-function $w = f(z)$ map $|z| < 1$ onto the complement of a Jordan arc slit C with one end point at infinity. We suppose that C is not a radial segment and that it contains at least one finite critical point of the corresponding $\boldsymbol{\Gamma}$-structure. Then $f(z)$ is the limit of $\mathfrak{D}_n$-functions $f^*(z)$ which map $|z| < 1$ onto the plane minus a single analytic slit with one end point at infinity and without critical points. The functions $f^*(z)$ satisfy the conditions of Lemmas XXVIII and XXX but not all of them can belong to K_n, for if they did, then the limit function $f(z)$ would also belong to K_n, which is contrary to Lemma XXVIII.

Let a function $w = f(z)$ belonging to K_n satisfy the $\mathfrak{D}_n$-equation

$$\left(\frac{z}{w}\frac{dw}{dz}\right)^2 P(w) = Q(z)$$

where

$$P(w) = \sum_{\nu=1}^{n-1} \frac{A_\nu}{w^\nu}.$$

There is the question whether the coefficient A_1 can be zero. However, whether the coefficient A_1 can be zero or not, this additional knowledge would not completely characterize K_n. If A_1 is not zero, then there is only one arc of the $\boldsymbol{\Gamma}$-structure issuing from the point at infinity.

It is of interest to note that the convex hull of the Koebe curve $(2e^{i\theta}, 3e^{2i\theta}, \cdots, ne^{i(n-1)\theta})$ does not contain the set V_n in case $n > 2$. To prove this we show that if ρ, $0 < \rho < 1$, is sufficiently small and if $\varphi \neq 0, \pm\pi, \pm 2\pi, \cdots$, then the linear form

$$F = 2 \operatorname{Re} [a_2 + \rho e^{i\varphi} a_n] \tag{10.3.16}$$

is not maximal at any point of the Koebe curve. Let

$$f(z) = z + a_2 z^2 + \cdots + a_n z^n + \cdots$$

be a function which maximizes F. In the notation of Chapter II we have $F_2 = 1$, $F_n = \rho e^{i\varphi}$, and $F_3 = F_4 = \cdots = F_{n-1} = 0$. By (2.3.2),

$$\begin{aligned} A_\nu &= \sum_{k=\nu+1}^{n} a_k^{(\nu+1)} F_k = a_n^{(\nu+1)} \rho e^{i\varphi}, \quad \nu = 2, 3, \cdots, n-1, \\ A_1 &= \sum_{k=2}^{n} a_k^{(2)} F_k = 1 + a_n^{(2)} \rho e^{i\varphi}. \end{aligned} \tag{10.3.17}$$

Since $|a_\nu| < e\nu$, $\nu = 1, 2, \cdots$, by Littlewood's inequality, we see that $A_1 \neq 0$ for $0 < \rho < \rho_0$ where we may choose $\rho_0 = 1/(e^2 n^3)$. Then

$$P(w) = \frac{1}{w} + \rho e^{i\varphi} \sum_{\nu=1}^{n-1} \frac{a_n^{(\nu+1)}}{w^\nu}, \tag{10.3.18}$$

and the integral

$$\zeta = \int (P(w))^{1/2} \frac{dw}{w} \tag{10.3.19}$$

has constant real part on the map of $|z| = 1$ by $w = f(z)$. Now suppose that the maximizing function $f(z)$ is of the form

$$f(z) = \frac{z}{(1 + e^{i\theta} z)^2} = z - 2e^{i\theta} z^2 + 3e^{2i\theta} z^3 - \cdots. \tag{10.3.20}$$

Then

$$a_n^{(\nu+1)} = (-1)^{n-\nu-1} \frac{(n + \nu)!}{(n - \nu - 1)!(2\nu + 1)!} e^{i(n-\nu-1)\theta}, \tag{10.3.21}$$

and the map of $|z| = 1$ by $w = f(z)$ is the line segment $w = re^{-i\theta}$, $1/4 \leqq r < \infty$. Since dw/w is real on this segment, we see from (10.3.19) that $P(re^{-i\theta}) < 0$ for $1/4 \leqq r < \infty$. From (10.3.21) we have

$$P(re^{-i\theta}) = \frac{e^{i\theta}}{r} + \rho \sum_{\nu=1}^{n-1} \frac{(n + \nu)!\, e^{i[(n-1)\theta+\varphi]}}{(n - \nu - 1)!(2\nu + 1)!} \frac{(-1)^{n-\nu-1}}{r^\nu}. \tag{10.3.22}$$

By taking r large it follows that the coefficient of $1/r$, namely

$$e^{i\theta} + \rho \frac{(n + 1)!\, e^{i[(n-1)\theta+\varphi]}}{(n - 2)!3!} (-1)^n, \tag{10.3.23}$$

is negative, and in particular its imaginary part is zero. Then the coefficient of $1/r^2$ is real; so $\sin [(n - 1)\theta + \varphi] = 0$. Since (10.3.23) is real, we have $\sin \theta = 0$. Hence $\theta = k\pi$, and then $\varphi = j\pi$. Since φ is not a multiple of π, we see that the maximum of the form (10.3.16) cannot occur at any point of the Koebe curve. In particular, the absolute maximum of the form (10.3.16) occurs at a point of K_n not on the Koebe curve.

10.4. It seems natural to expect that points $(a_2, a_3, \cdots, a_n)$ for which the corresponding function $f(z)$ satisfies more than one $\mathfrak{D}_n$-equation will lie on an edge or vertex of the surface while those points for which $f(z)$ satisfies only one $\mathfrak{D}_n$-equation will lie on a part of the surface which is in some sense more smooth. It is therefore of interest to investigate the class of functions $f(z)$ belonging to more than one $\mathfrak{D}_n$-equation. Theorem V shows that each such function is algebraic. The following is an example of an algebraic function satisfying two $\mathfrak{D}_n$-equations and it maps $|z| < 1$ onto the complement of a Jordan arc in the w-plane without finite singularities.

Let

$$\psi(w) = \frac{1}{w^{1/2}} \left\{ \alpha_0 + \frac{\alpha_1}{w} + \cdots + \frac{\alpha_\mu}{w^\mu} \right\}, \qquad \alpha_0 \neq 0.$$

Starting at $w = \infty$, prolong the locus $\operatorname{Re} \psi(w) = 0$ until it has mapping radius unity. By choosing the coefficients α_ν properly, this locus will not lie on a straight line through the origin. Let $w = f(z)$ map $|z| < 1$ onto the complement of the locus. On the locus, $\psi'(w)\cdot dw$ is pure imaginary, and, if m is a positive integer or zero, $\{\psi(w)\}^{2m}$ is real. Hence

$$\left[\psi(w)^{2m}\psi'(w)z\frac{dw}{dz}\right]^2 = \left(\frac{z}{w}\frac{dw}{dz}\right)^2 P(w)$$

is non-negative on $|z| = 1$, where

$$P(w) = \frac{A_{2m+1}}{w^{2m+1}} + \cdots + \frac{A_{k-1}}{w^{k-1}}, \qquad k = 2\{1 + \mu + m + 2m\mu\}.$$

By the Schwarz reflection principle

$$\left(\frac{z}{w}\frac{dw}{dz}\right)^2 P(w) = Q(z)$$

where $Q(z)$ is a Q-function. By choosing $m = 0$ and 1, $\mu = 1$, it follows that the function $f(z)$ satisfies several $\mathfrak{D}_n$-equations if $n \geqq 10$ where one of these equations has $A_1 \neq 0$. The locus $\operatorname{Re}\psi(w) = 0$ prolonged from infinity until it has mapping radius unity will be curved for proper choice of $\alpha_0, \alpha_1, \cdots, \alpha_\mu$, and will not pass through any finite zero of $P(w)$.

There are examples of functions satisfying several $\mathfrak{D}_n$-equations which map $|z| < 1$ onto the plane slit along straight lines. Consider the functions

$$f(z) = \frac{z}{\prod_{\nu=1}^{n-1}(1 - e^{i\theta_\nu}z)^{2\mu_\nu}} \tag{10.4.1}$$

depending on the $2n - 2$ real parameters $\theta_1, \theta_2, \cdots, \theta_{n-1}, \mu_1, \mu_2, \cdots, \mu_{n-1}$, where the θ_ν are arbitrary real parameters and the μ_ν satisfy

$$\mu_\nu = \frac{m_\nu}{k}, \qquad \sum_{\nu=1}^{n-1}\mu_\nu = 1. \tag{10.4.2}$$

Here m_ν and k are integers, $m_\nu \geqq 0$, $1 \leqq k \leqq n - 1$. A function (10.4.1) maps $|z| < 1$ onto the w-plane minus j $(1 \leqq j \leqq n - 1)$ rectilinear slits pointing toward $w = 0$ and adjacent slits form angles $2m_\nu\pi/k$ with each other. Given a function (10.4.1), let one of the slits lie on the ray $\arg(w) = \varphi$. If we choose $A_\nu = -|A_\nu|e^{i\nu\varphi}$, $A_\nu = 0$ unless $k \leqq \nu \leqq n - 1$ and $\nu \equiv 0 \pmod{k}$, and form the corresponding function $P(w)$, then the expression

$$\left(\frac{z}{w}\frac{dw}{dz}\right)^2 P(w), \qquad w = f(z),$$

will be non-negative on $|z| = 1$. The function $f(z)$ will then satisfy differential equations depending on $[(n - 1)/k]$ real parameters.

The functions (10.4.1), subject only to the conditions

$$\mu_\nu \geqq 0, \qquad \sum_{\nu=1}^{n-1} \mu_\nu \leqq 1,$$

are star-like. As the parameters vary, the corresponding point $(a_2, a_3, \cdots, a_n)$ sweeps out the coefficient region V_n^* of star-like schlicht functions and the boundary of V_n^* is characterized by the condition

$$\sum_{\nu=1}^{n-1} \mu_\nu = 1.$$

The intersection of the boundary of V_n^* with that of V_n is the subset defined by (10.4.2).

For if $f(z) = z + a_2 z^2 + a_3 z^3 + \cdots$ belongs to class $\mathfrak{S}$, then $f(z)$ is star-like if and only if the function $H(z)$,

$$(10.4.3) \qquad H(z) = \frac{zf'(z)}{f(z)} = 1 + \alpha_1 z + \alpha_2 z^2 + \cdots + \alpha_{n-1} z^{n-1} + \cdots,$$

has positive real part in $|z| < 1$. The coefficients $\alpha_1, \alpha_2, \cdots, \alpha_{n-1}$ define a point in euclidean space of $2n - 2$ real dimensions, and as $H(z) = 1 + \alpha_1 z + \cdots$ ranges over all functions which are regular in $|z| < 1$ and have positive real part there, the point $(\alpha_1, \alpha_2, \cdots, \alpha_{n-1})$ ranges over a set D_n^*. Relation (10.4.3) shows that $\alpha_1, \alpha_2, \cdots, \alpha_{n-1}$ are single-valued continuous functions of a_2, $a_3, \cdots, a_n$, while the integrated form,

$$f(z) = ze^{(\alpha_1 z + (1/2)\alpha_2 z^2 + \cdots + (1/n)\alpha_{n-1} z^{n-1} + \cdots},$$

shows that $a_2, a_3, \cdots, a_n$ are single-valued continuous functions of $\alpha_1, \alpha_2, \cdots,$ α_{n-1}. The correspondence between D_n^* and V_n^* is one-one and continuous, and boundary points correspond to boundary points. If $H(z) = 1 + \alpha_1 z + \alpha_2 z^2 + \cdots$ has positive real part, then

$$H(z) = 1 + 2 \sum_{\nu=1}^{n-1} \mu_\nu \frac{e^{i\theta_\nu} z}{1 - e^{i\theta_\nu} z} + b_n z^n + b_{n+1} z^{n+1} + \cdots$$

where $\Sigma\mu_\nu \geqq 0$, $\Sigma\mu_\nu \leqq 1$. The boundary of D_n^* is characterized by the condition $\Sigma\mu_\nu = 1$. If $f(z)$ belongs to class $\mathfrak{S}$ and is star-like, then

$$\frac{zf'(z)}{f(z)} = 1 + 2 \sum_{\nu=1}^{n-1} \mu_\nu \frac{e^{i\theta_\nu} z}{1 - e^{i\theta_\nu} z} + b_n z^n + \cdots;$$

so

$$f(z) = \frac{z}{\prod_{\nu=1}^{n-1} (1 - e^{i\theta_\nu} z)^{2\mu_\nu}} \{1 + c_n z^n + \cdots\}$$

and $f(z)$ has the same coefficients $a_2, a_3, \cdots, a_n$ as the function

$$\frac{z}{\prod_{\nu=1}^{n-1} (1 - e^{i\theta_\nu} z)^{2\mu_\nu}}.$$

The above examples show that there is a variety of algebraic $\mathfrak{D}_n$-functions which satisfy more than one $\mathfrak{D}_n$-equation. It is therefore of interest to show that under certain circumstances the algebraic function must be of simple type.

If $A_k \neq 0$ but $A_{k+1} = A_{k+2} = \cdots = A_{n-1} = 0$, we say that the $\mathfrak{D}_n$-equation is of degree $k + 1$. In that case $B_k \neq 0$, $B_{k+1} = B_{k+2} = \cdots = B_{n-1} = 0$. We say that two $\mathfrak{D}_n$-equations are different if one is not a multiple of the other.

Lemma XXXI. *Let $w = f(z)$ belong to class $\mathfrak{S}$ and satisfy a $\mathfrak{D}_n$-equation of degree $n = p + 1$ where p is a prime. Then it satisfies no other $\mathfrak{D}_n$-equation of degree equal to or less than $p + 1$ unless it is of the form*

$$f(z) = \frac{z}{(1 - e^{i\alpha}z)(1 - e^{i\beta}z)}.$$

Proof: If $w = f(z)$ satisfies two $\mathfrak{D}_n$-equations, each of degree n, let these equations be

$$\left(\frac{z}{w}\frac{dw}{dz}\right)^2 \sum_{\nu=1}^{n-1} \frac{A_\nu}{w^\nu} = \sum_{\nu=-(n-1)}^{n-1} \frac{B_\nu}{z^\nu}, \tag{10.4.3}$$

$$\left(\frac{z}{w}\frac{dw}{dz}\right)^2 \sum_{\nu=1}^{n-1} \frac{A'_\nu}{w^\nu} = \sum_{\nu=-(n-1)}^{n-1} \frac{B'_\nu}{z^\nu}. \tag{10.4.3$'$}$$

Let γ be such that $\gamma A_{n-1} - A'_{n-1} = 0$ where γ is not necessarily real. Multiplying (10.4.3) by γ and subtracting (10.4.3)$'$, we obtain

$$\left(\frac{z}{w}\frac{dw}{dz}\right)^2 \sum_{\nu=1}^{k-1} \frac{A''_\nu}{w^\nu} = \sum_{\nu=-q}^{k-1} \frac{B''_\nu}{z^\nu}, \qquad A''_{k-1} = B''_{k-1} \neq 0, \tag{10.4.4}$$

where $2 \leqq k \leqq n - 1 = p$.

If the two given $\mathfrak{D}_n$-equations are of different degrees, we divide one by the other, while if they are of the same degree n, we divide (10.4.4) by (10.4.3). In either case we obtain

$$w^{n-k}\frac{A''_1 w^{k-2} + \cdots + A''_{k-1}}{A_1 w^{n-2} + \cdots + A_{n-1}} = z^{n-k}\frac{B''_{-q}z^{q+k-1} + \cdots + B''_{k-1}}{B_{n-1}z^{2n-2} + \cdots + B_{n-1}}; \tag{10.4.5}$$

so w is an algebraic function of z and to each value of z there corresponds at most $n - 2$ values of w. From either of the given $\mathfrak{D}_n$-equations it follows that whenever z approaches zero, w must also approach zero. Taking the $(n - k)$th root of both sides of (10.4.5), we have

$$w\{\lambda_0 + \lambda_1 w + \cdots\} = ze^{i\tau}\{\mu_0 + \mu_1 z + \cdots\}, \ \lambda_0 \neq 0, \ \mu_0 \neq 0, \tag{10.4.6}$$

where τ is an integral multiple of $2\pi/(n - k)$. Thus in a neighborhood of the origin each branch of $w(z)$ is regular and has an expansion

$$w = \beta_1 z + \beta_2 z^2 + \cdots \tag{10.4.7}$$

where

$$\beta_1 = \mu_0 e^{i\tau}/\lambda_0.$$

Substituting the function (10.4.7) into (10.4.3), we have

$$A_{n-1} = B_{n-1}\beta_1^{n-1}.$$

Since $A_{n-1} = B_{n-1}$, it follows that $\beta_1^{n-1} = 1$. Similarly (10.4.4) shows that $\beta_1^{k-1} = 1$. The equations

$$\beta_1^{n-1} = \beta_1^{k-1} = 1$$

then show that $\beta_1 = 1$ since $k < n$ and $n - 1$ is a prime.

It therefore follows that

$$e^{i\tau} = \lambda_0/\mu_0$$

and from (10.4.6) we see that all branches of $w(z)$ coincide in a neighborhood of the origin. Then all branches of $w(z)$ coincide for all z and $w(z)$ is single-valued and therefore rational. Thus

$$w = z\frac{N_1(z)}{N_2(z)}$$

where $N_1(z)$ and $N_2(z)$ are polynomials without common factors and $N_1(0) = N_2(0) = 1$. It follows from (10.4.3) that as w approaches zero, z approaches zero or infinity. Hence $N_1(z)$ has no zero and is therefore a constant.

It is clear that $N_2(z)$ is not a constant since $w = z$ does not satisfy (10.4.3). If $N_2(z)$ is of first degree, then

$$w = \frac{z}{1 + az}.$$

Substituting into (10.4.3), we see that the left side of (10.4.3) approaches zero as z approaches infinity, while the right side becomes infinite. Thus $N_2(z)$ is of degree $r + 1$ where $r \geqq 1$. Then for large z

$$w = \frac{c}{z^r} + \frac{d}{z^{r+1}} + \cdots;$$

so from (10.4.3)

$$r^2 A_{n-1}\left(\frac{z^r}{c}\right)^{n-1}\{1 + o(1)\} = \bar{B}_{n-1}z^{n-1}\{1 + o(1)\}.$$

This shows that

$$(n - 1)r = n - 1, \qquad A_{n-1}r^2 = \bar{B}_{n-1}c^{n-1}.$$

Since $A_{n-1} = B_{n-1}$, it follows that

$$r = 1, \qquad |c| = 1.$$

Hence $N_2(z)$ is of precise degree 2 and

$$w = \frac{z}{1 + \lambda z + \mu z^2}$$

where $|\mu| = 1$. The product of the zeros of $1 + \lambda z + \mu z^2$ is of modulus unity and no zero can lie in $|z| < 1$ since $w(z)$ is regular there. Hence both zeros lie on $|z| = 1$ and we have

$$w = \frac{z}{(1 - e^{i\alpha}z)(1 - e^{i\beta}z)} .$$

This function maps $|z| < 1$ onto a domain whose entire boundary lies on a straight line through the origin in the w-plane.

Chapter XI

THE PORTION OF THE BOUNDARY OF V_n CORRESPONDING TO SINGLE ANALYTIC SLITS

11.1. Any $\mathfrak{D}_n$-function $w = f(z)$ maps $|z| < 1$ onto the w-plane minus a finite set of analytic Jordan arcs. The reciprocal $1/f(z)$ maps $|z| < 1$ onto the exterior of a set of arcs lying in a bounded region of the plane, and we have shown that each arc is analytic up to and including its end-points. In this chapter we consider the portion of the boundary of V_n corresponding to functions $w = f(z)$ which map $|z| < 1$ onto the exterior of a single arc. In special cases the function $w = f(z)$ may satisfy several $\mathfrak{D}_n$-equations

$$\left(\frac{w}{z}\frac{dw}{dz}\right)^2 P(w) = Q(z) \tag{11.1.1}$$

where

$$P(w) = \sum_{\nu=1}^{n-1} \frac{A_\nu}{w^\nu}, \qquad Q(z) = \sum_{\nu=-(n-1)}^{n-1} \frac{B_\nu}{z^\nu}. \tag{11.1.2}$$

We make the additional hypothesis, however, that at least one of these equations has the property that $P(w)$ does not vanish at a finite point of the bounding arc in the w-plane.

Thus let L_n be the set of boundary points of V_n which correspond to points of S_n having the following property: The identification of points on $|z| = 1$ defined by $Q(z)$ consists of precisely two arcs and their end points $e^{-i\alpha}$ and $e^{-i\beta}$. The function $Q(z)$ has a zero of precise order 2 at $e^{-i\alpha}$ and a zero of even order (possibly zero) at $e^{-i\beta}$ with no other zeros on $|z| = 1$. The distinguished cycle Σ is composed of the one point $e^{-i\beta}$. The $\mathfrak{D}_n$-functions belonging to points of L_n are precisely those described above and K_n belongs to L_n.

We shall first investigate the differential equations of the Löwner curves belonging to points of L_n. Let $w = f(z)$ be a $\mathfrak{D}_n$-function belonging to a point of L_n and satisfying the differential equation (11.1.1) where $P(w)$ does not vanish at any finite point of the map C of $|z| = 1$ by $w = f(z)$. According to Chapter IX, there is a family of functions

$$V(z, u) = z + b_2(u)z^2 + b_3(u)z^3 + \cdots \tag{11.1.3}$$

where

$$V(z, 1) = z, \qquad V(z, 0) = f(z). \tag{11.1.4}$$

Then $V(z, u)$ satisfies the equations

$$\left(\frac{z}{V}\frac{\partial V}{\partial z}\right)^2 T(V, u) = Q(z), \tag{11.1.5}$$

$$\frac{\partial V}{\partial u} = \frac{2e^{i\alpha} V^2}{1 - ue^{i\alpha} V}, \tag{11.1.6}$$

where

$$T(V, u) = \sum_{k=-(n-1)}^{n-1} \frac{C_k(u)}{V^k}, \qquad C_{-k}(u) = \overline{C_k(u)u^{2k}}. \tag{11.1.7}$$

The $C_k(u)$ satisfy the differential equations

$$\frac{dC_k}{du} = 2 \sum_{\nu=1}^{n-k-1} (k - \nu)C_{k+\nu} u^{\nu-1} e^{i\nu\alpha} \tag{11.1.8}$$

where the terminal conditions are

$$\begin{aligned} C_k(0) &= A_k, & k &= 1, 2, \cdots, n - 1; \\ C_k(0) &= 0, & k &= -n + 1, \cdots, 0; \end{aligned} \tag{11.1.9}$$

and

$$C_k(1) = B_k, \qquad -n + 1 \leqq k \leqq n - 1, \tag{11.1.10}$$

where the B_k are the coefficients of the given $Q(z)$.

These differential equations define a one-parameter set of coefficients $C_k(u)$, $k = -(n - 1), \cdots, n - 1$, which join the coefficients of $P(w)$ with those of $Q(z)$. Given either $P(w)$ or $Q(z)$ satisfying certain conditions, we shall use the method of successive approximation to construct the coefficients $C_k(u)$ leading from one to the other.

11.2. If the given function $Q(z)$ vanishes at two distinct points $e^{-i\alpha}$, $e^{-i\beta}$ of $|z| = 1$, then for $0 < u \leqq 1$ the function $T(z, u)$ will vanish at two distinct points $e^{-i\alpha(u)}/u$ and $e^{-i\beta(u)}/u$ of $|z| = 1/u$ and

$$\lim_{u \to 0} T(w, u) = P(w) = \sum_{\nu=1}^{n-1} \frac{A_\nu}{w^\nu}$$

will have $A_1 = 0$. Indeed, if $Q(z)$ has a zero at $e^{-i\beta}$ of multiplicity $2\gamma \geqq 0$, then $A_{\gamma+1} \neq 0$, $A_\nu = 0$ for $\nu \leqq \gamma$. Conversely, if $A_{\gamma+1} \neq 0$, $A_\nu = 0$ for $\nu \leqq \gamma$, and if $f(z) = V(z, 0)$ maps $|z| < 1$ onto the complement of a single analytic arc, then $Q(z)$ has a zero of precise order 2γ at $e^{-i\beta}$ and $T(z, u)$ has a zero of precise order 2γ at the point $e^{-i\beta(u)}/u$.

Now let

$$f(z, u) = z + a_2(u)z^2 + a_3(u)z^3 + \cdots \tag{11.2.1}$$

be the function defined in Chapter IX, **9.3,** which satisfies the $\mathfrak{D}_n$-equation

$$\left(\frac{z}{f}\frac{\partial f}{\partial z}\right)^2 P\left(\frac{f}{u}\right) = Q(z, u) \tag{11.2.2}$$

where

$$P(w) = \sum_{\nu=1}^{n-1} \frac{A_\nu}{w^\nu}, \qquad Q(uz, u) = T(z, u).$$

The function $f(z, u)$ also satisfies the equation

$$u\frac{\partial f}{\partial u} = f - z\frac{\partial f}{\partial z}\frac{1 + e^{i\alpha}z}{1 - e^{i\alpha}z}. \tag{11.2.3}$$

As u approaches zero,

$$f(z, u) \to \frac{z}{(1 + e^{i\alpha(0)}z)^2}, \qquad T(w, u) \to P(w),$$

where

$$A_{\gamma+1} = -|A_{\gamma+1}|e^{-i(\gamma+1)\alpha(0)}. \tag{11.2.4}$$

Here $A_\nu = 0$ for $\nu \leqq \gamma$, $A_{\gamma+1} \neq 0$. Given the function $P(w)$, we shall construct the function $T(z, u)$ for $0 \leqq u \leqq u^*$, where u^* is some positive number, such that

$$\lim_{u\to 0} T(w, u) = P(w).$$

To fix the ideas we suppose that $A_1 = A_2 = \cdots = A_\gamma = 0$, $A_{\gamma+1} \neq 0$, where $\gamma \geqq 1$. Although slightly more complicated, this case is degenerate in the sense that the number of free real parameters is diminished by 2γ. Thus let $P(w)$ be given such that $A_1 = A_2 = \cdots = A_\gamma = 0$, $A_{\gamma+1} \neq 0$, and let $\alpha(0)$ be one of the finitely many roots of the equation (11.2.4). Also, since $f(z, u)$ is independent of u in case $\gamma = n - 2$, suppose that $\gamma < n - 2$.

Let

$$\zeta_1(u), \zeta_2(u), \cdots, \zeta_{n-\gamma-2}(u)$$

be the zeros of $T(z, u)$ in $|z| < 1/u$. The multiplicity of each distinct zero of $T(z, u)$ is independent of u since each lies in $|z| < 1/u$. The differential equations for $\alpha(u)$, $\beta(u)$, $\zeta_1(u), \cdots, \zeta_{n-\gamma-2}(u)$ are

$$\frac{d\zeta_\nu}{du} = \frac{2e^{i\alpha}\zeta_\nu^2}{1 - e^{i\alpha}u\zeta_\nu}, \qquad \nu = 1, 2, \cdots, n - \gamma - 2, \tag{11.2.5}$$

$$\frac{d\beta}{du} = -\frac{1}{u}\cot\frac{\alpha - \beta}{2}, \tag{11.2.6}$$

$$\frac{d\alpha}{du} = \frac{\gamma}{u}\cot\frac{\alpha - \beta}{2} + 2\,\mathrm{Im}\sum_{\nu=1}^{n-\gamma-2}\frac{e^{i\alpha}\zeta_\nu}{1 - e^{i\alpha}u\zeta_\nu}. \tag{11.2.7}$$

Let

$$\varphi = \frac{\alpha - \beta}{2}, \qquad \psi = \alpha + \gamma\beta. \tag{11.2.8}$$

In place of equations (11.2.5), (11.2.6), and (11.2.7) we shall use the equations

$$\frac{d\zeta_\nu}{du} = \frac{2e^{i\alpha}\zeta_\nu^2}{1 - e^{i\alpha}u\zeta_\nu}, \qquad \nu = 1, 2, \cdots, n - \gamma - 2, \tag{11.2.9}$$

(11.2.10) $$\frac{d\psi}{du} = R(u),$$

(11.2.11) $$\frac{d}{du}\,(u^{(\gamma+1)/2}\cos\varphi) = -\frac{1}{2}u^{(\gamma+1)/2}R(u)\sin\varphi,$$

where

(11.2.12) $$R(u) = 2\,\mathrm{Im}\sum_{\nu=1}^{n-\gamma-2}\frac{e^{i\alpha}\zeta_\nu}{1-e^{i\alpha}u\,\zeta_\nu}.$$

If there are any solutions of these equations in an interval $0 < u \leqq u_0$ which have finite limits as u approaches zero, then (11.2.11) shows that

$$u^{(\gamma+1)/2}\cos\varphi = -\frac{1}{2}\int_0^u t^{(\gamma+1)/2}R(t)\sin\varphi\,dt = O(u^{(\gamma+3)/2});$$

so $\cos\varphi = O(u)$, and it follows from (11.2.11) that $d\varphi/du$ is bounded as u tends to zero.

Let there be given non-zero complex numbers $\zeta_1(0), \zeta_2(0), \cdots, \zeta_{n-\gamma-2}(0)$ and a real number $\alpha(0)$. If

(11.2.13) $$|\,\zeta_\nu(0)\,| \leqq M,$$

we show first that the solutions of equations (11.2.9)–(11.2.11) exist for $0 \leqq u \leqq u_1$, $u_1 = 1/[10(n-\gamma-2)M]$. Let

$$\zeta_\nu^{(0)}(u) = \zeta_\nu(0), \qquad \varphi^{(0)}(u) = \pi/2, \qquad \psi^{(0)}(u) = \psi(0).$$

If $\zeta_\nu^{(m-1)}(u)$, $\varphi^{(m-1)}(u)$, and $\psi^{(m-1)}(u)$ have been defined for some $m \geqq 1$ and

(11.2.14) $$|\,\zeta_\nu^{(m-1)}(u)\,| \leqq 2M, \qquad \nu = 1, 2, \cdots, n-\gamma-2,$$

for $0 \leqq u \leqq u_1$, let

(11.2.15) $$\zeta_\nu^{(m)}(u) = \zeta_\nu(0) + 2\int_0^u \frac{\exp\{i\alpha^{(m-1)}(t)\}\,\zeta_\nu^{(m-1)}(t)^2}{1-\exp\{i\alpha^{(m-1)}(t)\}\,t\zeta_\nu^{(m-1)}(t)}\,dt,$$

(11.2.16) $$\psi^{(m)}(u) = \psi(0) + \int_0^u R^{(m-1)}(t)\,dt,$$

(11.2.17) $$\cos\varphi^{(m)}(u) = -\frac{1}{2}\int_0^u (t/u)^{(\gamma+1)/2}R^{(m-1)}(t)\sin\varphi^{(m-1)}(t)\,dt,$$

where

(11.2.18) $$R^{(m-1)}(t) = 2\,\mathrm{Im}\sum_{\nu=1}^{n-\gamma-2}\frac{\exp\{i\alpha^{(m-1)}\}\zeta_\nu^{(m-1)}}{1-\exp\{i\alpha^{(m-1)}\}u\zeta_\nu^{(m-1)}}.$$

Then

$$| R^{(m-1)}(t) | \leqq 5M(n - \gamma - 2)$$

and so

$$(11.2.19) \qquad | \cos \varphi^{(m)}(u) | \leqq 5M \frac{n - \gamma - 2}{\gamma + 3} u \leqq \frac{1}{3}.$$

Moreover,

$$(11.2.20) \qquad \begin{aligned} | \zeta_\nu^{(m)}(u) | &\leqq 2M, \\ | \psi^{(m)}(u) - \psi(0) | &\leqq 5M(n - \gamma - 2)u \end{aligned}$$

in $0 \leqq u \leqq u_1$. Thus by induction all the approximating functions are defined and (11.2.20) is true for $m = 1, 2, \cdots$. The integrands of equations (11.2.15)–(11.2.17) have continuous first order partial derivatives with respect to φ, ψ, Re (ζ_1), Im (ζ_1), $\cdots$, Re $(\zeta_{n-\gamma-2})$, Im $(\zeta_{n-\gamma-2})$, provided that (11.2.14) is satisfied and $0 \leqq u \leqq u_1$. Because of inequality (11.2.19) the well known method of successive approximation may be used to show that these approximating functions converge to solutions of the equations (11.2.9)–(11.2.11) for $0 \leqq u \leqq u_1$ which assume the assigned initial values. These solutions may be shown to be unique.

Using the values $\zeta_1(u_1)$, $\zeta_2(u_2)$, $\cdots$, $\zeta_{n-\gamma-2}(u_1)$, $\psi(u_1)$, and $\varphi(u_1)$ as new initial values for the system of equations (11.2.9)–(11.2.11), we obtain a solution of these equations in an interval $u_1 \leqq u \leqq u_2$. This process may be repeated indefinitely so long as $u \mid \zeta_\nu \mid < 1$, for we note from (11.2.11) that when R is bounded, the derivative of $\cos \varphi$ is negative if $\cos \varphi$ is near $+1$, and is positive if $\cos \phi$ is near -1. Thus $\cos \varphi$ remains between -1 and $+1$. The differential equation (11.2.9), namely

$$(11.2.21) \qquad \frac{d \log (u\zeta_\nu)}{du} = \frac{1}{u} \frac{1 + e^{i\alpha} u\zeta_\nu}{1 - e^{i\alpha} u\zeta_\nu},$$

shows that $u \mid \zeta_\nu \mid$ is strictly increasing. Thus the solutions can be continued for arbitrarily large u so long as each of the functions $u \mid \zeta_\nu(u) \mid$ remains less than some constant λ, $\lambda < 1$. The real part of the right side of (11.2.21) is not less than

$$\frac{1}{u} \frac{1 - \lambda}{1 + \lambda};$$

so, as u becomes large, $u \mid \zeta_\nu(u) \mid$ tends to unity. In any case let u^* be the upper limit of u as max $(u \mid \zeta_1 \mid, \cdots, u \mid \zeta_{n-\gamma-2} \mid)$ tends to unity. The number u^* may be finite or infinite.

Thus, given any set of complex numbers $A_1, A_2, \cdots, A_{n-1}$, not all zero, where we no longer suppose that some of the leading terms vanish, and given a number $\alpha(0)$ which is a root of the equation (11.2.4), we obtain $\zeta_1(u)$, $\cdots$, $\zeta_{n-\gamma-2}(u)$, $\alpha(u)$, and, if $\gamma > 0$, a function $\beta(u)$. These functions satisfy equations

(11.2.5)–(11.2.7) and are unique for $0 \leqq u \leqq u^*$. We then obtain the unique function

$$Q(z, u) = | A_{n-1} | u^{n-1} \left[\frac{(z - e^{-i\alpha})^2}{-ze^{-i\alpha}}\right]\left[\frac{(z - e^{-i\beta})^2}{-ze^{-i\beta}}\right]^{\gamma} \tag{11.2.22}$$
$$\cdot \prod_{\nu=1}^{n-\gamma-2} | \zeta_\nu | \frac{(z - u\zeta_\nu)(z - 1/u\bar{\zeta}_\nu)}{-z\zeta_\nu} = \sum_{\nu=-(n-1)}^{n-1} \frac{B_\nu(u)}{z^\nu}$$

in the interval $0 \leqq u \leqq u^*$. The function $\alpha(u)$ being determined in this interval, the function $f(z, u)$ which satisfies (11.2.3) is also determined and it satisfies the differential equation (11.2.2). Its coefficients $a_k(u)$ are determined from the differential equation

$$\frac{d}{du}(u^{k-1} a_k) = -2u^{k-2} \sum_{\nu=1}^{k-1} \nu a_\nu(u) e^{i(k-\nu)\alpha(u)}. \tag{11.2.23}$$

The point $(a_2(u), a_3(u), \cdots, a_n(u))$ moves over a curve lying on the boundary of V_n as u varies from $u = 0$ to $u = u^*$. As $\mathfrak{A} = (A_1, A_2, \cdots, A_{n-1})$ moves over all possible non-null vectors and as $\alpha(0)$ takes each of the possible values allowed by (11.2.4), these curves sweep out the portion of the boundary of V_n whose points belong to functions $f(z)$ which map $| z | < 1$ onto the exteriors of single analytic slits with the properties described at the beginning of **11.1.** This is the surface L_n. In particular, these curves sweep out a boundary set which includes the supporting surface $\boldsymbol{K}_n$. We remark, however, that if these curves are prolonged farther, they will sweep out the whole boundary of V_n. These curves will in general have critical points or points of forking at the values u corresponding to the critical points of the $\boldsymbol{\Gamma}$-structure. Indeed, the condition that $\alpha(0)$ satisfy (11.2.4) is a reflection of the fact that the $\boldsymbol{\Gamma}$-structure has several analytic slits meeting at infinity.

The one-parameter family of functions

$$f(z) = \frac{z}{(1 - e^{i\theta} z)^2}, \qquad \theta \text{ real},$$

defines a curve $(2e^{i\theta}, 3e^{2i\theta}, \cdots, ne^{i(n-1)\theta})$ lying on the boundary of V_n, and we shall henceforth call this curve the Koebe curve Each of the curves traced by $(a_2(u), \cdots, a_n(u))$, $0 \leqq u \leqq u^*$, has its end point corresponding to $u = 0$ at a point of the Koebe curve. If $\mathfrak{A} = (A_1, A_2, \cdots, A_{n-1})$ is given, consider the set of loci $\boldsymbol{\Gamma}(\mathfrak{A})$ in the w-plane and let C be one of the analytic arcs of $\boldsymbol{\Gamma}(\mathfrak{A})$ emerging from $w = \infty$ and extending to a finite zero of $P(w)$ or to $w = 0$. Let $C(u)$ denote the subarc of C containing $w = \infty$ which has mapping radius $1/u$ and let $C^*(u)$ be the arc composed of points uw where w belongs to $C(u)$. Then $C^*(u)$ has mapping radius unity and $w = f(z, u)$ maps $| z | < 1$ onto the exterior of $C^*(u)$. If the finite end point of C is at a zero of $P(w)$, then u^* is finite and one end point of the curve $(a_2(u), \cdots, a_n(u))$ is at one of its points of forking. If the finite end point of C is at $w = 0$, then $u^* = \infty$ and the curve $(a_2(u), \cdots, a_n(u))$ has both its end points on the Koebe curve.

It is clear from the differential equation (11.2.2) that the arc $C^*(u)$ lies on the $\mathbf{\Gamma}$-structure $\mathbf{\Gamma}^*(u)$ defined by the vector $\mathfrak{A}(u) = (A_1 u, \cdots, A_{n-1} u^{n-1})$. Thus the two vectors $(A_1, A_2, \cdots, A_{n-1})$ and $(A_1 \tau, A_2 \tau^2, \cdots, A_{n-1} \tau^{n-1})$, where τ is any positive constant, generate the same curve on the boundary of V_n—or curves in case some of the leading coefficients A_ν are zero.

Integrating the equations (11.2.23), we obtain

$$a_2(u) = -\frac{2}{u} \int_0^u e^{i\alpha(t)}\, dt, \tag{11.2.24}$$

$$a_3(u) = -\frac{2}{u^2} \int_0^u t e^{2i\alpha(t)}\, dt + \frac{4}{u^2} \left(\int_0^u e^{i\alpha(t)}\, dt \right)^2, \tag{11.2.25}$$

$$\begin{aligned} a_4(u) = &-\frac{2}{u^3} \int_0^u t^2 e^{3i\alpha(t)}\, dt + \frac{12}{u^3} \int_{t_1=0}^u \int_{t_2=0}^u t_1 e^{2i\alpha(t_1)} e^{i\alpha(t_2)}\, dt_1\, dt_2 \\ &- \frac{4}{u^3} \int_{t_1=0}^u \int_{t_2=0}^{t_1} t_1 e^{2i\alpha(t_1)} e^{i\alpha(t_2)}\, dt_1\, dt_2 - \frac{8}{u^3} \left(\int_0^u e^{i\alpha(t)}\, dt \right)^3, \end{aligned} \tag{11.2.26}$$

and so on. By making a linear magnification of u we may suppose that for any given point on this curve $u = 1$. For example, if $t = uv$ and $\alpha(uv) = \alpha^*(v)$, the formula (11.2.24) becomes

$$a_2 = -2 \int_0^1 e^{i\alpha^*(v)}\, dv. \tag{11.2.24$'$}$$

The formulas thus obtained agree with the formulas (9.2.1) with $\alpha(v)$ replaced by $\alpha^*(v)$ and u set equal to zero.

Thus, given a point $(a_2, a_3, \cdots, a_n)$ on L_n, there is a curve $(a_2(u), a_3(u), \cdots, a_n(u))$ lying on L_n and joining the given point to a point of the Koebe curve and the given point corresponds to $u = 1$. The given point may also be joined to the origin $a_2 = 0, a_3 = 0, \cdots, a_n = 0$ of V_n by a curve $(b_2(u), b_3(u), \cdots, b_n(u))$, $0 \leqq u \leqq 1$, which lies entirely in the interior of V_n except for the end point at $u = 1$. Here $b_2(u), b_3(u), \cdots, b_n(u)$ are the coefficients of the function (11.1.3) which satisfies the equations (11.1.5) and (11.1.6).

11.3. We now investigate the solutions of the equation (11.1.1) in the case in which $Q(z)$ is given and has only one zero on $|z| = 1$ which is of precise order 2. The identification of points on $|z| = 1$ and the distinguished cycle $\mathbf{\Sigma}$ are then uniquely determined by Q. This case is slightly more restrictive than the one considered in the preceding paragraph. However, as $Q(z)$ varies over all possible Q-functions satisfying this condition, the corresponding point on the boundary of V_n covers all of L_n with the exception of a set of points depending on two fewer real parameters. We shall prove that the points on the boundary of V_n corresponding to such functions Q depend in an analytic way on the $2n - 3$ real parameters which define Q.

Given any $Q(z)$ with only one zero on $|z| = 1$ which is of precise order 2, it can be written in the form

$$Q(z) = \frac{(z - e^{-i\alpha})^2}{-ze^{-i\alpha}} \prod_{\nu=1}^{n-2} \left(1 + |z_\nu|^2 - z\bar{z}_\nu - \frac{z_\nu}{z}\right). \tag{11.3.1}$$

Since two functions Q are considered equal if one is a positive multiple of the other, each Q has one and only one representation in this form. Some of the z_ν may be zero, in which case the corresponding $Q(z)$ belongs to an integer less than n. From Chapter IX we know that there is a function

$$\begin{aligned} T(z, u) = \lambda(u) &\left(-\frac{e^{-i\alpha}}{z} + 2u - zu^2 e^{i\alpha}\right) \\ &\cdot \prod_{\nu=1}^{n-2} \left(1 - \frac{\zeta_\nu}{z} + u^2 |\zeta_\nu|^2 - u^2 z\bar{\zeta}_\nu\right) \end{aligned} \tag{11.3.2}$$

where $\lambda(u)$ is a positive function which tends to $|A_1| > 0$ as u approaches zero and tends to 1 as u tends to 1. Let $\zeta_1(u), \zeta_2(u), \cdots, \zeta_k(u)$ be the roots of $T(z, u)$ in $|z| < 1/u$, multiple roots being written multiply. In case $k < n - 2$, we define $\zeta_{k+1}(u) = \cdots = \zeta_{n-2}(u) = 0$. Let us write

$$\zeta_\nu(u) = \rho_\nu e^{i\varphi_\nu}, \qquad \rho_\nu = \rho_\nu(u), \qquad \varphi_\nu = \varphi_\nu(u), \qquad \nu = 1, 2, \cdots, n-2,$$

where, if $\rho_\nu = 0$, the number φ_ν is undefined, but we assign to it some value independent of u. Separating the equations (11.2.5), namely

$$\frac{d \log \zeta_\nu}{du} = \frac{2e^{i\alpha}\zeta_\nu}{1 - e^{i\alpha}u\zeta_\nu},$$

we obtain the system of equations:

$$\frac{d\rho_\nu}{du} = \frac{2\rho_\nu^2 [\cos(\alpha + \varphi_\nu) - u\rho_\nu]}{1 - 2u\rho_\nu \cos(\alpha + \varphi_\nu) + u^2\rho_\nu^2}, \tag{11.3.3}$$

$$\frac{d\varphi_\nu}{du} = \frac{2\rho_\nu \sin(\alpha + \varphi_\nu)}{1 - 2u\rho_\nu \cos(\alpha + \varphi_\nu) + u^2\rho_\nu^2}, \tag{11.3.4}$$

$$\frac{d\alpha}{du} = 2 \sum_{\nu=1}^{n-2} \rho_\nu \frac{\sin(\alpha + \varphi_\nu)}{1 - 2u\rho_\nu \cos(\alpha + \varphi_\nu) + u^2\rho_\nu^2}. \tag{11.3.5}$$

The initial values are $\rho_\nu(1)$, $\varphi_\nu(1)$, and $\alpha(1)$ and these numbers are defined by the given function $Q(z)$. The given initial values are real but, in order to prove that the solutions depend analytically on the initial values, we shall allow $\rho_\nu(1)$, $\varphi_\nu(1)$, and $\alpha(1)$ to be complex. Let $\Delta = \Delta(\delta, \eta, u)$ be the region

$$\begin{aligned} &|\operatorname{Im} \alpha| < \eta, \qquad |\rho_\nu| < \frac{1-\delta}{u}, \qquad |\operatorname{Im} \rho_\nu| < \frac{\eta}{u}, \\ &\qquad |\operatorname{Im} \varphi_\nu| < \eta, \qquad \nu = 1, 2, \cdots, n-2, \end{aligned} \tag{11.3.6}$$

where δ is any fixed positive number, $0 < \delta < 1$.

Let

$$R_\nu(u) = \frac{2\rho_\nu \cos(\alpha + \varphi_\nu) - 2u\rho_\nu^2}{1 - 2u\rho_\nu \cos(\alpha + \varphi_\nu) + u^2\rho_\nu^2} \tag{11.3.7}$$

$$= \frac{1}{u}\left\{\frac{1 - u^2\rho_\nu^2}{1 - 2u\rho_\nu \cos(\alpha + \varphi_\nu) + u^2\rho_\nu^2} - 1\right\},$$

$$S_\nu(u) = \frac{2\rho_\nu \sin(\alpha + \varphi_\nu)}{1 - 2u\rho_\nu \cos(\alpha + \varphi_\nu) + u^2\rho_\nu^2}, \tag{11.3.8}$$

and

$$y(u) = \sum_{\nu=1}^{n-2} \{|\rho_\nu \operatorname{Im} \alpha| + |\rho_\nu \operatorname{Im} \varphi_\nu| + |\operatorname{Im} \rho_\nu|\}. \tag{11.3.9}$$

From (11.3.3)–(11.3.5) we have

$$\frac{d\rho_\nu}{du} = \rho_\nu R_\nu(u), \tag{11.3.10}$$

$$\frac{d\varphi_\nu}{du} = S_\nu(u), \tag{11.3.11}$$

$$\frac{d\alpha}{du} = \sum_{\nu=1}^{n-2} S_\nu(u). \tag{11.3.12}$$

For each fixed value of u, $0 \leqq u \leqq 1$, the functions $R_\nu(u)$ and $S_\nu(u)$ are regular analytic functions of ρ_ν, φ_ν, and α in $\Delta(\delta, \eta, u)$, provided that η is small enough. When ρ_ν, φ_ν, and α are real and $-1 + \delta < u\rho < 1 - \delta$, the function $R_\nu(u)$ is greater than $(\delta/2 - 1)/u$. Choose η_0, $\eta_0 > 0$, so small that the real part of $uR_\nu(u)$ is greater than $(\delta/4) - 1$ in $\Delta_0(u) = \Delta(\delta, \eta_0, u)$ for all u, $0 \leqq u \leqq 1$, and $|1 - 2u\rho_\nu \cos(\alpha + \varphi_\nu) + u^2\rho_\nu^2| \geqq \delta^2/2$ in Δ_0. In $\Delta_0(u)$ we have

$$\operatorname{Re}(R_\nu) \geqq -b|\rho_\nu|, \tag{11.3.13}$$

where b is any number greater than some number b_0, $b_0 \geqq 1$, depending on $\Delta_0(u)$. We have also

$$|\operatorname{Im} R_\nu| \leqq Cy, \qquad |\operatorname{Im} S_\nu| \leqq Cy, \tag{11.3.14}$$

in $\Delta_0(u)$, where C is some positive constant.

Let

$$q = (eb)^{4/\delta}, \qquad u_0 = \frac{1}{bq}, \qquad \lambda = n(n-2)Cq, \qquad A = 3(n-2)q. \tag{11.3.15}$$

We may plainly suppose that $n > 2$ and hence by choosing b large that $\lambda > 1$, $A > 1$. Choose η_1 such that

$$\eta_1 < \frac{\eta_0}{A}e^{-\lambda}. \tag{11.3.16}$$

Given δ, $0 < \delta < 1$, let η_0 be chosen as above. Then choose η_1 such that (11.3.16) is satisfied and let the starting values $\rho_\nu(1)$, $\varphi_\nu(1)$, $\alpha(1)$ lie in $\Delta_1 = \Delta(\delta, \eta_1, 1)$. Let

$$\alpha^{(0)}(u) = \alpha(1), \qquad \rho_\nu^{(0)}(u) = \rho_\nu(1), \qquad \varphi_\nu^{(0)}(u) = \varphi_\nu(1).$$

Suppose that $\alpha^{(m-1)}(u)$, $\rho_\nu^{(m-1)}(u)$, and $\varphi_\nu^{(m-1)}(u)$ have been defined in the interval $0 < u \leqq 1$ for some $m \geqq 1$ and that for each u they lie in $\Delta_0(u)$. Suppose also that

$$y^{(m-1)} \leqq A\eta_1 e^{\lambda(1-u)}, \qquad 0 \leqq u \leqq 1, \tag{11.3.17}$$

$$| \rho_\nu^{(m-1)}(u) | \leqq q \, | \rho_\nu(1) |, \qquad 0 \leqq u \leqq u_0, \tag{11.3.18}$$

where $y^{(m-1)}$ is obtained by substituting the $(m-1)$st approximating functions in (11.3.9). We define for $\nu = 1, 2, \cdots, n-2$,

$$\rho_\nu^{(m)}(u) = \rho_\nu(1) \exp\left[-\int_u^1 R_\nu^{(m-1)}(t)\, dt \right], \tag{11.3.19}$$

$$\varphi_\nu^{(m)}(u) = \varphi_\nu(1) - \int_u^1 S_\nu^{(m-1)}(t)\, dt, \tag{11.3.20}$$

$$\alpha^{(m)}(u) = \alpha(1) - \int_u^1 \sum_{\nu=1}^{n-2} S_\nu^{(m-1)}(t)\, dt. \tag{11.3.21}$$

Since the $(m-1)$st approximating functions lie in $\Delta_0(u)$ and

$$\operatorname{Re}(R_\nu) \geqq \frac{1}{u}\left(\frac{\delta}{4} - 1\right)$$

in $\Delta_0(u)$, we have from (11.3.19)

$$| \rho_\nu^{(m)}(u) | \leqq \frac{| \rho_\nu(1) |}{u^{1-\delta/4}}, \qquad u_0 \leqq u \leqq 1.$$

For $0 < u \leqq u_0$,

$$| \rho_\nu^{(m)}(u) | \leqq \frac{| \rho_\nu(1) |}{u_0^{1-\delta/4}} \exp\left[bq \int_u^{u_0} dt \right] \leqq \frac{| \rho_\nu(1) |}{u_0^{1-\delta/4}} e^{bqu_0} \leqq q \, | \rho_\nu(1) |$$

by (11.3.15). Thus

$$\left| \frac{\rho_\nu^{(m)}(u)}{\rho_\nu(1)} \right| \leqq \begin{cases} \dfrac{1}{u^{1-\delta/4}} < q, & u_0 \leqq u \leqq 1, \\ q \qquad , & 0 \leqq u \leqq u_0. \end{cases} \tag{11.3.22}$$

Next, we have

$$\sum_{\nu=1}^{n-2} \{ | \rho_\nu^{(m)} \operatorname{Im} \alpha^{(m)} | + | \rho_\nu^{(m)} \operatorname{Im} \varphi_\nu^{(m)} | \} \tag{11.3.23}$$

$$\leqq 2(n-2)q\eta_1 + (n-1)(n-2)Cq \int_u^1 y^{(m-1)}\, dt.$$

Let $\rho_\nu(1) = |\rho_\nu(1)| \exp(i\theta_\nu)$ and let

$$g_\nu(u) = \operatorname{Im} \int_u^1 R_\nu^{(m-1)}(t)\, dt.$$

Then

(11.3.24) $$|g_\nu(u)| \leqq C \int_u^1 y^{(m-1)}\, dt.$$

We have

(11.3.25) $$\begin{aligned} |\operatorname{Im} \rho_\nu^{(m)}(u)| &\leqq q\,|\rho_\nu(1)|\,|\sin(\theta_\nu - g_\nu)| \leqq q\,|\rho_\nu(1)|\,(|\sin\theta_\nu| + |g_\nu|) \\ &\leqq q(|\operatorname{Im} \rho_\nu(1)| + |g_\nu|) \leqq q(\eta_1 + |g_\nu|). \end{aligned}$$

Thus

(11.3.26) $$\begin{aligned} y^{(m)} &\leqq 3(n-2)q\eta_1 + n(n-2)Cq \int_u^1 y^{(m-1)}\, dt \\ &\leqq A\eta_1 + \lambda A e^{\lambda} \eta_1 \int_u^1 e^{-\lambda t}\, dt \end{aligned}$$

by (11.3.15) and (11.3.17). Hence

(11.3.27) $$y^{(m)} \leqq A\eta_1 e^{\lambda(1-u)}, \qquad 0 \leqq u \leqq 1.$$

This completes the inductive proof that all the approximating solutions of the system of differential equations lie in $\Delta_0(u)$ for $0 \leqq u \leqq 1$ and satisfy the inequalities (11.3.17) and (11.3.18). Let $D = D(\eta_0, q)$ be the region

$$|\operatorname{Im} \alpha| < \eta_0, \qquad |\rho_\nu| \leqq q, \qquad |\operatorname{Im} \rho_\nu| < \eta_0,$$
$$|\operatorname{Im} \varphi_\nu| < \eta_0.$$

For each m the approximating functions lie in $\Delta_0(u)$ and, according to inequalities (11.3.16), (11.3.17), and (11.3.18), they also lie in D.

Let $\mathfrak{p}$ denote the point with coordinates $\rho_1, \varphi_1, \rho_2, \varphi_2, \cdots, \rho_{n-2}, \varphi_{n-2}, \alpha$, and define

$$\|\mathfrak{p}' - \mathfrak{p}''\| = |\alpha' - \alpha''| + \sum_{\nu=1}^{n-2} \{|\rho_\nu' - \rho_\nu''| + |\varphi_\nu' - \varphi_\nu''|\}.$$

In the common part of D and $\Delta_0(u)$, we have

(11.3.28) $$|R_\nu'(u) - R_\nu''(u)| \leqq M\,\|\mathfrak{p}' - \mathfrak{p}''\|,$$

(11.3.29) $$|S_\nu'(u) - S_\nu''(u)| \leqq M\,\|\mathfrak{p}' - \mathfrak{p}''\|,$$

where $R_\nu'(u)$ denotes the value of R_ν at the point $\mathfrak{p}' = (\rho_1', \varphi_1', \cdots, \rho_{n-2}', \varphi_{n-2}', \alpha')$ and similarly for $R_\nu''(u)$, $S_\nu'(u)$, $S_\nu''(u)$. The number M depends on δ, η_0, and q but is independent of u. Writing

$$\mathfrak{p}^{(m)} = (\rho_1^{(m)}, \varphi_1^{(m)}, \cdots, \rho_{n-2}^{(m)}, \varphi_{n-2}^{(m)}, \alpha^{(m)}),$$

suppose that for some m, $m \geqq 1$,

$$\| \mathfrak{p}^{(m)} - \mathfrak{p}^{(m-1)} \| \leqq B \frac{\gamma^{m-1}(1-u)^{m-1}}{(m-1)!}, \qquad 0 \leqq u \leqq 1, \tag{11.3.30}$$

where $\gamma = 2(n-2)(q+1)M$, and B is so chosen that the inequality is true for $m = 1$. Then, writing

$$x^{(m)} + iy^{(m)} = -\int_u^1 R_\nu^{(m)}(t)\, dt,$$

we have

$$\rho_\nu^{(m+1)} - \rho_\nu^{(m)} = \rho_\nu(1)\,\{(\exp x^{(m)} - \exp x^{(m-1)})(\exp iy^{(m)}) + (\exp x^{(m-1)})(\exp iy^{(m)} - \exp iy^{(m-1)})\}$$

$$= \rho_\nu(1)\,\{\exp x(x^{(m)} - x^{(m-1)})(\exp iy^{(m)}) + (\exp x^{(m-1)})(\exp iy^{(m)} - \exp iy^{(m-1)})\}$$

where x is some number between $x^{(m-1)}$ and $x^{(m)}$. By (11.3.18) we see that $\rho_\nu(1)e^x$ is bounded by q; so

$$| \rho_\nu^{(m+1)} - \rho_\nu^{(m)} | \leqq q\,\{|x^{(m)} - x^{(m-1)}| + |y^{(m)} - y^{(m-1)}|\} \tag{11.3.31}$$

$$\leqq \frac{2qMB\gamma^{m-1}}{m!}(1-u)^m.$$

By using (11.3.20) and (11.3.21) it follows that (11.3.30) is true with m replaced by $m+1$ and so it is true for all m. The approximating functions thus converge to solutions of the equations (11.3.3)–(11.3.5) and uniformly with respect to u, $\rho_\nu(1)$, $\varphi_\nu(1)$, $\alpha(1)$ for $0 \leqq u \leqq 1$ and $\rho(1)$, $\varphi_\nu(1)$, and $\alpha(1)$ in $\Delta_1 = \Delta(\delta, \eta_1, 1)$. These solutions are unique. Since $\rho_\nu^{(m)}(u)$, $\varphi_\nu^{(m)}(u)$, and $\alpha^{(m)}(u)$ are plainly analytic functions of $\rho_\nu(1)$, $\varphi_\nu(1)$, and $\alpha(1)$ for $m = 0$, it follows from (11.3.19), (11.3.20), and (11.3.21) that for each fixed u the mth approximating functions are analytic functions of the starting values. Owing to the uniformity of the convergence of the approximating functions, it follows that the solutions $\rho_\nu(u)$, $\varphi_\nu(u)$, and $\alpha(u)$ are analytic functions of the starting values throughout $\Delta_1 = \Delta(\delta, \eta_1, 1)$ for each fixed u in $0 \leqq u \leqq 1$. The solutions and their first order partial derivatives with respect to $\rho_\nu(1)$, $\varphi_\nu(1)$, and $\alpha(1)$ are bounded by a constant K which is independent of u.

For each fixed u, $0 \leqq u \leqq 1$, the function $\alpha(u)$ is an analytic function of the parameters $\rho_\nu(1)$, $\varphi_\nu(1)$, $\alpha(1)$ defined by the given Q-function $Q(z)$; so by Chapter IX the coefficients a_2, a_3, $\cdots$ of the $\mathfrak{D}_n$-function $f(z)$ determined by Q are analytic functions of these parameters. The real and imaginary parts of the roots of $P(w)$ are also analytic functions of $\rho_\nu(1)$, $\varphi_\nu(1)$, and $\alpha(1)$, where $w = f(z)$ satisfies the $\mathfrak{D}_n$-equation

$$\left(\frac{z}{w}\frac{dw}{dz}\right)^2 P(w) = Q(z).$$

For each fixed z in $|z| < 1$, the function

$$f(z) = z + a_2z^2 + a_3z^3 + \cdots$$

is analytic in the parameters $\rho_\nu(1)$, $\varphi_\nu(1)$, and $\alpha(1)$. To investigate the analytic dependence of the values taken by $f(z)$ at fixed points z on $|z| = 1$, we consider instead the function

$$s = F(z) = \frac{1}{f(z)} = \frac{1}{z} + b_0 + b_1z + \cdots$$

which satisfies

$$\left(\frac{z}{s}\frac{ds}{dz}\right)^2 P\left(\frac{1}{s}\right) = Q(z). \tag{11.3.32}$$

The image of $|z| = 1$ by $s = F(z)$ is contained in a domain D which is interior to the circle $|s| < 8$ and which contains no zero of $P(1/s)$ inside or on its boundary. If $s_0 = F(z_0)$, where $|z_0| < 1$ but s_0 is interior to D and near the image of $|z| = 1$ by F, let

$$\zeta = \int_{s_0}^{s} \left(P\left(\frac{1}{s}\right)\right)^{1/2} \frac{ds}{s} = \int_{z_0}^{z} (Q(z))^{1/2} \frac{dz}{z}. \tag{11.3.33}$$

As the parameters $\rho_\nu(1)$, $\varphi_\nu(1)$, and $\alpha(1)$ vary in a small neighborhood with z_0 fixed, the value s_0 as well as the real and imaginary parts of the coefficients of $P(w)$ depends analytically on the parameters. It readily follows that for each fixed z in $0 < |z| < 1 + \epsilon$, where ϵ is some positive number, the function $F(z)$ depends analytically on the parameters.

Finally, if $0 < u < 1$, then the functions $\zeta_\nu(u)$ and $\exp(-i\alpha(u))/u$ are zeros of $T(z, u)$ while $z_\nu(u) = u\zeta_\nu(u)$ and $e^{-i\alpha(u)}$ are the zeros of $Q(z, u)$. Using the formula (11.1.5), we obtain a one-parameter family of functions $V(z, u)$ whose coefficients $b_2(u), b_3(u), \cdots, b_n(u)$ trace a curve from the boundary of V_n to the origin $b_2 = 0, \cdots, b_n = 0$. The coefficients $a_2(u), a_3(u), \cdots, a_n(u)$ of the function $f(z, u)$ which satisfies equation (11.2.2) where $P(w) = T(w, 0)$ define a curve on the boundary of V_n joining the point $a_2(1), a_3(1), \cdots, a_n(1)$ corresponding to the given $Q(z)$ with some point on the Koebe curve. For each fixed u the points $(b_2(u), b_3(u), \cdots, b_n(u))$ and $(a_2(u), a_3(u), \cdots, a_n(u))$ depend analytically on the parameters.

Instead of considering the system of differential equations for the roots $e^{-i\alpha}/u, \zeta_1, \zeta_2, \cdots, \zeta_{n-2}$, we may construct the solutions using the differential equations for the coefficients $C_k(u)$. The coefficients $C_k(u)$ satisfy the equations

$$\frac{dC_k}{du} = 2\sum_{\nu=1}^{n-k-1} (k - \nu)C_{k+\nu}\, u^{\nu-1}\, e^{i\nu\alpha} \tag{11.3.34}$$

for $k = -(n-1), \cdots, n-1$, and $e^{i\alpha}$ satisfies the equation

$$\sum_{\nu=1}^{n-1} \nu u^\nu \{C_\nu e^{i\nu\alpha} - \bar{C}_\nu e^{-i\nu\alpha}\} = 0. \tag{11.3.35}$$

Differentiating this equation, we obtain, using (11.3.34) and simplifying,

$$g \frac{d\alpha}{du} = g_1 \tag{11.3.36}$$

where

$$g = -\sum_{\nu=1}^{n-1} \nu^2 u^{\nu-1} \{C_\nu e^{i\nu\alpha} + \bar{C}_\nu e^{-i\nu\alpha}\}, \tag{11.3.37}$$

$$\begin{aligned} g_1 &= \frac{1}{3iu} \sum_{\nu=1}^{n-1} \nu(\nu^2 + 2)u^{\nu-1}\{C_\nu e^{i\nu\alpha} - \bar{C}_\nu e^{-i\nu\alpha}\} \\ &= \frac{1}{3i} \sum_{\nu=2}^{n-1} \nu(\nu^2 - 1)u^{\nu-2}\{C_\nu e^{i\nu\alpha} - \bar{C}_\nu e^{-i\nu\alpha}\} \end{aligned} \tag{11.3.38}$$

by (11.3.35). We use equations (11.3.34) for $k = 1, 2, \cdots, n - 1$ and equation (11.3.36). The initial conditions which we consider are

$$\alpha(1) = \alpha, \qquad C_k(1) = B_k, \tag{11.3.39}$$

where the B_k, $k = 1, 2, \cdots, n - 1$, and α are defined by a Q-function

$$Q(z) = \sum_{\nu=-(n-1)}^{n-1} \frac{B_\nu}{z^\nu} \tag{11.3.40}$$

which has a zero of precise order 2 at the point $e^{-i\alpha}$ on $|z| = 1$ and no other zeros on $|z| = 1$. We have shown that there is a solution of the equations (11.3.3), (11.3.4), and (11.3.5) under the corresponding initial conditions, and it follows that there is a solution of equations (11.3.34) and (11.3.36). Given a function $Q(z)$ satisfying the above conditions, there is an ϵ such that for all Q-functions $Q^*(z)$ with coefficients B^* which satisfy

$$\sum_{\nu=-(n-1)}^{n-1} |B_\nu^* - B_\nu|^2 < \epsilon^2, \tag{11.3.41}$$

the corresponding solutions $C_\nu^*(u)$ and $\alpha^*(u)$ are uniformly bounded in u, $0 \leqq u \leqq 1$, and $g_1^*(u)/g(u)$ is also bounded uniformly in u. The well-known method of successive approximation shows that these solutions have continuous derivatives with respect to the real and imaginary parts of the B_ν^* provided that (11.3.41) is satisfied. Then the function $1/f^*(z)$ defined by $Q^*(z)$ can be differentiated with respect to the real and imaginary parts of the B_ν^* uniformly in z on $|z| = 1$.

11.4. Let M_n be the set of boundary points of V_n whose points belong to functions $f(z)$ which satisfy one and only one $\mathfrak{D}_n$-equation and in this equation the function $Q(z)$ vanishes at only one point of $|z| = 1$ where it has a zero of precise order 2. Two $\mathfrak{D}_n$-equations are considered equal if one is a positive multiple of the other. An alternative definition of M_n is that a function belonging to one of its points satisfies only one $\mathfrak{D}_n$-equation, the function $P(w)$ has the property that its coefficient A_1 is not zero, and it has no zero at a finite point of the map of $|z| = 1$ by $w = f(z)$. Clearly no point of the Koebe curve belongs to M_n if $n > 2$.

An example given in Chapter X shows that if $n \geqq 10$, then there is a Q-function which vanishes at only one point of $|z| = 1$, where it has a zero of precise order 2, but which defines a point not in M_n.

THEOREM VII. *The set M_n is not empty, it is open, and its closure contains the supporting surface K_n. There is a one-one bicontinuous correspondence between points of M_n and a class of Q-functions vanishing at only one point of $|z| = 1$ where there is a zero of precise order 2. If $\mathfrak{a}$ is any point of M_n and $f(z)$ the function belonging to $\mathfrak{a}$, let $Q(z)$ be the corresponding Q-function and let $e^{-i\alpha}$ be the point on $|z| = 1$ where $Q(z)$ vanishes. Let*

$$\psi(z) = \sum_{\nu=-(n-1)}^{n-1} \frac{\beta_\nu}{z^\nu}, \qquad \beta_{-\nu} = \bar{\beta}_\nu,$$

have a zero of first or higher order at $z = e^{-i\alpha}$, and let

$$\chi(z) = \frac{1}{2}\int_{e^{-i\alpha}}^{z} \frac{\psi(v)}{(Q(v))^{1/2}} \frac{dv}{v}$$

where $(Q(v))^{1/2}$ is analytic on $|v| = 1$ except for a discontinuity at the point $e^{-i\beta}$ which maps into infinity by $f(z)$. As v passes counterclockwise through $e^{-i\alpha}$, we suppose that $(Q(v))^{1/2}$ passes from negative to positive values. Then the function $f^*(z)$ belonging to any point $\mathfrak{a}^*$ of M_n in a sufficiently small neighborhood $U(\mathfrak{a}, \epsilon)$ of $\mathfrak{a}$ is given by the formula

$$f^*(z) = f(z) - \epsilon \frac{f(z)^2}{2\pi i} \int_{|t|=1} \left(\frac{tf'(t)}{f(t)}\right)^2 \frac{\chi(t)}{(Q(t))^{1/2}[f(z) - f(t)]} \frac{dt}{t} + o(\epsilon)$$

where the integration is counterclockwise around $|t| = 1$ beginning and ending at $e^{-i\beta}$. Here the function $(Q(t))^{1/2}$ is to be interpreted as above. The term $o(\epsilon)$ holds uniformly in $U(\mathfrak{a}, \epsilon)$ if ϵ is a sufficiently small positive number. As $\mathfrak{a}^*$ sweeps over all of $U(\mathfrak{a}, \epsilon)$, the coefficients $\beta_1, \beta_2, \cdots, \beta_{n-1}$ sweep out a full neighborhood of the origin $\beta_1 = 0, \beta_2 = 0, \cdots, \beta_{n-1} = 0$.

If a function $w = f(z)$ belonging to a point on the boundary of V_n satisfies several $\mathfrak{D}_n$-equations, then it satisfies at least one in which the Q-function has as many zeros on $|z| = 1$ as any Q-function belonging to $f(z)$. We call such a $\mathfrak{D}_n$-equation maximal (relative to $f(z)$ and the point $\mathfrak{a}$); if the function $Q(z)$ of a $\mathfrak{D}_n$-equation which is maximal with respect to $f(z)$ has $2j$ zeros on $|z| = 1$, we say that the function $f(z)$ and the point $\mathfrak{a}$ have index j. Two Q-functions will be considered equal if their ratio is a constant, which is necessarily positive. We therefore suppose in the present section, unless the contrary is stated, that each Q-function

$$Q(z) = \sum_{\nu=-(n-1)}^{n-1} \frac{B_\nu}{z^\nu}$$

is normalized by the condition

$$\sum_{\nu=-(n-1)}^{n-1} | B_\nu |^2 = 1.$$

In order to study the closure of the set $\boldsymbol{M}_n$ it will be convenient to introduce a slightly larger set $\boldsymbol{M}_n^*$ on the boundary of V_n. Let $\mathfrak{a}$ be a point on the boundary of V_n and let $f(z)$ be the function of class $\mathfrak{S}$ which belongs to $\mathfrak{a}$. We shall say that the point $\mathfrak{a}$ belongs to the set $\boldsymbol{M}_n^*$ if the maximal $\mathfrak{D}_n$-equation

$$\left(\frac{z}{w}\frac{dw}{dz}\right)^2 P(w) = Q(z) \tag{11.4.1}$$

satisfied by $w = f(z)$ is unique and if:

1. The unit circumference $| z | = 1$ is the sum of two identified arcs U and L and their end points $e^{-i\alpha}$ and $e^{-i\beta}$.
2. The distinguished cycle $\boldsymbol{\Sigma}$ which maps into the point at infinity in the w-plane is the single point $z = e^{-i\beta}$.
3. $Q(z)$ has a zero of precise order 2 at $z = e^{-i\alpha}$, no zero at $z = e^{-i\beta}$, and at most one of each pair of identified points of $| z | = 1$ is a zero of $Q(z)$.

The function $w = f(z)$ belonging to a point of $\boldsymbol{M}_n^*$ then maps $| z | < 1$ onto the complement of a Jordan arc in the w-plane with at most a finite number of critical points, and with one end point at $w = \infty$. The function

$$P(w) = \sum_{\nu=1}^{n-1} \frac{A_\nu}{w^\nu}$$

belonging to a maximal $\mathfrak{D}_n$-equation satisfied by $w = f(z)$ does not vanish at the finite tip of this Jordan arc, and the point at infinity is of order -1 since $A_1 \neq 0$.

Let $\mathfrak{a}$ be a point on the boundary of V_n with index j and let the function $w = f(z)$ of class $\mathfrak{S}$ belonging to $\mathfrak{a}$ map $| z | < 1$ onto the exterior of a single Jordan arc in the w-plane with one end point at $w = \infty$. Let

$$\left(\frac{z}{w}\frac{dw}{dz}\right)^2 P(w) = Q(z)$$

be a maximal $\mathfrak{D}_n$-equation which is satisfied by $w = f(z)$ and let

$$z_1 = e^{-i\alpha}, z_2, z_3, \cdots, z_j$$

be the zeros of $(Q(z))^{1/2}$ on $| z | = 1$, and

$$z_{j+1}, z_{j+2}, \cdots, z_m$$

the zeros of $Q(z)$ in $| z | < 1$. Here $m \leqq n - 1$, and $z = e^{-i\alpha}$ is the point which maps into the finite tip of the Jordan arc in the w-plane. The function $Q(z)$ can be written in the form

$$Q(z) = C \prod_{\nu=1}^{j} \frac{(z - z_\nu)^2}{-zz_\nu} \prod_{\nu=j+1}^{m} \frac{(z - z_\nu)(z - 1/\bar{z}_\nu)}{-zz_\nu} = \frac{(z - z_1)^2}{-zz_1} S(z)$$

where C is some positive constant and $S(z) \geqq 0$ on $|z| = 1$. Then $|z| = 1$ is the sum of two identified arcs U and L and their end points; and if z' belonging to U and z'' belonging to L are identified, then

$$\int_{z''}^{z'} \left(\frac{(z - e^{-i\alpha})^2}{-ze^{-i\alpha}}\right)^{1/2} \cdot |(S(z))^{1/2}| \frac{dz}{z} = 0$$

where the integration is counterclockwise from z'' through $z = e^{-i\alpha}$ to z'. Let $\epsilon > 0$ be given and let α_1 satisfy $|\alpha - \alpha_1| < \epsilon$. Define a Q-function $Q_1(z)$ by the relation

$$Q_1(z) = C_1 \frac{(z - e^{-i\alpha_1})^2}{-ze^{-i\alpha_1}} \prod_{\nu=2}^{j} \frac{(z - z_\nu)^2}{-zz_\nu} \prod_{\nu=j+1}^{m} \frac{(z - z_\nu)(z - 1/\bar{z}_\nu)}{-zz_\nu}$$

where C_1 is a positive constant chosen to normalize $Q_1(z)$. Let $|z| = 1$ be the sum of two arcs U_1 and L_1 and their end points $z = \exp(-i\alpha_1)$ and $z = \exp(-i\beta_1)$, the arcs being identified by the relation

$$\int_{z_1''}^{z_1'} \left(\frac{(z - e^{-i\alpha_1})^2}{-ze^{-i\alpha_1}}\right)^{1/2} \cdot |(S(z))^{1/2}| \frac{dz}{z} = 0. \tag{11.4.2}$$

Here z_1'' belongs to L_1 and z_1' to U_1. The identification may also be expressed by the relation

$$\int_{\theta_1''}^{\theta_1'} [e^{i(\theta+\alpha_1)/2} - e^{-i(\theta+\alpha_1)/2}] R(\theta)\, d\theta = 0 \tag{11.4.3}$$

where $R(\theta)$ is non-negative and is independent of α_1. This is an equation of the form

$$pe^{i\alpha_1/2} - \bar{p}e^{-i\alpha_1/2} = 0 \tag{11.4.4}$$

where $p \neq 0$ since R is non-negative and $\exp(i\theta/2)$ lies in a half-plane over the range of integration. The point $\exp(-i\beta_1)$ is determined from (11.4.3) by the relation $\theta_1' = \theta_1'' + 2\pi$. For fixed z_1' and z_1'' the solution α_1 of (11.4.4) is unique modulo 2π. Thus we can choose α_1, $0 < |\alpha_1 - \alpha| < \epsilon$, such that $Q_1(z)$ has a zero of precise order 2 at $z = \exp(-i\alpha_1)$, no zero at $z = \exp(-i\beta_1)$, and such that no zero of $Q_1(z)$ on L_1 is identified with a zero of $Q_1(z)$ on U_1.

Now let $f_1(z)$ be a function of class $\mathfrak{S}$ which realizes the identification defined by the metric $|(Q_1(z))^{1/2}\, dz/z|$ where $|z| = 1$ is the sum of two identified arcs U_1, L_1 and their end points $\exp(-i\alpha_1)$ and $\exp(-i\beta_1)$, and the distinguished cycle $\mathbf{\Sigma}$ is the single point $z = \exp(-i\beta_1)$. Then $w = f_1(z)$ satisfies the $\mathfrak{D}_n$-equation

$$\left(\frac{z}{w}\frac{dw}{dz}\right)^2 P_1(w) = Q_1(z).$$

As ϵ approaches zero, the identification defined by $Q_1(z)$ approaches that defined by $Q(z)$, and $f_1(z)$ approaches $f(z)$ uniformly in every closed set in $|z| < 1$. It follows that the point $\mathfrak{a}_1 = (a_2^{(1)}, a_3^{(1)}, \cdots, a_n^{(1)})$ to which $f_1(z)$ belongs must approach the point $\mathfrak{a} = (a_2, a_3, \cdots, a_n)$ to which $f(z)$ belongs. Thus, given any $\delta > 0$, there is an $\epsilon > 0$ such that

$$\sum_{\nu=2}^{n} | a_\nu^{(1)} - a_\nu |^2 < \delta^2.$$

If the function $w = f_1(z)$ satisfies any other $\mathfrak{D}_n$-equation

$$\left(\frac{z}{w}\frac{dw}{dz}\right)^2 P_2(w) = Q_2(z),$$

then since $f_1(z)$ realizes the identification defined by $Q_\nu(z)$ for $\nu = 1, 2$, it follows that

$$\frac{Q_1(z_1'')}{Q_2(z_1'')} = \frac{Q_1(z_1')}{Q_2(z_1')}. \tag{11.4.5}$$

If z_ν, $z_\nu \neq \exp(-i\alpha_1)$, is a zero of $Q_1(z)$ on $| z | = 1$ of precise order $2k$, $k \geqq 1$, then $Q_1(z)$ is not zero at the identified point according to the above construction. If $Q_2(z)$ has a zero of precise order 2λ at z_ν and of precise order 2μ at the point identified with z_ν, then relation (11.4.5) shows that $\lambda = k + \mu$. The statement $\lambda = k + \mu$ remains true for $k = 0$ if $Q_1(z)$ does not vanish at the point identified with z_ν. This shows in particular that $\lambda \geqq k$. We write

$$Q_1(z) = \frac{(z - e^{-i\alpha_1})^2}{-ze^{-i\alpha_1}} \left\{\prod_{\nu=2}^{j} \frac{(z - z_\nu)^2}{-zz_\nu}\right\} N_1(z),$$

$$Q_2(z) = \frac{(z - e^{-i\alpha_1})^2}{-ze^{-i\alpha_1}} \left\{\prod_{\nu=2}^{j} \frac{(z - z_\nu)^2}{-zz_\nu}\right\} N_2(z),$$

where $N_1(z)$ and $N_2(z)$ are of the form

$$N_1(z) = \sum_{\nu=-n+1+j}^{n-1-j} \frac{B_\nu^{(1)}}{z^\nu}, \qquad N_2(z) = \sum_{\nu=-n+1+j}^{n-1-j} \frac{B_\nu^{(2)}}{z^\nu},$$

and $N_1(z) > 0$ on $| z | = 1$, $N_2(z) \geqq 0$ on $| z | = 1$. Let η be the least upper bound of all ρ for which the function

$$N_1(z) - \rho N_2(z)$$

is non-negative on $| z | = 1$. It is clear that $0 < \eta < \infty$. Then the function $N_1 - \eta N_2$ is non-negative on $| z | = 1$ and has at least one zero there. Then let

$$Q_3(z) = C_3\{Q_1(z) - \eta Q_2(z)\} = \sum_{\nu=-(n-1)}^{n-1} \frac{B_\nu^{(3)}}{z^\nu}$$

where C_3 is a constant so chosen that

$$\sum_{\nu=-(n-1)}^{n-1} | B_\nu^{(3)} |^2 = 1.$$

It is clear that $Q_3(z)$ has more than $2j$ zeros on $|z| = 1$ and is non-negative there. If we write

$$P_3(w) = C_3\{P_1(w) - \eta P_2(w)\},$$

then $w = f_1(z)$ satisfies the $\mathfrak{D}_n$-equation

$$\left(\frac{z}{w}\frac{dw}{dz}\right)^2 P_3(w) = Q_3(z);$$

so $f_1(z)$ has index greater than j. We have thus shown that if the function $f_1(z)$ satisfies more than one $\mathfrak{D}_n$-equation, then it has index greater than j. In the particular case in which $j = n - 1$, the function $w = f_1(z)$ satisfies a unique $\mathfrak{D}_n$-equation. A consequence of this argument is that if a function $f(z)$ of class $\mathfrak{S}$ belonging to some point on the boundary of V_n has index 1, then it satisfies only one $\mathfrak{D}_n$-equation. For if it should satisfy two of them, then it would satisfy some $\mathfrak{D}_n$-equation whose Q-function had at least four zeros on $|z| = 1$.

Now let $\mathfrak{a} = (a_2, a_3, \cdots, a_n)$ be a point of $\boldsymbol{K}_n$ with index j. Then the function $w = f(z)$ belonging to $\mathfrak{a}$ maps $|z| < 1$ onto the complement of a Jordan arc in the w-plane with one end point at $w = \infty$. Given any $\delta > 0$, there is a function $w = f_1(z)$ belonging to a point $\mathfrak{a}_1 = (a_2^{(1)}, a_3^{(1)}, \cdots, a_n^{(1)})$ on the boundary of $\boldsymbol{V}_n$ such that

$$\sum_{\nu=2}^{n} |a_\nu^{(1)} - a_\nu|^2 < \delta^2,$$

and $w = f_1(z)$ maps $|z| < 1$ onto the complement of a single Jordan arc in the w-plane with one end point at $w = \infty$. Moreover, $w = f_1(z)$ satisfies a $\mathfrak{D}_n$-equation

$$\left(\frac{z}{w}\frac{dw}{dz}\right)^2 P_1(w) = Q_1(z)$$

in which $Q_1(z)$ has a zero of precise order 2 at $z = \exp(-i\alpha_1)$ but does not vanish at the point $z = \exp(-i\beta_1)$ which maps into $w = \infty$, and no zero of $Q_1(z)$ on $|z| = 1$ is identified with a zero except for the point $\exp(-i\alpha_1)$ which is identified with itself. Then either $w = f_1(z)$ satisfies a unique $\mathfrak{D}_n$-equation or it has index greater than j. In the latter case we repeat the argument, obtaining a function $w = f_2(z)$ which either satisfies a unique $\mathfrak{D}_n$-equation or has index greater than $j + 1$. Ultimately we obtain a function $w = f^*(z)$ which satisfies a unique $\mathfrak{D}_n$-equation and so belongs to $\boldsymbol{M}_n^*$. It follows that the closure of $\boldsymbol{M}_n^*$ contains the supporting surface $\boldsymbol{K}_n$.

Now let $\mathfrak{a} = (a_1, a_2, \cdots, a_n)$ be a point of $\boldsymbol{M}_n^*$ with index j, and let $f(z)$ belong to $\mathfrak{a}$. Then $w = f(z)$ satisfies a $\mathfrak{D}_n$-equation

$$\left(\frac{z}{w}\frac{dw}{dz}\right)^2 P(w) = Q(z) \tag{11.4.6}$$

where $Q(z)$ has $2j$ zeros on $|z| = 1$. If $j = 1$, then $\mathfrak{a}$ belongs to $\boldsymbol{M}_n$. We now show that there is a $\delta > 0$ such that all points $\mathfrak{a}^* = (a_2^*, a_3^*, \cdots, a_n^*)$ on the boundary of V_n which satisfy

$$\sum_{\nu=2}^{n} |a_\nu^* - a_\nu|^2 < \delta^2 \tag{11.4.7}$$

have index j or less. For let $\mathfrak{a}_1, \mathfrak{a}_2, \cdots$ be a sequence of points on the boundary of V_n with limit $\mathfrak{a}$, and let $f_\nu(z)$ belong to $\mathfrak{a}_\nu$ and $f(z)$ belong to $\mathfrak{a}$. Since $f_\nu(z)$ belongs to class $\mathfrak{S}$, it follows that there is a subsequence for which $f_\nu(z)$ converges to a limit function of class $\mathfrak{S}$. The limit belongs to the point $\mathfrak{a}$; so it is the function $f(z)$. It readily follows that the entire sequence $f_1(z), f_2(z), \cdots$ converges to $f(z)$. If $\mathfrak{a}_\nu$ has index j_ν, let $w = f_\nu(z)$ satisfy the $\mathfrak{D}_n$-equation

$$\left(\frac{z}{w}\frac{dw}{dz}\right)^2 P_\nu(w) = Q_\nu(z)$$

where $Q_\nu(z)$ has $2j_\nu$ zeros on $|z| = 1$. The coefficients of $Q_\nu(z)$ are uniformly bounded; so the coefficients of $P_\nu(w)$ must be uniformly bounded. Thus there is a subsequence of integers such that the functions $P_\nu(w)$ and $Q_\nu(z)$ converge to limits $P^\Delta(w)$ and $Q^\Delta(z)$ respectively. Since $f_\nu(z)$ converges to $f(z)$, it follows that $w = f(z)$ satisfies the $\mathfrak{D}_n$-equation

$$\left(\frac{z}{w}\frac{dw}{dz}\right)^2 P^\Delta(w) = Q^\Delta(z).$$

This shows that there is a $\delta > 0$ such that if (11.4.7) is true, then $\mathfrak{a}^*$ has index j or less. For if not, the sequence $\mathfrak{a}_1, \mathfrak{a}_2, \cdots$ could be chosen such that $Q_\nu(z)$ would have $2j_\nu \geqq 2j + 2$ zeros on $|z| = 1$ and this would imply that $Q^\Delta(z)$ has at least $2j + 2$ zeros on $|z| = 1$, a contradiction.

Let the function $w = f^*(z)$ belonging to $\mathfrak{a}^*$ satisfy the $\mathfrak{D}_n$-equation

$$\left(\frac{z}{w}\frac{dw}{dz}\right)^2 P^*(w) = Q^*(z)$$

where

$$Q^*(z) = \sum_{\nu=-(n-1)}^{n-1} \frac{B_\nu^*}{z^\nu}.$$

Given $\epsilon > 0$, there is a $\delta > 0$ such that if (11.4.7) is true, then $Q^*(z)$ has $2j$ or fewer zeros on $|z| = 1$, and those $Q^*(z)$ which have $2j$ zeros on $|z| = 1$ satisfy

$$\sum_{\nu=-(n-1)}^{n-1} |B_\nu^* - B_\nu|^2 < \epsilon^2. \tag{11.4.8}$$

The first part of this statement has been proved. To prove the second part, let the sequence $\mathfrak{a}_1, \mathfrak{a}_2, \cdots$ be such that $Q_\nu(z)$ has $2j$ zeros on $|z| = 1$. Then $Q^\Delta(z)$ has at least $2j$ zeros on $|z| = 1$, and it can have no more since $\mathfrak{a}$ has index j. Since $\mathfrak{a}$ belongs to $\boldsymbol{M}_n^*$, its maximal $\mathfrak{D}_n$-equation is unique, and this implies that $Q^\Delta(z) = Q(z)$. The statement follows.

We now show that each point of M_n^* lies in the closure of M_n. Let $\mathfrak{a} = (a_2, a_3, \cdots, a_n)$ be a point of M_n^* and let it have index j. If $j = 1$, then $\mathfrak{a}$ belongs to M_n; so suppose that $j > 1$. The function $w = f(z)$ belonging to $\mathfrak{a}$ satisfies the maximal $\mathfrak{D}_n$-equation

$$\left(\frac{z}{w}\frac{dw}{dz}\right)^2 P(w) = Q(z)$$

where $Q(z)$ has precisely $2j$ zeros on $|z| = 1$. The unit circumference $|z| = 1$ is the sum of two identified arcs U and L and their end points $z = e^{-i\alpha}$ and $z = e^{-i\beta}$, and the distinguished cycle Σ which maps into $w = \infty$ is the single point $z = e^{-i\beta}$. Let

$$z_1 = e^{-i\alpha}, z_2, z_3, \cdots, z_j$$

be the zeros of $(Q(z))^{1/2}$ which lie on $|z| = 1$ and let

$$z_{j+1}, z_{j+2}, \cdots, z_m,$$

where $m \leqq n - 1$, be the zeros of $Q(z)$ that lie in $|z| < 1$. Then, since $\mathfrak{a}$ belongs to M_n^*, the zeros $z_2, z_3, \cdots, z_j$ are interior points of U or L; and, although several of them may coincide, there is no pair which is identified by $w = f(z)$. Let ρ be some number satisfying $0 < \rho < 1$, and let $Q_1(z)$ be defined by

$$Q_1(z) = C_1(\rho)\frac{(z - z_1)^2}{-zz_1}\frac{(z - \rho z_2)(z - 1/\rho\bar{z}_2)}{-zz_2}\prod_{\nu=3}^{n}\frac{(z - z_\nu)(z - 1/\bar{z}_\nu)}{-zz_\nu}$$

where $C_1(\rho) > 0$ is chosen such that $Q_1(z)$ is a normalized Q-function. Let $|z| = 1$ be the sum of two arcs U_1 and L_1 identified in the metric defined by Q_1, and let their end points be $z = \exp(-i\alpha)$ and $z = \exp(-i\beta_1)$. Let $w = f_1(z)$ be the function of class $\mathfrak{S}$ which realizes this identification and maps $z = \exp(-i\beta_1)$ into $w = \infty$. Let $\mathfrak{a}_1 = (a_2^{(1)}, a_3^{(1)}, \cdots, a_n^{(1)})$ be the point on the boundary of V_n belonging to $f_1(z)$ and, given $\epsilon_1 > 0$, choose ρ, $0 < \rho < 1$, such that

$$\sum_{\nu=2}^{n} |a_\nu - a_\nu^{(1)}|^2 < \epsilon_1^2,$$

and such that none of the zeros $z_3, z_4, \cdots, z_j$ of $(Q_1(z))^{1/2}$ on $|z| = 1$ are identified with another zero or lie at the point $z = \exp(-i\beta_1)$. There exists such a ρ since, as ρ tends to 1, the identification defined by Q_1 tends toward that defined by Q, and f_1 tends toward f. If $\epsilon_1 > 0$ is sufficiently small, then (according to the above remarks) the point $\mathfrak{a}_1$ has index j or less. But $w = f_1(z)$ satisfies a $\mathfrak{D}_n$-equation of the form

$$\left(\frac{z}{w}\frac{dw}{dz}\right)^2 P_1(w) = Q_1(z) \tag{11.4.9}$$

and $Q_1(z)$ has $2j - 2$ zeros on $|z| = 1$; so $f_1(z)$ has index j or $j - 1$. If $w = f_1(z)$ satisfies another $\mathfrak{D}_n$-equation

$$\left(\frac{z}{w}\frac{dw}{dz}\right)^2 P_2(w) = Q_2(z), \tag{11.4.10}$$

then, by an argument previously used, it satisfies a $\mathfrak{D}_n$-equation

$$\left(\frac{z}{w}\frac{dw}{dz}\right)^2 P_3(w) = Q_3(z)$$

where

$$Q_3(z) = \gamma(Q_1(z) - \eta Q_2(z)), \qquad P_3(w) = \gamma(P_1(w) - \eta P_2(w)),$$

and $(Q_3(z))^{1/2}$ has the zeros $z_1, z_3, z_4, \cdots, z_j$ on $|z| = 1$ and at least one other zero on $|z| = 1$. In this case $\mathfrak{a}_1$ has index j and the zeros of $(Q_3(z))^{1/2}$ on $|z| = 1$ are precisely the set $z_1, z_3, z_4, \cdots, z_j, z^\Delta$, where z^Δ is some number of modulus 1. It has also been shown that if $Q_1(z)$ has a zero of precise order $2k$, $k \geqq 0$, at some point τ of U_1 or L_1 and if $Q_3(z)$ has a zero of precise order 2λ at τ and of precise order 2μ at the point identified with τ where $\lambda \geqq 0$, $\mu \geqq 0$, then $\lambda = k + \mu$. This shows that if $(Q_3(z))^{1/2}$ has a zero in U_1 or L_1 other than one of the zeros $z_3, z_4, \cdots, z_j$, then it must also have an additional zero at the identified point. But this is impossible since z^Δ is the only possible additional zero in U_1 or L_1. This shows that if $w = f_1(z)$ satisfies (11.4.10), then the additional zero z^Δ on $|z| = 1$ must lie at one of the points $z = \exp(-i\alpha)$, $z = \exp(-i\beta_1)$. As ϵ_1 approaches zero, we see from (11.4.8) that $Q_3(z)$ approaches $Q(z)$; so the zeros of $Q_3(z)$ approach the zeros of $Q(z)$. But $Q(z)$ has a zero of precise order 2 at $z = e^{-i\alpha}$ and has no zero at $z = \exp(-i\beta_1)$. Since α is fixed as ϵ_1 approaches zero and $\exp(-i\beta_1)$ approaches $e^{-i\beta}$, it follows that z^Δ must approach $z = e^{-i\alpha}$ or $z = e^{-i\beta}$, and this is impossible.

Thus if $\epsilon_1 > 0$ is sufficiently small, the function $w = f_1(z)$ belonging to $\mathfrak{a}_1$ satisfies a unique maximal $\mathfrak{D}_n$-equation. Thus $\mathfrak{a}_1$ belongs to $\boldsymbol{M}_n^*$ and has index $j - 1$. If $j > 2$, there is a point $\mathfrak{a}_2$ of $\boldsymbol{M}_n^*$ near $\mathfrak{a}_1$ with index $j - 2$. Repeating the argument, we see that there is a point $\mathfrak{a}_{j-1}$ of $\boldsymbol{M}_n^*$ near $\mathfrak{a}$ which has index 1. This point belongs to $\boldsymbol{M}_n$, and it follows that $\boldsymbol{M}_n^*$ lies in the closure of $\boldsymbol{M}_n$. That is, each point of $\boldsymbol{M}_n^*$ is either a point of $\boldsymbol{M}_n$ or a limit point of points of $\boldsymbol{M}_n$.

We have shown that $\boldsymbol{K}_n$ lies in the closure of $\boldsymbol{M}_n^*$ and that $\boldsymbol{M}_n^*$ lies in the closure of $\boldsymbol{M}_n$. It follows that the supporting surface $\boldsymbol{K}_n$ lies in the closure of $\boldsymbol{M}_n$. This shows in particular that the set $\boldsymbol{M}_n$ exists. If $n > 2$, the set $\boldsymbol{M}_n$ does not contain the Koebe curve; and we know in this case that there are points of $\boldsymbol{K}_n$ not on the Koebe curve. The argument that has been used shows that $\boldsymbol{M}_n$ is an open set. For if $\mathfrak{a}$ is a point of $\boldsymbol{M}_n$, then it has index 1, and we have shown that there is a neighborhood G consisting of all points $\mathfrak{a}^* = (a_2^*, \cdots, a_n^*)$ on the boundary of $\boldsymbol{V}_n$ for which (11.4.7) is true such that each point of G has index 1. We have also shown that if a point has index 1, then it satisfies only one $\mathfrak{D}_n$-equation. Thus each point of G belongs to $\boldsymbol{M}_n$, and $\boldsymbol{M}_n$ is open.

Given any point $\mathfrak{a} = (a_2, a_3, \cdots, a_n)$ of $\boldsymbol{M}_n$, let

$$Q(z) = \gamma \frac{(z - e^{-i\alpha})^2}{-ze^{-i\alpha}} \prod_{\nu=1}^{n-} \left(1 + \rho_\nu^2 - z\rho_\nu e^{-i\varphi_\nu} - \frac{\rho_\nu e^{i\varphi_\nu}}{z}\right)$$

be the corresponding Q-function. If the coefficient B_{n-1} of $1/z^{n-1}$ is zero, then some of the numbers ρ_ν are zero. Let γ, $\gamma > 0$, be chosen such that $Q(z)$ is normalized. Then γ is bounded above and below by positive constants since $0 \leqq \rho_\nu < 1$. There is a one–one correspondence between points of $\boldsymbol{M}_n$ and a subset of Q-functions. No Q-function outside this subset gives rise to a point of $\boldsymbol{M}_n$. The discussion given in **11.3** and **11.4** shows that the correspondence is bicontinuous. There is a topological mapping of $\boldsymbol{M}_n$ onto an open subset of points $(B_1, B_2, \cdots, B_{n-1})$ on the sphere

$$\sum_{\nu=1}^{n-1} | B_\nu |^2 = 1.$$

In **11.2** it was shown that, given any set of constants $A_1, A_2, \cdots, A_{n-1}$ not all zero, there is a curve extending from a point on the Koebe curve and lying on the boundary of V_n. There is a number u^*, $u^* = u^*(A_1, A_2, \cdots, A_{n-1})$, such that for $0 < u < u^*$ the function $w = f(z)$ belonging to the point satisfies a $\mathfrak{D}_n$-equation of the form

$$\left(\frac{z}{w}\frac{dw}{dz}\right)^2 P\left(\frac{w}{u}\right) = Q(z, u) \tag{11.4.11}$$

where

$$P(w) = \sum_{\nu=1}^{n-1} \frac{A_\nu}{w^\nu}.$$

We show that if any point of one of these curves lies on $\boldsymbol{M}_n$, then all points of the curve for which $0 < u < u^*$ also lie on $\boldsymbol{M}_n$. The end points $u = 0$ and $u = u^*$ of the curve obviously cannot belong to $\boldsymbol{M}_n$ for $n > 2$. Let the curve corresponding to $A_1, A_2, \cdots, A_{n-1}$ contain a point belonging to $\boldsymbol{M}_n$, and let this point correspond to the value $u = u_1$, $0 < u_1 < u^*$. Then $A_1 \neq 0$. Let $\boldsymbol{\Gamma}_w$ denote the $\boldsymbol{\Gamma}$-structure defined by $A_1, A_2, \cdots, A_{n-1}$, and let C be the arc of $\boldsymbol{\Gamma}_w$ which has one end point at $w = \infty$. Let $C(u)$ be the subarc of C which is the image of $|z| = 1$ by the function $(1/u) f(z, u)$, $0 < u < u^*$. If the point corresponding to $u = u_2$, $0 < u_2 < u^*$, does not belong to $\boldsymbol{M}_n$, then the function $w = f(z, u_2)$ satisfies a $\mathfrak{D}_n$-equation

$$\left(\frac{z}{w}\frac{dw}{dz}\right)^2 P^*(w) = Q^*(z)$$

where $Q^*(z)$ has zeros on $|z| = 1$ of total multiplicity exceeding 2. It also satisfies the $\mathfrak{D}_n$-equation

$$\left(\frac{z}{w}\frac{dw}{dz}\right)^2 P\left(\frac{w}{u_2}\right) = Q(z, u_2)$$

where $P(w/u_2)/P^*(w)$ is not a constant. If ϵ, $\epsilon > 0$, is sufficiently small, then

$$\left(\frac{z}{w}\frac{dw}{dz}\right)^2 \left[P\left(\frac{w}{u_2}\right) - \epsilon P^*(w)\right] = Q(z, u_2) - \epsilon Q^*(z)$$

is a $\mathfrak{D}_n$-equation since the right side is non-negative on $|z| = 1$. On some subarc of C the function

$$s = \int (P(w) - \epsilon P^*(u_2 w))^{1/2} \frac{dw}{w}$$

has a constant real part and, given any u, $0 < u < u^*$, there is an $\epsilon > 0$ such that s is a regular function of w in a neighborhood of $C(u)$ except at $w = \infty$. If

$$\zeta = \int (P(w))^{1/2} \frac{dw}{w},$$

then the arc $C(u)$ is mapped onto a line segment $\mathrm{Re}(\zeta) = $ constant which we denote by L. The function $s(w(\zeta))$ is regular in a neighborhood of L. Since it has constant real part on a subarc of L, it must have constant real part on the whole of L. This implies that $w = f(z, u_1)$ satisfies the $\mathfrak{D}_n$-equation

$$\left(\frac{z}{w}\frac{dw}{dz}\right)^2 \left[P\left(\frac{w}{u_1}\right) - \epsilon P^*\left(\frac{u_2 w}{u_1}\right)\right] = Q^{**}(z)$$

for some Q-function $Q^{**}(z)$. Since $w = f(z, u_1)$ also satisfies the $\mathfrak{D}_n$-equation (11.4.11), we have a contradiction.

11.5. It remains to prove the variational formula of Theorem VII. Any given Q-function may always be written in the form

$$Q(z) = \frac{(z - e^{-i\alpha})^2}{-ze^{-i\alpha}} \prod_{\nu=1}^{n-2} \left(1 + |z_\nu|^2 - \frac{z_\nu}{z} - \bar{z}_\nu z\right). \tag{11.5.1}$$

We suppose that $|z_\nu| < 1$, $\nu = 1, 2, \cdots, n - 2$. Let $Q(z)$ belong to a point $\mathfrak{a}$ of $\boldsymbol{M}_n$, and let

$$Q^*(z) = \frac{(z - e^{-i\alpha^*})^2}{-ze^{-i\alpha^*}} \prod_{\nu=1}^{n-2} \left(1 + |z_\nu^*|^2 - \frac{z_\nu^*}{z} - \bar{z}_\nu^* z\right) \tag{11.5.2}$$

be another Q-function such that

$$|\alpha^* - \alpha|^2 + \sum_{\nu=1}^{n-2} |z_\nu^* - z_\nu|^2 = \epsilon^2. \tag{11.5.3}$$

There is an ϵ_0, $\epsilon_0 > 0$, such that the set of points $\mathfrak{a}^*$ defined by all $Q^*(z)$ which satisfy (11.5.3) with $0 \leqq \epsilon \leqq \epsilon_0$ belongs to $\boldsymbol{M}_n$, and there is a one–one correspondence between the points $\mathfrak{a}^*$ and the functions $Q^*(z)$. Let

$$\chi(z) = \frac{Q^*(z) - Q(z)}{\epsilon}, \tag{11.5.4}$$

where the function $\chi(z)$ depends on α^*, z_ν^*, $\nu = 1, 2, \cdots, n - 2$. We have

$$\chi(z) = \psi(z) + \epsilon T(z), \tag{11.5.5}$$

where

$$\psi(z) = \sum_{\nu=-(n-1)}^{n-1} \frac{\beta_\nu}{z^\nu}, \qquad T(z) = \sum_{\nu=-(n-1)}^{n-1} \frac{\tau_\nu}{z^\nu}. \tag{11.5.6}$$

The functions $\psi(z)$ and $T(z)$ are real on $|z| = 1$ and there is a positive number C depending on Q such that

$$\sum_{\nu=-(n-1)}^{n-1} |\beta_\nu|^2 < C, \qquad \sum_{\nu=-(n-1)}^{n-1} |\tau_\nu|^2 < C. \tag{11.5.7}$$

If $Q(z)$ has a zero of precise order $\gamma > 0$ at a point z_0, then for any function $Q^*(z)$ satisfying (11.5.3) for small ϵ, the function $\psi(z)$ may be chosen without loss of generality to have a zero of order $\gamma - 1$ or more at the point z_0. For, writing $z_\nu = x_\nu + iy_\nu$, $\nu = 1, 2, \cdots, n-2$, and expanding in a Taylor series, we obtain

$$\begin{aligned} Q^*(z) = Q(z)\Big\{1 &+ \left[\frac{i(z + e^{-i\alpha})}{z - e^{-i\alpha}}\right](\alpha^* - \alpha) \\ &+ \sum_{\nu=1}^{n-2}\left(\frac{2x_\nu - 1/z - z}{1 + |z_\nu|^2 - z_\nu/z - \bar{z}_\nu z}\right)(x_\nu^* - x_\nu) \\ &+ \sum_{\nu=1}^{n-2}\left(\frac{2y_\nu - i/z + iz}{1 + |z_\nu|^2 - z_\nu/z - \bar{z}_\nu z}\right)(y_\nu^* - y_\nu)\Big\} + O(\epsilon^2), \end{aligned} \tag{11.5.8}$$

where the term $O(\epsilon^2)$ holds uniformly for all $Q^*(z)$ satisfying (11.5.3) with $0 < \epsilon \leqq \epsilon_0$. We may take

$$\begin{aligned} \psi(z) = Q(z)\Big\{&\left[\frac{i(z + e^{-i\alpha})}{z - e^{-i\alpha}}\right]\frac{\alpha^* - \alpha}{\epsilon} \\ &+ \sum_{\nu=1}^{n-2}\left(\frac{2x_\nu - 1/z - z}{1 + |z_\nu|^2 - z_\nu/z - \bar{z}_\nu z}\right)\frac{x_\nu^* - x_\nu}{\epsilon} \\ &+ \sum_{\nu=1}^{n-2}\left(\frac{2y_\nu - i/z + iz}{1 + |z_\nu|^2 - z_\nu/z - \bar{z}_\nu z}\right)\frac{y_\nu^* - y_\nu}{\epsilon}\Big\}. \end{aligned} \tag{11.5.9}$$

If $Q(z)$ has a root of multiplicity $\gamma \geqq 1$ at a point z_1, let $z_1 = z_2 = \cdots = z_\gamma$. Then $\psi(z)$ has a zero of order at least $\gamma - 1$ at the point z_1. In the case where $\alpha^* = \alpha$ and $Q(z)$ has at least one zero in $|z| < 1$ of order greater than 1, the function $\psi(z)$ may vanish identically. Now we have shown that every $Q(z)$ can be written in the form (11.5.5) where at each zero of $Q(z)$ of order γ the function $\psi(z)$ has a zero of order $\gamma - 1$ at least. Conversely, if $\psi(z)$ is any function of the form (11.5.6) where $\beta_{-\nu} = \bar{\beta}_\nu$ and which has this property concerning the order of its zeros, then there is a number $C = C(\epsilon, \psi)$ such that

$$Q^*(z) = Q(z) + \epsilon\psi(z) + \epsilon^2 T(z) \tag{11.5.10}$$

is a Q-function having only one zero on $|z| = 1$ and this of precise order 2. In fact, the function $T(z)$ may be taken equal to a constant depending on ϵ.

However, it may be more convenient to use the coefficients of the Q-functions as parameters. Given a Q-function

$$Q(z) = \sum_{\nu=-(n-1)}^{n-1} \frac{B_\nu}{z^\nu}, \qquad B_{-\nu} = \bar{B}_\nu, \tag{11.5.11}$$

which corresponds to a point $\mathfrak{a}$ of $\boldsymbol{M}_n$, let

$$Q^*(z) = \sum_{\nu=-(n-1)}^{n-1} \frac{B_\nu^*}{z^\nu}, \qquad B_{-\nu}^* = \bar{B}_\nu^*, \tag{11.5.12}$$

be a Q-function such that

$$\sum_{\nu=1}^{n-1} | B_\nu^* - B_\nu |^2 = \epsilon^2. \tag{11.5.13}$$

There is an ϵ_0, $\epsilon_0 > 0$, such that the set of points $\mathfrak{a}^*$ defined by all $Q^*(z)$ satisfying (11.5.13) belongs to $\boldsymbol{M}_n$. We shall show that $Q^*(z)$ may be written in the form

$$Q^*(z) = Q(z) + \epsilon\psi(z) + Q^*(e^{-i\alpha}) \tag{11.5.14}$$

where

$$\psi(z) = \sum_{\nu=-(n-1)}^{n-1} \frac{\beta_\nu}{z^\nu}, \qquad \beta_{-\nu} = \bar{\beta}_\nu, \tag{11.5.15}$$

$$Q^*(e^{-i\alpha}) = O(\epsilon^2).$$

We observe that the function $\psi(z)$ vanishes at $e^{-i\alpha}$. From (11.5.13) it follows that $| Q^{*\prime}(z) - Q'(z) | < C\epsilon$ in a neighborhood of $| z | = 1$, where C is some positive constant. Also on $| z - e^{-i\alpha} | = \rho\epsilon$, we have

$$| Q'(z) | > | Q''(e^{-i\alpha}) | \rho\epsilon/2;$$

hence if ρ is first chosen large, then ϵ small, we see from Rouché's theorem that

$$Q^{*\prime}(z) = Q'(z) + (Q^{*\prime} - Q')$$

has a simple zero in $| z - e^{-i\alpha} | < \rho\epsilon$. Since $Q^{*\prime}(z)$ has only one zero in a neighborhood of $e^{-i\alpha}$, it follows that $| \alpha^* - \alpha | = O(\epsilon)$ and so $Q^*(e^{-i\alpha}) = O(\epsilon^2)$. Conversely, if $\psi(z)$ is any function of the form (11.5.15) which vanishes at $e^{-i\alpha}$, then there is a $C = C(\epsilon, \psi)$ such that the function $Q^*(z)$ defined by

$$Q^*(z) = Q(z) + \epsilon\psi(z) + \epsilon^2 C \tag{11.5.16}$$

is a Q-function with only one zero on $| z | = 1$, this zero being of precise order 2.

Let $Q(z)$ be a Q-function belonging to a point of $\boldsymbol{M}_n$. Let $e^{-i\alpha}$ be the point on $| z | = 1$ where Q vanishes, and let $e^{-i\beta}$ be the point on $| z | = 1$ which maps into $w = \infty$ by the corresponding function $w = f(z)$. Suppose that $Q^*(z)$ is a near-by Q-function, and let $\psi(z)$ be defined in one of the above two ways. Let $\exp(-i\alpha^*)$ be the point on $| z | = 1$ where $Q^*(z)$ vanishes, and let $\exp(-i\beta^*)$ be the point which maps into infinity by the corresponding function $f^*(z)$. Then

$$\alpha^* - \alpha = \rho\epsilon \tag{11.5.17}$$

where ρ is bounded as ϵ approaches zero. The identification of points on $|z| = 1$ by $Q(z)$ is defined by

$$\int_{e^{-i\alpha}}^{z'} (Q(t))^{1/2} \frac{dt}{t} = \int_{e^{-i\alpha}}^{z''} (Q(t))^{1/2} \frac{dt}{t} \tag{11.5.18}$$

where z' lies on the upper arc U generated by a point moving from $e^{-i\alpha}$ counterclockwise to $e^{-i\beta}$ and where z'' lies on the lower arc L generated by a point moving from $e^{-i\alpha}$ clockwise to $e^{-i\beta}$. The function $(Q(t))^{1/2}$ is positive on U and negative on L. The formula (11.5.18) may be written

$$\int_{z''}^{z'} (Q(t))^{1/2} \frac{dt}{t} = 0 \tag{11.5.19}$$

where the integration is counterclockwise along $|z| = 1$ from z'' to z' through the point $e^{-i\alpha}$. Since $Q(t)$ has a double zero at $e^{-i\alpha}$, we see that $(Q(t))^{1/2}$ is analytic in a neighborhood of $e^{-i\alpha}$. According to our convention regarding the sign of $(Q(t))^{1/2}$ on $|z| = 1$, we see that $(Q(t))^{1/2}$ has a discontinuity at $e^{-i\beta}$. The formula for $Q^*(z)$ corresponding to (11.5.19) is

$$\int_{\zeta''}^{\zeta'} (Q^*(t))^{1/2} \frac{dt}{t} = 0. \tag{11.5.20}$$

Taking $\zeta' = \zeta'' = \exp(-i\beta^*)$, we obtain

$$\oint_{e^{-i\beta^*}} (Q^*(t))^{1/2} \frac{dt}{t} = 0 \tag{11.5.21}$$

where the integration is in the counterclockwise sense. We plainly have

$$\oint_{e^{-i\beta^*}} (Q^*(t))^{1/2} \frac{dt}{t} = \oint_{e^{-i\beta}} (Q^*(t))^{1/2} \frac{dt}{t} + 2\int_{e^{-i\beta}}^{e^{-i\beta^*}} (Q^*(t))^{1/2} \frac{dt}{t} = 0 \tag{11.5.22}$$

where $(Q^*(t))^{1/2} > 0$ on the small arc joining $e^{-i\beta}$ and $e^{-i\beta^*}$. Thus

$$2i(\beta^* - \beta)(Q(e^{-i\beta}))^{1/2} = \oint_{e^{-i\beta}} (Q^*(t))^{1/2} \frac{dt}{t} + O(\beta^* - \beta)^2 + O(\epsilon^2) \tag{11.5.23}$$

where $(Q(e^{-i\beta}))^{1/2} > 0$. Deforming the contour of integration slightly in the neighborhood of $e^{-i\alpha}$ and expanding, we have

$$\oint_{e^{-i\beta}} (Q(t))^{1/2} \left\{ 1 + \frac{\epsilon}{2} \frac{\psi(t)}{Q(t)} + O(\epsilon^2) \right\} \frac{dt}{t} \tag{11.5.24}$$
$$= \frac{\epsilon}{2} \oint_{e^{-i\beta}} (Q(t))^{1/2} \frac{\psi(t)}{Q(t)} \frac{dt}{t} + O(\epsilon^2).$$

It therefore follows that

$$\beta^* - \beta = d \cdot \epsilon, \tag{11.5.25}$$

where

$$d = \frac{1}{4i(Q(e^{-i\beta}))^{1/2}} \oint_{e^{-i\beta}} (Q(t))^{1/2} \frac{\psi(t)}{Q(t)} \frac{dt}{t} + O(\epsilon),$$

and $Q(e^{-i\beta}) > 0$.

We shall map $|z| < 1$ onto $|\zeta| < 1$ by a bilinear transformation

$$\zeta = \zeta(z) = e^{i\lambda} \frac{z - \gamma}{1 - \bar{\gamma}z} \tag{11.5.26}$$

in such a way that $z = e^{-i\alpha}$ goes into $\zeta = e^{-i\alpha^*}$ and $z = e^{-i\beta}$ into $\zeta = e^{-i\beta^*}$. The transformation (11.5.26) is not uniquely defined by these conditions and there is a one-parameter family of such transformations. To obtain this family of mappings, let

$$s_1 = \frac{1 + e^{i\alpha} z}{1 - e^{i\alpha} z}, \qquad s_2 = \frac{1 + e^{i\alpha^*} \zeta}{1 - e^{i\alpha^*} \zeta}. \tag{11.5.27}$$

Then $|z| < 1$ is mapped onto $\mathrm{Re}(s_1) > 0$ and $|\zeta| < 1$ is mapped onto $\mathrm{Re}(s_2) > 0$. The mapping from s_1 to s_2 is of the form

$$s_2 = as_1 + ib \tag{11.5.28}$$

where $a > 0$ and b is real. If $z = e^{-i\beta}$, then $\zeta = e^{-i\beta^*}$; so

$$s_1 = (1 + e^{i(\alpha-\beta)})/(1 - e^{i(\alpha-\beta)}) \text{ and } s_2 = (1 + e^{i(\alpha^*-\beta^*)})/(1 - e^{i(\alpha^*-\beta^*)}).$$

Hence

$$\frac{1 + e^{i(\alpha^*-\beta^*)}}{1 - e^{i(\alpha^*-\beta^*)}} = a \frac{1 + e^{i(\alpha-\beta)}}{1 - e^{i(\alpha-\beta)}} + ib. \tag{11.5.29}$$

If we compose the mappings (11.5.26) and (11.5.27), we obtain

$$\zeta = e^{i(\alpha-\alpha^*)} \frac{a - ib + 1}{a + ib + 1} \frac{z + \dfrac{a + ib - 1}{a - ib + 1} e^{-i\alpha}}{1 + \dfrac{a - ib - 1}{a + ib + 1} e^{i\alpha} \cdot z}.$$

Eliminating ib with the aid of (11.5.29) and writing

$$R = a \frac{\sin((\alpha^* - \beta^*)/2)}{\sin((\alpha - \beta)/2)} > 0,$$

we find that

$$e^{i\lambda} = e^{i(\alpha-\alpha^*+\beta-\beta^*)/2} \frac{R - e^{i(\alpha-\beta+\alpha^*-\beta^*)/2}}{1 - Re^{i(\alpha-\beta+\alpha^*-\beta^*)/2}}, \tag{11.5.30}$$

$$\gamma = e^{-i\alpha} \frac{1 - Re^{i((\alpha-\beta)/2-(\alpha^*-\beta^*)/2)}}{1 - Re^{-i((\alpha-\beta)/2+(\alpha^*-\beta^*)/2)}}. \tag{11.5.31}$$

The parameter a is at our disposal, and so we choose a to make $R = 1$. By (11.5.17) and (11.5.25)

$$\alpha^* - \alpha = \rho\epsilon, \qquad \beta^* - \beta = d\epsilon,$$

and so

$$e^{i\lambda} = e^{-i\epsilon(\rho+d)/2}\,\epsilon = 1 - i\,\frac{\rho + d}{2}\,\epsilon + o(\epsilon), \tag{11.5.32}$$

$$\gamma = \frac{e^{-i(\alpha+\beta)/2}}{4\sin((\alpha - \beta)/2)}\,(\rho - d)\epsilon + o(\epsilon). \tag{11.5.33}$$

Thus

$$\zeta = z + \{M - Nz - \bar{M}z^2\}\,\epsilon + o(\epsilon), \tag{11.5.34}$$

where

$$M = (d - \rho)\,\frac{e^{-i(\alpha+\beta)/2}}{4\sin((\alpha - \beta)/2)}, \qquad N = i\,\frac{\rho + d}{2}. \tag{11.5.35}$$

Here the $o(\epsilon)$ holds uniformly in $|z| \leqq 1$. Let

$$F(z) = \frac{1}{f(z)} = \frac{1}{z} + \cdots, \qquad F^*(\zeta) = \frac{1}{f^*(\zeta)} = \frac{1}{\zeta} + \cdots, \tag{11.5.36}$$

where $f(z)$ belongs to the point $\mathfrak{a}$ and $f^*(\zeta)$ belongs to the point $\mathfrak{a}^*$. Then $w = F(z)$ maps $|z| < 1$ onto the exterior of a single analytic Jordan arc J with one end point at $w = 0$ and lying in the circle $|w| < 4$. The point $z = e^{-i\beta}$ maps into the end point at $w = 0$ and $z = e^{-i\alpha}$ maps into the other end point. Likewise $w^* = F^*(\zeta)$ maps $|\zeta| < 1$ onto the exterior of a single analytic Jordan arc J^* with one end point at $w^* = 0$ and lying in $|w^*| < 4$; the point $\zeta = e^{-i\beta^*}$ maps into $w^* = 0$ and $\zeta = e^{-i\alpha^*}$ maps into the other end point. The function

$$F^{**}(z) = \frac{F^*(\zeta(z))}{1 - F^*(\zeta(z))/F^*(\zeta(0))} \tag{11.5.37}$$

maps $|z| < 1$ onto the exterior of a single analytic Jordan arc with one end point at the origin, $z = e^{-i\beta}$ going into this end point and $e^{-i\alpha}$ into the other. In the ring $1/2 \leqq z \leqq 1$, we have

$$F^{**}(z) = F^*(z) + \{M[F(z)^2 + F'(z)] - NzF'(z) - \bar{M}z^2F'(z)\}\,\epsilon + o(\epsilon)$$

where in the coefficient of ϵ the function F^* and its derivative have been replaced by F and its derivative. Multiplying $F^{**}(z)$ by a suitable constant, we obtain a function $H(z)$ which has a pole of residue unity at $z = 0$:

$$\begin{aligned} H(z) = F^*(z) + \epsilon\,\{&M[F'(z) + F(z)^2 + za_2F(z)] \\ &- N[zF'(z) + F(z)] - \bar{M}z^2F'(z)\} \\ &+ o(\epsilon) \end{aligned} \tag{11.5.38}$$

where

$$F(z) = 1/f(z) = 1/z - a_2 + \cdots.$$

The $o(\epsilon)$, which has been shown to hold uniformly in the ring $1/2 \leqq |z| \leqq 1$, now holds uniformly in $|z| \leqq 1$. The function $w_1 = H(z)$ maps $|z| < 1$ onto the exterior of an analytic Jordan arc J_1 with one end point at $w_1 = 0$, and $z = e^{-i\beta}$ maps into this end point, $z = e^{-i\alpha}$ into the other end point. The arc J_1 lies in $|w| < 4$ and is obtained from J^* by a bilinear transformation with the origin preserved.

The function $F^*(\zeta)$ identifies the points ζ' and ζ''' on $|\zeta| = 1$, that is, $F^*(\zeta') = F^*(\zeta''')$. Let ζ' be the image of z' on the upper arc U of $|z| = 1$ under the mapping (11.5.26), and let ζ''' be the image of z''' under this mapping. Then $H(z)$ identifies the points z' and z'''.

We shall now compute $z''' - z''$. The points ζ' and ζ''' are identified by $Q^*(\zeta)$,

$$\int_{\zeta'''}^{\zeta'} (Q^*(\tau))^{1/2} \frac{d\tau}{\tau} = 0 \tag{11.5.39}$$

where the integration is counterclockwise along $|\zeta| = 1$ from ζ''' through $\exp(-i\alpha^*)$ to ζ'. Since $(Q^*(\tau))^{1/2}$ is negative from ζ''' to $\exp(-i\alpha^*)$ and positive from $\exp(-i\alpha^*)$ to ζ', it is analytic in a domain containing the arc from ζ''' to ζ'. Recalling that

$$Q^*(z) = Q(z) + \epsilon\psi + O(\epsilon^2),$$

we note that in a circle of radius δ with center at $e^{-i\alpha}$, where δ is a positive number depending on $Q(z)$, the function

$$(Q^*(z))^{1/2} - (Q(z))^{1/2} - \frac{\epsilon\psi(z)}{2(Q(z))^{1/2}} \tag{11.5.40}$$

is regular. Since this function is $O(\epsilon^2)$ on the circumference of this circle, it is uniformly $O(\epsilon^2)$ throughout the interior of this circle. At any point of $|z| = 1$ outside this circle, the function (11.5.40) is clearly uniformly $O(\epsilon^2)$.

Under the linear transformation

$$\zeta = e^{i\lambda} \frac{z - \gamma}{1 - \bar{\gamma} z}, \qquad z = e^{-i\lambda} \frac{\zeta + \gamma e^{i\lambda}}{1 + \bar{\gamma} e^{-i\lambda} \zeta},$$

the points z' and z''' map into the points ζ' and ζ''' respectively. Let

$$\tau = e^{i\lambda} \frac{v - \gamma}{1 - \bar{\gamma} v}.$$

By (11.5.39) we have

$$0 = \int_{\zeta'''}^{\zeta'} (Q^*(\tau))^{1/2} \frac{d\tau}{\tau} = \int_{z'''}^{z'} \left(Q^* \left(e^{i\lambda} \frac{v - \gamma}{1 - \bar{\gamma} v} \right) \right)^{1/2} \frac{1 - |\gamma|^2}{(v - \gamma)(1 - \bar{\gamma} v)} \, dv$$

$$= \int_{z'''}^{z'} (Q(v))^{1/2} \frac{dv}{v} + \epsilon[K(z') - K(z'')] + o\{\epsilon \, | z' - e^{-i\alpha} |\}$$

where

(11.5.41) $$K(z) = \frac{1}{2}\int_{e^{-i\alpha}}^{z} (Q(v))^{1/2}\left\{\frac{\psi(v)}{Q(v)} + G(v)\right\}\frac{dv}{v},$$

(11.5.42) $$G(v) = \frac{Q'(v)}{Q(v)}[M - Nv - \bar{M}v^2] - 2\left[\bar{M}v + \frac{M}{v}\right].$$

The function $(Q(v))^{1/2}$ in the integrand of $K(z')$ is positive on the upper arc U and is negative on the lower arc L; so $(Q(v))^{1/2}$ is analytic on $|z| = 1$ except at the point $e^{-i\beta}$ where it has a discontinuity. Integrating by parts, we have

$$\int_{e^{-i\alpha}}^{z}\frac{Q'(v)}{Q(v)}(Q(v))^{1/2}v^k\frac{dv}{v} = 2z^{k-1}(Q(z))^{1/2} - 2(k-1)\int_{e^{-i\alpha}}^{z}(Q(v))^{1/2}v^{k-2}\,dv.$$

Taking $k = 0, 1, 2$, we obtain

$$\frac{1}{2}\int_{e^{-i\alpha}}^{z}(Q(v))^{1/2}G(v)\frac{dv}{v} = (Q(z))^{1/2}\left(\frac{M}{z} - N - \bar{M}z\right).$$

Hence

(11.5.43) $$K(z) = \frac{1}{2}\int_{e^{-i\alpha}}^{z}\frac{\psi(v)}{(Q(v))^{1/2}}\frac{dv}{v} + (Q(z))^{1/2}\left(\frac{M}{z} - N - \bar{M}z\right).$$

Comparing with (11.5.19), we obtain

(11.5.44) $$z''' - z'' = \epsilon\frac{z''}{(Q(z''))^{1/2}}\{K(z') - K(z'')\} + o(\epsilon).$$

The function $w_1(w)$, where $w = F(z)$, $w_1 = H(z)$, maps the exterior of J_1 onto the exterior of J. In this mapping the point at infinity goes into the point at infinity, and the end points of the arcs correspond to one another. For $|w| > 4$, we have

(11.5.45) $$w_1(w) = w + b_0 + b_1/w + \cdots.$$

Let w be a point not on J, and let C be a rectifiable closed Jordan curve containing J in its interior and w in its exterior. Then, computing the residues in the exterior of C, we obtain

(11.5.46) $$\begin{aligned} w_1(w) &= w - \frac{1}{2\pi i}\int_C w_1(t)\left\{\frac{1}{t-w} - \frac{1}{t}\right\}dt \\ &= w - \frac{w}{2\pi i}\int_C w_1(t)\frac{dt}{t(t-w)}, \end{aligned}$$

where the integration is counterclockwise around C. If z' is a point on the upper arc U of $|z| = 1$ which is traversed by a point moving counterclockwise from $e^{-i\alpha}$ to $e^{-i\beta}$ and z'' is the point on the lower arc L which is identified with z' under the identification defined by $Q(z)$, then writing

$$t' = F(z'), \qquad t'' = F(z''),$$

we see, since $F(z') = F(z'')$, that t' and t'' are points on the opposite edges of the arc J. In formula (11.5.44) we may shrink the contour C onto the two edges of J. If we let

$$p(t') = w_1(t') - w_1(t''), \tag{11.5.47}$$

then formula (11.5.46) becomes

$$w_1(w) = w - \frac{w}{2\pi i}\int_0^T p(t')\,\frac{dt'}{t'(t'-w)} \tag{11.5.48}$$

where the integration is along the right edge of J from 0 to the end point at T. The points $w_1(t')$ and $w_1(t'')$ lie on opposite edges of J_1 but are not in general coincident. It is clear that

$$w_1(t') = H(z'), \qquad w_1(t'') = H(z'').$$

Let w_1''' be the point on J_1 which coincides with $w_1(t')$ in value but lies on the opposite edge of J_1. Then

$$w_1''' = H(z''');$$

so $H(z''') = H(z')$, where z''' lies on the lower arc L of $|z| = 1$ and the difference $z''' - z''$ is given by formula (11.5.44).

We have

$$\begin{aligned} p(t') &= w_1(t') - w_1(t'') = H(z') - H(z'') \\ &= H(z') - H(z''') + H(z''') - H(z'') \\ &= (z''' - z'')H'(z'') + o(\epsilon) = \epsilon\,\frac{z''}{(Q(z''))^{1/2}}\,\{K(z') - K(z'')\}\,F'(z'') \\ &\quad + o(\epsilon) \end{aligned}$$

by (11.5.44). Thus if s is any point of $|s| < 1$, $s \neq 0$, we have by (11.5.48)

$$\begin{aligned} H(s) = F(s) - \epsilon\,\frac{F(s)}{2\pi i}\int_U \frac{z''}{(Q(z''))^{1/2}}\,&\{K(z') - K(z'')\}F'(z'') \\ &\cdot\frac{F'(z')\,dz'}{F(z')(F(z') - F(s))} + o(\epsilon), \end{aligned} \tag{11.5.49}$$

where the integration is clockwise along the upper arc U from $e^{-i\beta}$ to $e^{-i\alpha}$ and z'' is the point of L identified with z'. Since $F(z') = F(z'')$, we have from (11.5.19)

$$\frac{z'F'(z')}{(Q(z'))^{1/2}} = \frac{z''F'(z'')}{(Q(z''))^{1/2}}. \tag{11.5.50}$$

From (11.5.49) we see that

$$H(s) = F(s) + \epsilon\,\frac{F(s)}{2\pi i}\int_{|z|=1} \frac{z}{(Q(z))^{1/2}}\,K(z)\,\frac{F'(z)^2\,dz}{F(z)[F(z) - F(s)]} + o(\epsilon), \tag{11.5.51}$$

where the integration is counterclockwise starting from $e^{-i\beta}$. Then from (11.5.38),

$$
\begin{aligned}
F^*(s) = F(s) &+ \epsilon \frac{F(s)}{2\pi i} \int_{|z|=1} \frac{z}{(Q(z))^{1/2}} K(z) \frac{F'(z)^2\, dz}{F(z)[F(z) - F(s)]} \\
&+ \epsilon\,\{-M[F'(s) + F(s)^2 + 2a_2 F(s)] + N[sF'(s) + F(s)] \\
&+ \bar{M}s^2F'(s)\} + o(\epsilon).
\end{aligned} \tag{11.5.52}
$$

We now show that several of the terms in the equation (11.5.52) cancel. With the expression (11.5.43) in mind, we note that

$$
\begin{aligned}
\frac{1}{2\pi i} \int_{|z|=1} (M - Nz - \bar{M}z^2) &\frac{F'(z)^2\, dz}{F(z)[F(z) - F(s)]} \\
&= \frac{F'(s)}{F(s)} (M - Ns - \bar{M}s^2) + M[2a_2 + F(s)] - N.
\end{aligned}
$$

Thus we obtain from (11.5.52)

$$
F^*(s) = F(s) + \epsilon \frac{F(s)}{2\pi i} \int_{|z|=1} \frac{z}{(Q(z))^{1/2}} \chi(z) \frac{F'(z)^2\, dz}{F(z)[F(z) - F(s)]} + o(\epsilon), \tag{11.5.53}
$$

where

$$
\chi(z) = \frac{1}{2} \int_{e^{-i\alpha}}^{z} \frac{\psi(v)}{(Q(v))^{1/2}} \frac{dv}{v}.
$$

Since $F^*(s) = 1/f^*(s)$, $F(s) = 1/f(s)$, we have finally

$$
f^*(s) = f(s) - \epsilon \frac{f(s)^2}{2\pi i} \int_{|z|=1} \left(\frac{zf'(z)}{f(z)}\right)^2 \frac{\chi(z)}{(Q(z))^{1/2}[f(s) - f(z)]} \frac{dz}{z} + o(\epsilon). \tag{11.5.54}
$$

This completes the proof of Theorem VII.

Chapter XII

PARAMETRIZATION OF THE BOUNDARY OF V_n

12.1. In this chapter we study the space S_n defined in Chapter VIII and show that it can be dissected into finitely many portions $\Omega_1, \Omega_2, \cdots, \Omega_N$, $N = N(n)$, such that the points of Ω_ν are expressible in terms of a set of real parameters $\omega_1, \omega_2, \cdots, \omega_k$, $k = k(\nu, n) \leqq 2n - 3$. This is accomplished by considering the images of $|z| = 1$ by functions $w = f(z)$ belonging to boundary points of V_n. The image always defines a tree, and the purpose of the present chapter is to describe the tree in terms of finitely many parameters. The problem is not completely topological because a characterization of the tree depends on the lengths of its various arcs or branches in terms of a certain metric. The topological theory of trees is well understood, but for the sake of completeness we give a formula for the number of topologically different trees having a fixed number of tips. An analogous formula for the number of trees having a fixed number of branches was given by Cayley in 1857. However, the parametrization of trees whose branch lengths are defined by means of a certain metric does not seem to be given explicitly in the literature.

Let Q denote a function of the form

$$Q(z) = \sum_{\nu=-(n-1)}^{n-1} \frac{B_\nu}{z^\nu}, \qquad B_{-\nu} = \bar{B}_\nu, \tag{12.1.1}$$

which is non-negative on $|z| = 1$ with at least one zero there. Then $Q = Q_{\mathfrak{B}}(z)$ depends on the $n - 1$ complex numbers $B_1, B_2, \cdots, B_{n-1}$. However, two functions Q are considered equal if their ratio is a positive constant and so Q depends on $2n - 3$ real parameters. We shall normalize in this chapter by assuming that

$$\| \mathfrak{B} \| = \left\{ \sum_{\nu=1}^{n-1} |B_\nu|^2 \right\}^{1/2} = 1. \tag{12.1.2}$$

We assume that the identification of arcs on $|z| = 1$ satisfies the conditions (i), (ii), and (iii) of Chapter VIII and that there is a distinguished cycle $\boldsymbol{\Sigma}$ satisfying condition (v). We also assume without loss of generality that the identification is canonical, that is, the points of every cycle containing precisely two terms except possibly $\boldsymbol{\Sigma}$ are interior points of identified arcs. Then Q, the identification of points on $|z| = 1$, and the cycle $\boldsymbol{\Sigma}$ define a point in the space S_n discussed in Chapter VIII.

By Theorem V there is a function $w = f(z)$ which maps $|z| < 1$ onto the plane minus finitely many piecewise analytic Jordan arcs. Each Jordan arc is the image of a pair of identified arcs on $|z| = 1$, the two arcs of the pair mapping into the two edges of the Jordan arc with identified points brought together.

The set of these arcs forms a tree and the point at infinity will be called the root of the tree. By an inversion we obtain a tree whose root lies at the origin.

The image of a pair of identified arcs on $|z| = 1$ is a branch of the tree and the image of a cycle of vertices on $|z| = 1$ is a knot. By our assumption that the identification is canonical, there is no cycle containing two vertices except possibly Σ and so the knots are of three kinds: terminal knots or free tips of branches, which correspond to a cycle of one term; points where three or more branches meet, which correspond to cycles of three or more points; the root of the tree which corresponds to Σ. There may be one, two, or more branches issuing from the root.

Let τ ($\tau \geqq 1$) denote the number of terminal knots; if the root is a terminal knot, we do not count it. We define an internal knot of the tree to be any knot which is neither a tip nor the root. Let μ be the number of internal knots and let m_j ($1 \leqq j \leqq \mu$) be the number of branches issuing from the jth internal knot. Clearly $m_j \geqq 3$. Let m_0 be the number of branches or stalks issuing from the root: $m_0 = 1$ if the root is terminal, $m_0 \geqq 2$ otherwise. Finally let ν denote the total number of branches on the tree. We have

$$\nu = \tau + \mu,$$

that is, the number of branches is equal to the number of tips plus the number of internal knots. Also, counting the number of branches which issue from each knot, we obtain twice the total number of branches. Hence

$$2\nu = \tau + m_0 + \sum_{j=1}^{\mu} m_j$$

and so

$$\tau = m_0 + \sum_{j=1}^{\mu} (m_j - 2).$$

We define

$$\lambda_j = m_j - 2, \qquad j = 1, 2, \cdots, \mu,$$

to be the multiplicity of an internal knot and we have

$$\tau = m_0 + \sum_{j=1}^{\mu} \lambda_j . \tag{12.1.3}$$

Since each terminal tip of the tree is the image of a point on $|z| = 1$ where Q vanishes, we see that $\tau \leqq n - 1$. Since $\lambda_j \geqq 1$, we have by (12.1.3)

$$\mu + 1 \leqq m_0 + \sum_{j=1}^{\mu} \lambda_j = \tau \leqq n - 1,$$

that is,

$$\mu \leqq n - 2.$$

Thus

$$\nu = \tau + \mu \leqq 2n - 3.$$

Since each branch of the tree corresponds to a pair of identified arcs on $|z| = 1$, we see that the total number 2ν of arcs on $|z| = 1$ does not exceed $4n - 6$ and so the number of vertices on $|z| = 1$ does not exceed $4n - 6$.

12.2. We now devise a system for describing the topological structure of a tree, and we say that a knot of the tree is of (topological) height h if a path starting at the knot traverses at least h branches in order to reach the root. Starting at a knot of height h, one and only one path on the tree joins the knot to the root and traverses h branches. If there were two, the tree would contain a Jordan curve. The root itself is of height zero.

Imagine that the root of the tree is placed at the origin of the plane and let the remainder of the tree lie in the upper half-plane. Starting at a point on the positive axis near the root, we turn counterclockwise round a small circle with center at the origin until we first meet a stalk issuing from the root, and then proceed along this stalk until we encounter a knot. Starting at a point on the stalk just below the knot, we turn counterclockwise until we meet another branch of the tree, and we then proceed along it, and so on. Each knot encountered in this way has a height one greater than the preceding until we meet a terminal knot. This is a consequence of the fact that the tree contains no Jordan curve. At a terminal knot we turn completely around counterclockwise until we meet the opposite edge of the same terminal branch. Proceeding along this edge, we meet next the knot which has a height one less than the terminal knot and is the same knot which we met immediately preceding the terminal knot. We proceed in this fashion completely around the tree until we arrive finally at a point near the root such that, turning counterclockwise around a small circle with center at the root, we reach the negative axis without encountering a further branch of the tree.

As the whole tree is traversed in this manner, starting from the root, write down in order the height of each knot as it is met, giving a sequence of numbers

$$h_0, h_1, \cdots, h_i, \qquad h_0 = h_i = 0. \tag{12.2.1}$$

If a number of this sequence has the property that it is less than the two adjacent numbers, we call it a b-number. We also consider h_0 and h_i to be b-numbers. The τ terminal knots define $\tau - 1$ interterminal openings and there are two additional openings, one between the positive axis and the tree, the other between the negative axis and the tree. The b-numbers correspond to these $\tau + 1$ openings; in fact, a b-number is the height of the crotch at the bottom of the corresponding opening. Let

$$b_0, b_1, \cdots, b_{\tau-1}, b_\tau,$$

where $b_0 = b_\tau = 0$, be the sequence of b-numbers arranged in the same order as they occur in (12.2.1). The b-sequence is characterized by the following property:

(P) The first and last numbers b_0 and b_τ are zero. Given an integer q, $1 \leqq q \leqq \tau$, let b_p, $p < q$, be the last predecessor of b_q which does not exceed b_q in magnitude and write

$$x = \max (b_{p+1}, b_{p+2}, \cdots, b_q).$$

Then $(b_{p+1}, b_{p+2}, \cdots, b_q)$ contains all integers $b_q, b_q + 1, \cdots, x$.

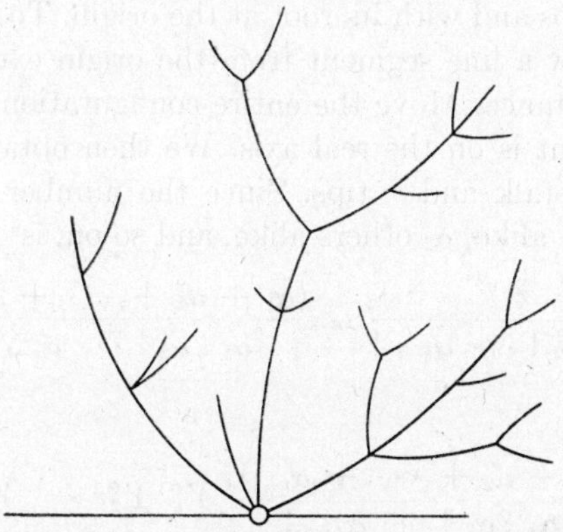

FIG. 5

The property (P) is an arithmetical characterization of the structure of the tree. A little reflection shows that there is a one-one correspondence between sequences (b) with the property (P) and trees in the upper half-plane with roots at the origin as described above. For example, the sequence

$$b_0 = 0, 2, 1, 2, 3, 1, 2, 0, 1, 3, 4, 4, 2, 3, 1, 0, 0, 1, 1, 2, b_\tau = 0$$

corresponds to the tree shown in Figure 5. Thus two trees in the half-plane are topologically equal if and only if their b-sequences are identical.

In the case of trees in the full plane we identify the terms b_0 and b_τ in the b-sequence and delete either one of these two terms. We then say that two trees are equal if and only if the b-sequence of one tree is obtainable from the b-sequence of the other by a cyclic permutation.

Removal of the root divides a tree in the plane into r subtrees, $r \geqq 1$. The number r is the number of stalks issuing from the root. If the r subtrees are not identical, a complete tree in the plane corresponds to exactly r different complete trees in the half-plane because of the cyclic rotation. However, if the r subtrees are all alike, the tree in the plane corresponds to only one tree in the half-plane since the cyclic rotations give the same tree in the half-plane.

Let Y_τ denote the number of different trees in the plane or half-plane which have one stalk ($r = 1$) and τ terminal knots. We observe that the trees with one stalk and τ terminal knots can be formed from the trees with one stalk and a smaller number of terminal knots in the following manner. Assume that $\tau \geqq 2$ and let

$$\tau = \alpha_1 + 2\alpha_2 + \cdots + (\tau - 1)\alpha_{\tau-1}, \qquad r = \alpha_1 + \alpha_2 + \cdots + \alpha_{\tau-1},$$

where $\alpha_1, \alpha_2, \cdots, \alpha_{\tau-1}$ are non-negative integers. Then $r \geqq 2$. Place α_1 trees with one stalk and one tip each in the upper half-plane, then among these trees place α_2 trees with one stalk and two tips each, and so on. Here each tree is assumed to have its root at the origin. This gives a tree in the upper half-plane with r stalks and τ tips and with its root at the origin. To obtain a tree with one stalk and τ tips, draw a line segment from the origin extending into the lower half-plane a finite distance. Move the entire configuration upward until the free tip of the line segment is on the real axis. We then obtain a tree in the upper half-plane with one stalk and τ tips. Since the number of combinations of r things of which α_1 are alike, α_2 others alike, and so on, is

$$\frac{r\,!}{\alpha_1\,!\,\alpha_2\,!\cdots\alpha_{\tau-1}\,!} = \frac{(\alpha_1 + \alpha_2 + \cdots + \alpha_{\tau-1})!}{\alpha_1\,!\,\alpha_2\,!\cdots\alpha_{\tau-1}\,!},$$

there are

$$\frac{(\alpha_1 + \alpha_2 + \cdots + \alpha_{\tau-1})!}{\alpha_1\,!\,\alpha_2\,!\cdots\alpha_{\tau-1}\,!} Y_1^{\alpha_1} Y_2^{\alpha_2} \cdots Y_{\tau-1}^{\alpha_{\tau-1}}$$

ways of forming a tree with one stalk and τ tips from α_1 trees with one stalk and one tip, α_2 trees with one stalk and two tips, $\cdots$, $\alpha_{\tau-1}$ trees with one stalk and $\tau - 1$ tips. Define

$$Y_\tau^{(r)} = \sum \frac{(\alpha_1 + \alpha_2 + \cdots + \alpha_\tau)!}{\alpha_1\,!\,\alpha_2\,!\cdots\alpha_\tau\,!} Y_1^{\alpha_1} Y_2^{\alpha_2} \cdots Y_\tau^{\alpha_\tau} \tag{12.2.2}$$

where the summation is over all integers $\alpha_1, \alpha_2, \cdots, \alpha_\tau$ satisfying

$$\alpha_1 \geqq 0, \quad \alpha_2 \geqq 0, \quad \cdots, \quad \alpha_\tau \geqq 0;$$
$$\alpha_1 + 2\alpha_2 + \cdots + \tau\alpha_\tau = \tau; \qquad \alpha_1 + \alpha_2 + \cdots + \alpha_\tau = r.$$

Observing that $\alpha_\tau = 1$ if $r = 1$, we see from (12.2.2) that

$$Y_\tau^{(1)} = Y_\tau\,;$$

so $Y_\tau^{(1)}$ is the number of different trees in the plane or half-plane with one stalk and τ terminal knots. If $r > 1$, then $\alpha_\tau = 0$ and the preceding discussion shows that $Y_\tau^{(r)}$ is the number of different trees in the upper half-plane with r stalks and τ terminal tips. For by (12.2.2) $Y_\tau^{(r)}$ is the number of different ways of forming a tree in the half-plane with r stalks and τ terminal tips from r trees, each with one stalk and fewer than τ terminal tips. Then

$$Y_\tau = Y_\tau^{(1)} = Y_\tau^{(2)} + Y_\tau^{(3)} + \cdots + Y_\tau^{(\tau)}, \qquad \tau \geqq 2. \tag{12.2.3}$$

The preceding discussion shows that

$$\left(\sum_{\tau=1}^{\infty} Y_\tau x^\tau\right)^r = \sum_{\tau=r}^{\infty} Y_\tau^{(r)} x^\tau$$

for $r = 1, 2, 3, \cdots$. This is a formal rule for computing the $Y_\tau^{(r)}$ from the Y_τ since these series converge only for $x = 0$. Conversely the Y_τ can be computed from the $Y_\tau^{(r)}$ by using (12.2.3). Thus we obtain the formal relation

(12.2.4)
$$\begin{aligned}\left(1 - \sum_{\tau=r}^{\infty} Y_\tau x^\tau\right)^{-1} &= 1 + \sum_{r=1}^{\infty}\sum_{\tau=r}^{\infty} Y_\tau^{(r)} x^\tau \\ &= 1 + Y_1 x + 2\{Y_2 x^2 + Y_3 x^3 + \cdots\}\end{aligned}$$

where $Y_1 = 1$.

Since $Y_\tau^{(r)}$ is the number of different trees in the half-plane having r stalks and τ tips, $Y_\tau^{(r)}$ is equal to the total number of b-sequences

(12.2.5) $$b_0, b_1, \cdots, b_\tau$$

containing exactly $r + 1$ zeros. The total number of trees in the half-plane having τ tips or the total number of b-sequences (12.2.5) is

$$H_\tau = Y_\tau^{(1)} + Y_\tau^{(2)} + \cdots + Y_\tau^{(\tau)} = 2Y_\tau, \qquad \tau \geqq 2.$$

When $\tau = 1$, $H_\tau = 1$.

Finally, let $T_\tau^{(r)}$ be the number of different trees in the full plane having r stalks and τ tips. If r is not a divisor of τ, it is plain that

$$T_\tau^{(r)} = \frac{1}{r} Y_\tau^{(r)}.$$

On the other hand, if r is a divisor of τ, $\tau = tr$, a tree in the plane may be composed of r equal subtrees, each having t tips and one stalk, and such a tree corresponds to just one tree in the half-plane. Hence if r is a divisor of τ,

$$T_\tau^{(r)} = \frac{1}{r} Y_\tau^{(r)} + \frac{r-1}{r} Y_{\tau/r}.$$

The total number of different trees in the plane with precisely τ tips is

$$N_\tau = T_\tau^{(1)} + T_\tau^{(2)} + \cdots + T_\tau^{(\tau)}.$$

From (12.2.4) we obtain

$$Y_2 x^2 + Y_3 x^3 + Y_4 x^4 + \cdots = \left(\sum_1^\infty Y_\tau x^\tau\right)^2 + \left(\sum_1^\infty Y_\tau x^\tau\right)^3 + \left(\sum_1^\infty Y_\tau x^\tau\right)^4 + \cdots.$$

Thus

$$Y_1 = 1, \qquad Y_2 = Y_1^2 = 1, \qquad Y_3 = Y_1^3 + 2Y_1Y_2 = 3,$$
$$Y_4 = Y_1^4 + 3Y_1^2Y_2 + Y_2^2 + 2Y_1Y_3 = 11, \cdots$$

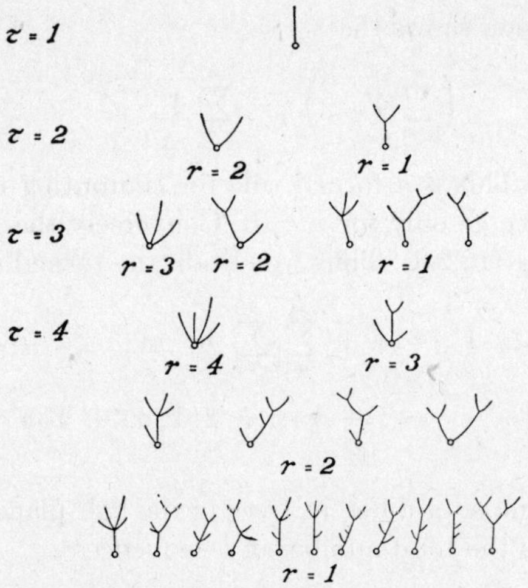

FIG. 6

and

$$Y_1^{(1)} = Y_1 = 1: \quad H_1 = 1;$$

$$Y_2^{(1)} = Y_2 = 1, \quad Y_2^{(2)} = Y_1^2 = 1: \quad H_2 = 2;$$

$$Y_3^{(1)} = Y_3 = 3, \quad Y_3^{(2)} = 2Y_1Y_2 = 2, \quad Y_3^{(3)} = Y_1^3 = 1: \quad H_3 = 6;$$

$$Y_4^{(1)} = Y_4 = 11, \quad Y_4^{(2)} = Y_2^2 + 2Y_1Y_3 = 7,$$

$$Y_4^{(3)} = 3Y_1^2Y_2 = 3, \quad Y_4^{(4)} = Y_1^4 = 1: \quad H_4 = 22;$$

$$T_1^{(1)} = Y_1^{(1)} = 1: \quad N_1 = 1;$$

$$T_2^{(1)} = Y_2^{(1)} = 1, \quad T_2^{(2)} = Y_2^{(2)}/2 + Y_1/2 = 1: \quad N_2 = 2;$$

$$T_3^{(1)} = Y_3^{(1)} = 3, \quad T_3^{(2)} = Y_3^{(2)}/2 = 1,$$

$$T_3^{(3)} = Y_3^{(3)}/3 + 2Y_1/3 = 1: \quad N_3 = 5;$$

$$T_4^{(1)} = Y_4^{(1)} = 11, \quad T_4^{(2)} = Y_4^{(2)}/2 + Y_2/2 = 4,$$

$$T_4^{(3)} = Y_4^{(3)}/3 = 1, \quad T_4^{(4)} = Y_4^{(4)}/4 + 3Y_1/4 = 1: \quad N_4 = 17.$$

The trees in the plane up to $\tau = 4$ are shown in Figure 6.

Let $f(z)$ be a $\mathfrak{D}_n$-function belonging to a point of $\boldsymbol{S}_n$. Then $w = f(z)$ transforms $|z| = 1$ into a set of arcs forming a tree and the root is the point at $w = \infty$. Since $Q(z)$ has at most $n - 1$ zeros on $|z| = 1$, this tree has at most $n - 1$ tips. Thus the image of $|z| = 1$ by $\mathfrak{D}_n$-functions is a tree in the plane having τ tips, $1 \leqq \tau \leqq n - 1$, and there are

$$\sum_{\nu=1}^{n-1} N_\nu$$

different trees of this kind. For example, the complete set of $\mathfrak{D}_5$-functions transforms $|z| = 1$ into the 25 trees shown in Figure 6.

12.3. Let $|z| \leqq 1$ be mapped onto the plane in such a way that any pair of identified arcs I'_ν, I''_ν maps into a branch of the tree whose length is equal to

$$\int_{I'_\nu} |dZ| = \int_{I''_\nu} |dZ|, \qquad |dZ| = \left| (Q(z))^{1/2} \frac{dz}{z} \right|. \tag{12.3.1}$$

This defines the length of any branch of the tree. In a similar way we define the distance between any two points of the same branch.

There is a shortest path or geodesic in this metric joining any two points of the tree and the geodesic must be a Jordan arc. Since the tree contains no closed Jordan curve, the geodesic is unique. A geodesic on the tree is characterized by the property that it is a Jordan arc.

To characterize a tree metrically as well as topologically, we define the (metric) altitude of a knot to be the distance from the knot to the root. Now place the tree in the upper half-plane and traverse the tree in the way described above (starting at a point on the positive axis to the right of the root). Let

$$\chi_1, \chi_2, \cdots, \chi_l, \qquad \chi_0 = \chi_l = 0,$$

be the altitudes of the knots written in the order in which they are met. Adjacent numbers are distinct. If a number of this sequence is less than the two adjacent numbers or is equal to zero, we call it a β-number. In complete analogy with the b-sequence we obtain the β-sequence,

$$\beta_0, \beta_1, \cdots, \beta_{\tau-1}, \beta_\tau,$$

where $\beta_0 = \beta_\tau = 0$.

From the β-sequence we may construct the b-sequence (which characterizes the tree topologically) as follows. If $\beta_i = 0$, we define $b_i = 0$. If $\beta_i > 0$, let β_{i_1} $(i_1 < i)$ be the first β-number to the left of β_i which is less than β_i. If $\beta_{i_1} > 0$, let β_{i_2} $(i_2 < i_1)$ be the first β-number to the left of β_{i_1} which is less than β_{i_1}, and so on. We stop when we obtain a number $\beta_{i_\mu} = 0$. Similarly, let β_{j_1} $(j_1 > i)$ be the first β-number to the right of β_i which is less than β_i. If $\beta_{j_1} > 0$, we repeat the process until we find a number $\beta_{j_\nu} = 0$. We obtain a sequence

$$\beta_{i_\mu} = 0, \qquad \beta_{i_{\mu-1}}, \cdots, \beta_{i_1}, \beta_i, \beta_{j_1}, \beta_{j_2}, \cdots, \beta_{j_\nu} = 0, \tag{12.3.2}$$

where

$$0 = \beta_{i_\mu} < \beta_{i_{\mu-1}} < \cdots < \beta_{i_1} < \beta_i, \qquad \beta_i > \beta_{j_1} > \cdots > \beta_{j_{\nu-1}} > \beta_{j_\nu} = 0.$$

Let $x_0, x_1, \cdots, x_k$, $0 = x_0 < x_1 < x_2 < \cdots < x_k = \beta_i$, $k \leqq \mu + \nu - 1$, be the different numerical values assumed by the numbers in the sequence (12.3.2). For each value x_m there is at least one knot whose altitude is x_m, and a geodesic connecting the knot β_i with the root must clearly pass through at least one knot whose altitude is x_m. Since the geodesic passes over the least number of branches between the knot β_i and the root, this least number is k and we define $b_i = k$. The sequence

$$b_0 = 0, b_1, \cdots, b_{\tau-1}, b_\tau = 0,$$

defined in this way is therefore the b-sequence of the tree.

Given any sequence of non-negative numbers

$$\beta_0, \beta_1, \cdots, \beta_{\tau-1}, \beta_\tau,$$

where $\beta_0 = \beta_\tau = 0$, we show that the b-sequence formed from it by the above rule has the property (P) characterizing b-sequences. If β_i, $1 \leqq i \leqq \tau - 1$, is greater than zero, we denote by S_i the sequence (12.3.2) used in defining b_i. Let S_{i_1} and S_{j_1} be the sequences corresponding to the numbers β_{i_1} and β_{j_1} of (12.3.2). Then we have

$$S_{i_1}: \beta_{i_\mu}, \beta_{i_{\mu-1}}, \cdots, \beta_{i_1}, \beta_{j_p}, \beta_{j_{p+1}}, \cdots, \beta_{j_\nu};$$

$$S_{j_1}: \beta_{i_\mu}, \beta_{i_{\mu-1}}, \cdots, \beta_{i_q}, \beta_{j_1}, \beta_{j_2}, \cdots, \beta_{j_\nu}.$$

Hence $b_{i_1} \leqq b_i - 1 < b_i$, $b_{j_1} \leqq b_i - 1 < b_i$. If $\beta_{i_1} > \beta_{j_1}$, then $p = 1$ and $b_{i_1} = b_i - 1$. If $\beta_{i_1} = \beta_{j_1}$, then $p = 2$ but in this case we also have $b_{i_1} = b_i - 1$. Similarly $b_{j_1} = b_i - 1$ if $\beta_{j_1} \geqq \beta_{i_1}$. We call b_{i_1} and b_{j_1} the left and right numbers belonging to b_i, and we see that one of these two numbers is always equal to $b_i - 1$. If $i_1 < i - 1$, each sequence S_j, $i_1 + 1 \leqq j \leqq i$, has the form

$$S_j: \beta_{i_\mu}, \beta_{i_{\mu-1}}, \cdots, \beta_{i_1}, \cdots, \beta_j, \cdots, \beta_{j_1}, \beta_{j_2}, \cdots, \beta_{j_\nu}$$

and so $b_j \geqq b_i$ $(j = i_1 + 1, \cdots, i - 1)$.

Now let b_q, $1 \leqq q \leqq \tau$, be defined by

$$S_q: \beta_{q_\mu}, \beta_{q_{\mu-1}}, \cdots, \beta_{q_1}, \beta_q, \beta_{r_1}, \beta_{r_2}, \cdots, \beta_{r_\nu}$$

and let b_p be the last predecessor of b_q in the b-sequence which does not exceed b_q. Then $q_1 \leqq p < q$. If $p = q - 1$, there is nothing to prove; so assume that $p < q - 1$ and let

$$x = \max(b_{p+1}, \cdots, b_q).$$

Let b_r, $p + 1 \leqq r \leqq q$, have the value x. If $x = b_q$, again there is nothing to prove. If $x > b_q$, then $x > b_p$ since $b_q \geqq b_p$; so $p < r < q$. Either the left or the right number belonging to b_r has the value $x - 1$, say $b_{r_1} = x - 1$. If $x - 1 > b_q$, then $x - 1 > b_p$; so $p < r_1 < q$ and we may repeat the process. Thus the sequence

$$b_{p+1}, \cdots, b_q$$

contains all integers b_q, $b_q + 1$, $\cdots$, x and the b-sequence has the property (P).

For example, the β-sequence

$$0, 1, 1, 1, 1, 0.5, 0.2, 0.5, 0.01, 0.2, 0.7, 0.9, 0$$

gives the b-sequence

$$0, 4, 4, 4, 4, 3, 2, 3, 1, 2, 3, 4, 0$$

which corresponds to the tree shown in Figure 7.

12.4. It will be convenient from now on to interpret the subscript on a β-number modulo τ, $\beta_{k+\tau} = \beta_k$. The numbers β_0 and β_τ ($\beta_0 = \beta_\tau = 0$) are then the

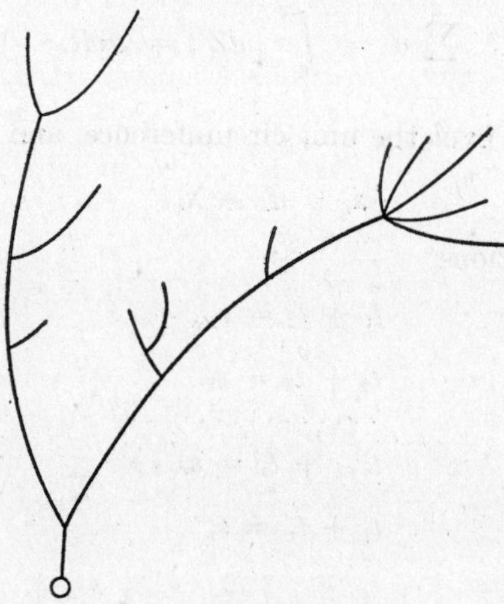

FIG. 7

same and the two numbers adjacent to $\beta_0 = \beta_\tau$ are $\beta_{\tau-1}$ and $\beta_{\tau+1} = \beta_1$. A number in the β-sequence

$$\beta_0, \beta_1, \beta_2, \cdots, \beta_\tau$$

which is equal to or less than each of the two adjacent numbers will be called a minimal number, and a number which is equal to or greater than each of the two adjacent numbers will be called a maximal number. If $\tau = 1$, define $\lambda_1 = 0$; if $\tau > 1$, let

$$\lambda_i = \begin{cases} \beta_{i+1} + \beta_{i-1} - 2\beta_i, & \text{if } \beta_i \text{ is minimal,} \\ 0, \quad \text{if } \beta_i \text{ is maximal,} \\ \beta_{i+1} - \beta_i, & \text{if } \beta_{i-1} < \beta_i < \beta_{i+1}, \\ \beta_{i-1} - \beta_i, & \text{if } \beta_{i-1} > \beta_i > \beta_{i+1}. \end{cases} \tag{12.4.1}$$

Let the tree be placed in the upper half-plane with root at the origin. As we traverse the tree starting at a point on the positive axis, let $t_1, t_2, \cdots, t_\tau$ be the lengths of the terminal branches in the order in which they are met. The tips of the terminal branches correspond to zeros $z_1, z_2, \cdots, z_\tau$ of $Q(z)$ on $|z| = 1$, and we suppose that the tip of the terminal branch of length t_i corresponds to z_i. Since β_i is the altitude of the crotch at the bottom of the interterminal space between the branches t_i and t_{i+1}, we see that λ_i is the distance traversed on the tree after leaving the terminal branch t_i until meeting the next terminal branch t_{i+1}. Write

(12.4.2) $$d_i = \int_{z_i}^{z_{i+1}} |dZ|, \qquad |dZ| = \left|(Q(z))^{1/2}\frac{dz}{z}\right|,$$

(12.4.3) $$\sum_{i=1}^{\tau} d_i = \int_0^{2\pi} |dZ| = 2\pi R,$$

where the integration is over the unit circumference, and let

$$s_i = d_i - \lambda_i .$$

Then we have the equations

$$\begin{aligned} t_1 + t_2 &= s_1 , \\ t_2 + t_3 &= s_2 , \\ \cdots\cdots&\cdots\cdots, \\ t_{\tau-1} + t_\tau &= s_{\tau-1} , \\ t_1 + t_\tau &= s_\tau \end{aligned}$$

whose determinant is

$$D = \begin{vmatrix} 1 & 1 & 0 & 0 & \cdot & \cdot & \cdot & 0 \\ 0 & 1 & 1 & 0 & \cdot & \cdot & \cdot & 0 \\ \cdot & \cdot & \cdot & \cdot & \cdot & \cdot & \cdot & \cdot \\ 0 & 0 & \cdot & \cdot & \cdot & 0 & 1 & 1 \\ 1 & 0 & \cdot & \cdot & \cdot & 0 & 0 & 1 \end{vmatrix} = 1 + (-1)^{\tau-1}, \qquad \tau \geqq 2.$$

If τ is odd, D is not zero and the s_i determine the t_i. In this case

(12.4.4) $$\begin{aligned} s_1 - s_2 + s_3 - s_4 + \cdots + s_\tau &= 2t_1 , \\ s_1 + s_2 - s_3 + s_4 - \cdots - s_\tau &= 2t_2 , \\ -s_1 + s_2 + s_3 - s_4 + \cdots + s_\tau &= 2t_3 , \\ \cdots\cdots\cdots\cdots&\cdots\cdots, \\ -s_1 + s_2 - \cdots + s_{\tau-1} + s_\tau &= 2t_\tau . \end{aligned}$$

When τ is even, the augmented matrix must have rank $\tau - 1$ and this gives the condition

$$s_1 - s_2 + s_3 - s_4 + \cdots - s_\tau = 0.$$

If this condition is satisfied, we may specify t_1, $t_1 = s_0$ say, and the remaining terminal lengths are given by

$$
\begin{aligned}
&s_1 - s_0 = t_2, \\
&s_2 - s_1 + s_0 = t_3, \\
(12.4.5)\qquad &s_3 - s_2 + s_1 - s_0 = t_4, \\
&\cdots\cdots\cdots\cdots\cdots\cdots, \\
&s_{\tau-1} - s_{\tau-2} + \cdots + s_1 - s_0 = t_\tau .
\end{aligned}
$$

Since $t_i > 0$, $i = 1, 2, \cdots, \tau$, the equations (12.4.4) and (12.4.5) give necessary conditions on the s_i. However, these conditions are somewhat troublesome and so it is better to prescribe the terminal lengths t_i directly e shall now describe one method of doing this.

12.5. A number β_i may be both minimal and maximal, in which case $\lambda_i = 0$. On the other hand, if $\lambda_i = 0$, then β_i is maximal. For suppose that β_i is minimal and that

$$\lambda_i = \beta_{i+1} + \beta_{i-1} - 2\beta_i = (\beta_{i+1} - \beta_i) + (\beta_{i-1} - \beta_i) = 0.$$

Since $\beta_{i+1} - \beta_i$ and $\beta_{i-1} - \beta_i$ have the same sign, both are zero and so β_i is also a maximal number. We shall henceforth call β_i maximal if $\lambda_i = 0$ and we shall call β_i minimal if and only if it is minimal with $\lambda_i > 0$. A β-sequence contains at least one maximal number, and so at least one of the numbers

$$\lambda_1, \lambda_2, \cdots, \lambda_\tau$$

is equal to zero. If all numbers λ_i are zero, then all β_i are zero; so consider the case in which at least one λ_i is not zero. Then the zeros in the λ-sequence divide it into intervals $\Lambda_1, \Lambda_2, \cdots, \Lambda_p$, $p \geqq 1$, where

$$\Lambda_j = (\lambda_1^{(j)}, \lambda_2^{(j)}, \cdots, \lambda_{\mu_j}^{(j)}), \qquad \lambda_{ij}^{(j)} > 0, \quad i = 1, 2, \cdots, \mu_j,$$

the numbers λ between two intervals Λ being all equal to zero. We assume that the intervals Λ are ordered in the way in which they occur in the λ-sequence.

We say that the numbers β_i and λ_i correspond to one another and that the β-numbers corresponding to $\lambda_1^{(j)}, \lambda_2^{(j)}, \cdots, \lambda_{\mu_j}^{(j)}$ belong to Λ_j. For simplicity let Λ_j be the sequence

$$\lambda_i, \lambda_{i+1}, \cdots, \lambda_k .$$

Then $\lambda_{i-1} = \lambda_{k+1} = 0$; so β_{i-1} and β_{k+1} are maximal. Thus the β's corresponding to Λ_j and λ_{i-1}, λ_{k+1} can be written

$$\beta_{i-1} \geqq \beta_i, \qquad \beta_{i+1}, \cdots, \beta_k \leqq \beta_{k+1} .$$

It is then clear that among $\beta_i, \beta_{i+1}, \cdots, \beta_k$ there is at least one, say β_ν, which is as small as any; and, if there are several, let ν, $\nu \leqq k$, be as large as possible. If $\nu < k$, then $\beta_{\nu-1} \geqq \beta_\nu$, $\beta_{\nu+1} > \beta_\nu$; so β_ν is minimal. If $\nu = k$, then $\beta_{k-1} \geqq \beta_k$,

$\beta_{k+1} \geqq \beta_k$. Since $\lambda_k \neq 0$, either $\beta_{k-1} > \beta_k$, $\beta_{k+1} > \beta_k$, or both; so β_k is minimal. Thus among the β-numbers belonging to Λ_j there is at least one minimal number. On the other hand, between any two non-adjacent minimal numbers β there is at least one maximal number and Λ_j cannot contain a maximal number since $\lambda_m \neq 0$, $m = i, i + 1, \cdots, k$. Hence among the β's belonging to Λ_j there is either precisely one minimal β or precisely two minimal β's which are necessarily adjacent. Thus the β-numbers corresponding to $\lambda_{i-1}, \lambda_i, \cdots, \lambda_k, \lambda_{k+1}$ are of one of the four possible forms:

$$\beta_{i-1} = \beta_i < \cdots < \beta_k < \beta_{k+1};$$

$$\beta_{i-1} > \beta_i > \cdots > \beta_\nu < \beta_{\nu+1} < \cdots < \beta_k < \beta_{k+1}, \quad i \leqq \nu \leqq k;$$

$$\beta_{i-1} > \beta_i > \cdots > \beta_\nu = \beta_{\nu+1} < \cdots < \beta_k < \beta_{k+1}, \quad i \leqq \nu \leqq k - 1;$$

$$\beta_{i-1} > \beta_i > \cdots > \beta_k = \beta_{k+1}.$$

Now the β-numbers, τ in number, correspond one-one to the interterminal spaces between the tips of the tree and determine the altitudes of the knots at the bottoms of the interterminal spaces. Several β-numbers may correspond to the same knot of the tree, however. Since two adjacent minimal β-numbers are equal, they correspond to the same knot. We may then say that each interval Λ_j contains one and only one minimal point of the tree.

Write

$$D_j = \lambda_1^{(j)} + \lambda_2^{(j)} + \cdots + \lambda_{\mu_j}^{(j)}.$$

Let β_j' be the value of the maximal β-number immediately preceding Λ_j. If Λ_j is the sequence $\lambda_i, \lambda_{i+1}, \cdots, \lambda_k$ mentioned above, then $\beta_j' = \beta_{i-1}$. Let x_j, $0 \leqq x_j \leqq D_j$, be the decrement in altitude between the knot of altitude β_j' and the minimal point of Λ_j. Given β_j', x_j, and the λ-numbers of Λ_j, the β-numbers belonging to Λ_j are determined.

Given a Q-function $Q(z)$, let it vanish at h distinct points on $|z| = 1$, $1 \leqq h \leqq n - 1$. Choose any τ points $z_1, z_2, \cdots, z_\tau$ of this set, $1 \leqq \tau \leqq h$, and let them be numbered in clockwise fashion around the unit circumference. Then the positive numbers $d_1, d_2, \cdots, d_\tau$ are defined by (12.4.2) and they satisfy (12.4.3). Choose τ positive numbers $t_1, t_2, \cdots, t_\tau$ such that

$$t_i + t_{i+1} \leqq d_i \qquad (i = 1, 2, \cdots, \tau)$$

with equality for at least one i. Here the subscripts are to be interpreted modulo τ; so $t_{\tau+1} = t_1$. Writing $s_i = t_i + t_{i+1}$, let

$$\lambda_i = d_i - s_i.$$

We then obtain a λ-sequence

$$\lambda_1, \lambda_2, \cdots, \lambda_\tau$$

which contains at least one zero. If all numbers λ in this sequence are zero, we define $\beta_i = \lambda_i = 0$ $(i = 1, 2, \cdots, \tau)$. Otherwise form the intervals Λ_j, $j = 1, 2, \cdots, p$. If Λ_j is the sequence

$$\lambda_1^{(j)}, \lambda_2^{(j)}, \cdots, \lambda_{\mu_j}^{(j)},$$

let

$$D_j = \lambda_1^{(j)} + \lambda_2^{(j)} + \cdots + \lambda_{\mu_j}^{(j)}.$$

Choose numbers x_j, $0 \leqq x_j \leqq D_j$, subject to the condition

$$x_1 + x_2 + \cdots + x_p = \pi R - (t_1 + t_2 + \cdots + t_\tau). \tag{12.5.1}$$

Starting with Λ_1, assign to the β-number immediately preceding Λ_1 (which is maximal) the fictitious value α_1. The remaining fictitious β-values $\alpha_2, \alpha_3, \cdots, \alpha_\tau$ are then determined in terms of the $\lambda_1, \lambda_2, \cdots, \lambda_\tau, x_1, x_2, \cdots, x_p$, and the assigned value α_1. Let

$$m = \min(\alpha_1, \alpha_2, \cdots, \alpha_\tau)$$

and define

$$\beta_i = \alpha_i - m, \qquad i = 1, 2, \cdots, \tau.$$

We obtain a β-sequence, and the β-numbers which are zero correspond to the root. However, the β-sequence obtained in this way may be a cyclic permutation of the β-sequence defined in **12.3.**

EXAMPLE: $d_1 = 1, \quad d_2 = .9, \quad d_3 = 1.8, \quad d_4 = .7,$

$$2\pi R = d_1 + d_2 + d_3 + d_4 = 4.4.$$

Choose $t_1 = .4$, $t_2 = .3$, $t_3 = .6$, $t_4 = .3$. Then

$$s_1 = .7, \quad s_2 = .9, \quad s_3 = .9, \quad s_4 = .7,$$

$$\lambda_1 = .3, \quad \lambda_2 = 0, \quad \lambda_3 = .9, \quad \lambda_4 = 0,$$

$$\Lambda_1 = (\lambda_3), \quad \Lambda_2 = (\lambda_1),$$

$$D_1 = .9, \quad D_2 = .3,$$

$$x_1 + x_2 = 2.2 - (.4 + .3 + .6 + .3) = .6.$$

Choose $x_1 = .4$, $x_2 = .2$. Assuming $\alpha_1 = 1$, we obtain

$$\alpha_1 = 1, \alpha_2 = 1 - x_1 = .6, \alpha_3 = \alpha_2 + D_1 - x_1 = 1.1, \alpha_4 = \alpha_3 - x_2 = .9,$$

$$\alpha_5 = \alpha_1 = \alpha_4 + D_2 - x_2 = 1,$$

$$m = \min(\alpha_1, \alpha_2, \alpha_3, \alpha_4) = .6,$$

$$\beta_1 = .4, \quad \beta_2 = 0, \quad \beta_3 = .5, \quad \beta_4 = .3.$$

By a cyclic rearrangement the β-sequence may be written

$$0, \quad .5, \quad .3, \quad .4, \quad 0,$$

and the corresponding b-sequence then is

$$0, \quad 2, \quad 1, \quad 2, \quad 0.$$

The tree is shown in Figure 8; encircled numbers indicate the branch lengths.

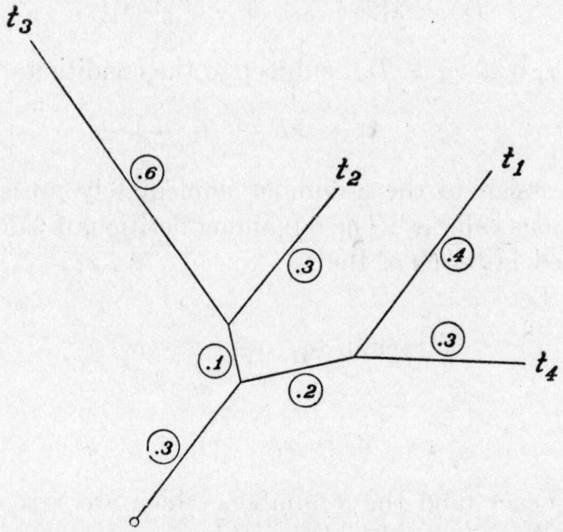

FIG. 8

12.6. The function Q is determined by the vector $\mathfrak{B} = (B_1, B_2, \cdots, B_{n-1})$,

$$\|\mathfrak{B}\| = \left(\sum_{\nu=1}^{n-1} |B_\nu|^2\right)^{1/2} = 1,$$

and so it depends on $2n - 3$ real parameters. However, if Q has ν zeros on $|z| = 1$, $1 \leqq \nu \leqq n - 1$, and if we vary Q in such a way that ν zeros remain on $|z| = 1$, then Q depends on $2n - \nu - 2$ real parameters. If the λ-sequence contains p zeros, no two of which are adjacent, we obtain p intervals Λ_j. In this case there are $\tau - p$ parameters among $t_1, t_2, \cdots, t_\tau$, and $p - 1$ parameters among $x_1, x_2, \cdots, x_p$, giving $\tau - 1$ parameters in all. Since consecutive zeros in the λ-sequence count as one zero in forming the intervals Λ_j, we see that if two zeros in the λ-sequence are adjacent, then the number of parameters is $\tau - 2$, and so on. Thus the points of S_n depend generally on

$$(2n - \tau - 2) + (\tau - 1) = 2n - 3$$

real parameters.

We remark that the identification of points on $|z| = 1$ is uniquely determined by $Q(z)$ if and only if $\nu = 1$, in which case $\tau = 1$.

12.7. Given a point of S_n, let

$$z_1, z_2, \cdots, z_\nu, \qquad \nu \geqq 1,$$

be the distinct zeros of Q on $|z| = 1$ enumerated in a clockwise fashion around $|z| = 1$, and let $2\sigma_\nu$ be the order of z_ν, $\sigma_\nu > 0$. The τ tips of the terminal branches of the tree correspond to τ of these zeros, $1 \leqq \tau \leqq m$. We may indicate this information by a two-rowed matrix

$$\begin{pmatrix} \sigma_1 & \sigma_2 & \cdots & \sigma_m \\ 0 & 1 & \cdots & 0 \end{pmatrix}.$$

The first row gives the half-orders of the zeros of Q on $|z| = 1$ and the second row, which consists of 0's and 1's, indicates the zeros which map into tips of the tree. If the number below σ_ν is 0, this zero does not map into a tip whereas if the number is 1, it does. In order to exhibit the topological structure of the tree, it is better to replace each 1 in the second row by the b-number which corresponds to the crotch of the tree between this tip and the following one. For each point P of $\mathbf{S}_n$ we thus obtain a two-rowed matrix

$$\mathfrak{M}_P = \begin{pmatrix} \sigma_1 & \sigma_2 & \cdots & \sigma_m \\ 0 & b_1 & \cdots & 0 \end{pmatrix}. \tag{12.7.1}$$

We consider two such matrices equal if one can be obtained from the other by a cyclic rearrangement of the columns.

We say that a point P of $\mathbf{S}_n$ is of order k, $k \geqq 2$, if the total number of zeros of the corresponding function Q, multiplicities counted, is $2k - 2$.

Any zero z_ν of Q on $|z| = 1$ is mapped into some point of the tree, which may be a tip, a knot, the root, or an interior point of a branch. If the order of z_ν is $2\sigma_\nu$, we refer to the corresponding point on the tree as the point σ_ν. Let P_1, P_2 be two points of $\mathbf{S}_n$. If $\mathfrak{M}_{P_1} = \mathfrak{M}_{P_2}$, then by making a cyclic rearrangement of the columns of $\mathfrak{M}_{P_2}$ the two matrices $\mathfrak{M}_{P_1}$, $\mathfrak{M}_{P_2}$ can be made identical and in particular $\sigma_\nu^{(1)} = \sigma_\nu^{(2)}$ ($\nu = 1, 2, \cdots, m$). We say that two points P_1, P_2 of $\mathbf{S}_n$ are similar if P_1 and P_2 are of the same order, if $\mathfrak{M}_{P_1} = \mathfrak{M}_{P_2}$, and if there is a topological mapping of the tree at P_1 onto the tree at P_2 which carries each point $\sigma_\nu^{(1)}$ into the corresponding point $\sigma_\nu^{(2)}$. The whole of $\mathbf{S}_n$ is thus divided into classes of similar points, and these classes are finite in number. We denote them by $\mathbf{\Omega}_1, \mathbf{\Omega}_2, \cdots, \mathbf{\Omega}_N$, $N = N(n)$. If the points of $\mathbf{\Omega}_\nu$ are of order k, we say that $\mathbf{\Omega}_\nu$ is of order k.

The points P of any one of these regions $\mathbf{\Omega}_\nu$ depend on a set of real parameters $\omega_1, \omega_2, \cdots, \omega_k$. Certain of these parameters define Q and the others define the tree. The number and nature of these parameters change as we pass from one $\mathbf{\Omega}_\nu$ to another. We have seen that the number $k = k(\nu)$ of parameters defining $\mathbf{\Omega}_\nu$ cannot exceed $2n - 3$, but for some regions $\mathbf{\Omega}_\nu$ the number of parameters is less than $2n - 3$. The domains $\mathbf{\Omega}_\nu$ map into corresponding domains $\mathbf{\Pi}_\nu$ on the boundary of V_n, and the parametrization of the $\mathbf{\Omega}_\nu$ is thus carried over to the $\mathbf{\Pi}_\nu$, $\nu = 1, 2, \cdots, N$.

We remark that if $\mathbf{\Omega}_\nu$ is of order k, the functions $w = f(z)$ belonging to $\mathbf{\Pi}_\nu$ are $\mathfrak{D}_k$-functions. If $k < n$, the points $(a_2, a_3, \cdots, a_k)$ belong to the boundary of V_k.

Chapter XIII

THE REGION V_3

13.1. In this chapter we find the parametric equations for the boundary of V_3.

The differential equations for functions $w = f(z)$ belonging to the boundary of V_3 are of the form

$$\left(\frac{z}{w}\frac{dw}{dz}\right)^2 P(w) = Q(z)$$

where

$$P(w) = \frac{A_1}{w} + \frac{A_2}{w^2}, \qquad Q(z) = B_0 + \frac{B_1}{z} + \bar{B}_1 z + \frac{B_2}{z^2} + \bar{B}_2 z^2.$$

The parametric equations involve only elementary functions, for in the case $n = 3$ the functions $(P(w))^{1/2}/w$ and $(Q(z))^{1/2}/z$ are integrable in terms of elementary functions.

The boundary is composed of two hypersurfaces $\mathbf{\Pi}_1$ and $\mathbf{\Pi}_2$ plus their intersection. Strictly speaking, in the terminology of Chapter XII the boundary of V_3 (apart from points of order 2) is composed of five parts $\mathbf{\Pi}_1, \mathbf{\Pi}_2, \cdots, \mathbf{\Pi}_5$: $\mathbf{\Pi}_1$ and $\mathbf{\Pi}_2$ are hypersurfaces of dimension 3, $\mathbf{\Pi}_3$ and $\mathbf{\Pi}_4$ are surfaces, and $\mathbf{\Pi}_5$ is a curve. The corresponding portions $\mathbf{\Omega}_1, \mathbf{\Omega}_2, \mathbf{\Omega}_3, \mathbf{\Omega}_4$, and $\mathbf{\Omega}_5$ of S_3 may be described as follows. At a point of $\mathbf{\Omega}_1$, the function Q (which has four zeros in all, multiplicities counted) has just the one double zero on $|z| = 1$ and the identification of points on $|z| = 1$ yields a tree with one branch. At a point of $\mathbf{\Omega}_2$, Q has two double zeros on $|z| = 1$ and so has no zero in $|z| < 1$. The tree is composed of three branches, a stalk issuing from the root and two terminal branches. At a point of $\mathbf{\Omega}_3$, Q has two double zeros on $|z| = 1$ but the tree is composed of only one branch while at a point of $\mathbf{\Omega}_4$, the function Q has two double zeros on $|z| = 1$ and the tree consists of two stalks issuing from the root. Finally at a point of $\mathbf{\Omega}_5$, Q has a quadruple zero on $|z| = 1$ and no other zeros; the tree then consists of one branch.

Because a theorem would be rather long, we summarize the results proved in this chapter in the remainder of the present section and begin the proof in **13.2.** The functions $w = f(z)$ belonging to the boundary of V_3 map $|z| < 1$ onto the w-plane minus a set of arcs forming a tree of one of the above types and the root is the point at $w = \infty$. At a point of $\mathbf{\Pi}_1$, $w = f(z)$ maps $|z| < 1$ onto the w-plane minus a single analytic slit extending to $w = \infty$ while at a point of $\mathbf{\Pi}_2$, the slit consists of a straight line arg (w) = constant extending from $w = \infty$ to some finite point where the straight line bifurcates into two curved terminal slits which make angles $2\pi/3$ with the rectilinear portion. Therefore the slit is forked. At points of $\mathbf{\Pi}_3$, the slits differ from those corresponding to $\mathbf{\Pi}_2$ in that one of the two terminal branches is missing. A slit belonging to a function $f(z)$

of Π_4 lies on a straight line through the origin. Finally at a point of Π_5, the function has the form

$$w = \frac{z}{(1 + e^{i\theta} z)^2}, \qquad \theta \text{ real.}$$

It therefore maps $|z| < 1$ onto the plane minus a rectilinear slit extending from $w = e^{-i\theta}/4$ to $w = \infty$.

Since it turns out that the sum $\Pi_2 + \Pi_4$ is an analytic hypersurface which intersects Π_1 along Π_3, we ignored Π_4 in describing the boundary of V_3 in Chapter I. In any case, it is clear that we have only to find Π_1 and Π_2; the remainder of the boundary is of lower dimensionality and may be obtained by passing to the limit.

Because of the rotational property described in Chapter I, we may suppose that $a_2 \geqq 0$ or $a_3 \geqq 0$. If $a_2 \geqq 0$, the parametric formulas to be obtained are as follows:

(1) Suppose that $0 < \rho < 1$, $-\pi/2 \leqq \varphi \leqq \pi/2$, and let

$$r^2 = 1 + 6\rho^2 + \rho^4 + 4\rho(1 + \rho^2) \cos 2\varphi, \qquad r > 0,$$

$$\tan \alpha = \frac{2\rho \sin 2\varphi}{1 + 2\rho \cos 2\varphi + \rho^2}, \qquad \left(-\frac{\pi}{2} < \alpha < \frac{\pi}{2}\right),$$

$$C_1 = \frac{(1 + \rho)^2}{2\rho} \cos \varphi, \qquad C_2 = \frac{(1 - \rho)^2}{2\rho} \sin \varphi,$$

$$\lambda = C_1 \log \frac{r}{(1 + \rho)^2} + C_2\, \alpha - 2 \cos \varphi, \qquad \mu = C_2 \log \frac{r}{(1 - \rho)^2} - C_1\, \alpha + 2 \sin \varphi.$$

Then Π_1 is defined by

$$\begin{aligned} a_2 &= (\lambda^2 + \mu^2)^{1/2}, \\ a_3 &= \lambda^2 + \mu^2 + (C_1 + iC_2)(\lambda - i\mu) + \left(\rho + \frac{1}{\rho} + \cos 2\varphi\right) \frac{\lambda - i\mu}{\lambda + i\mu}. \end{aligned} \tag{13.1.1}$$

(2) Suppose that $0 < \varphi \leqq \pi/2$ and let

$$-2(\sin \varphi - \varphi \cos \varphi) \leqq \mu \leqq 2(\sin \varphi - \varphi \cos \varphi),$$

$$\lambda = 2 \cos \varphi \{\log (\cos \varphi) - 1\}.$$

Then Π_2 is defined by

$$\begin{aligned} a_2 &= (\lambda^2 + \mu^2)^{1/2}, \\ a_3 &= \lambda^2 + \mu^2 + 2(\cos \varphi)(\lambda - i\mu) + (2 \cos^2 \varphi + 1) \frac{\lambda - i\mu}{\lambda + i\mu}. \end{aligned} \tag{13.1.2}$$

In that section of V_3 for which $a_2 \geqq 0$, the surface Π_4 appears as a curve and it is given by (13.1.2) with $\varphi = \pi/2$; that is, it is the parabolic arc

$$a_3 = a_2^2 - 1, \qquad 0 \leqq a_2 \leqq 2. \tag{13.1.3}$$

The intersection Π_3 of Π_1 and Π_2 is obtained by letting ρ tend to 1 in (13.1.1) or by taking

$$\mu = \pm 2(\sin \varphi - \varphi \cos \varphi)$$

in (13.1.2).

For any fixed value of φ, let ρ tend to zero in the formulas (13.1.1). It is readily verified that (a_2, a_3) tends to the point (2, 3). Thus for fixed φ the formulas (13.1.1) define a curve as ρ varies, and the curve begins at a point of Π_3 for $\rho = 1$ and tends to the point (2, 3) as ρ tends to 0. When φ is near 0, the curve lies in a small neighborhood of $a_2 = 2$, $a_3 = 3$; and when $\varphi = 0$, the curve degenerates to this point. In other words when $\varphi = 0$, formulas (13.1.1) give $a_2 = 2$, $a_3 = 3$ for any value of ρ, $0 < \rho < 1$. Hence Π_1 is a continuous but not a one-one image of the parameter space and in the strict sense of differential geometry consists of two surfaces Π_1^+, Π_1^- which intersect at the point $a_2 = 2$, $a_3 = 3$. The surface Π_1^+ corresponds to $0 < \varphi < \pi/2$ and Π_1^- to $-\pi/2 < \varphi < 0$. If we no longer restrict a_2 to be real, Π_1 consists (strictly speaking) of two 3-dimensional hypersurfaces which meet along the curve $(2e^{i\theta}, 3e^{2i\theta})$, θ real.

Now consider the section of Π_2 for which $a_2 \geqq 0$, and for a fixed value of φ let μ vary. We obtain a curve which extends from a point (a_2, a_3) on the intersection Π_3 of Π_1 and Π_2 to the point $(a_2, \bar{a}_3)$ of Π_3 which is the image of (a_2, a_3) with respect to the plane $\text{Im}(a_3) = 0$. This curve crosses the plane $\text{Im}(a_3) = 0$ when $\mu = 0$. When φ is near zero, this curve lies in a small neighborhood of the point (2, 3); and when $\varphi = 0$, it degenerates to this point (since $\mu = 0$ when $\varphi = 0$). The curve $\mu = 0$, $0 \leqq \varphi \leqq \pi/2$, is given by

$$\begin{aligned} a_2 &= |\lambda|, \\ a_3 &= \lambda^2 + 2(\cos\varphi)\lambda + 2\cos^2\varphi + 1 \end{aligned} \tag{13.1.4}$$

and it extends from the point (2, 3) to the point (0, 1) as φ varies from 0 to $\pi/2$. The curves (13.1.3) and (13.1.4) together with their images in the plane $a_2 = 0$ bound the 2-dimensional region of (a_2, a_3) when both a_2, a_3 are real.

Plate I, pp. xii–xiii, shows the portion of V_3 for which $a_2 \geqq 0$, $\text{Im}(a_3) \geqq 0$. This portion is one-half of the entire cross-section for which $\text{Im}(a_2) = 0$. The surface Π_1 is shown in yellow and the surface Π_2 is shown in blue. Table I of the Appendix gives points on the boundary of this portion of V_3: Part I of this table gives points on Π_1 and Part II gives points on Π_2. The corresponding values of the parameters are indicated.

By a rotation we can make a_3 real, $a_3 \geqq 0$. When $a_3 \geqq 0$, let the point (a_2, a_3) be denoted by (b_2, b_3). Points (b_2, b_3) may be obtained from points (a_2, a_3) by using the transformation

$$\begin{aligned} b_2 &= \frac{a_2}{2^{1/2}}\left\{\left(1 + \frac{\text{Re}(a_3)}{|a_3|}\right)^{1/2} \pm i\left(1 - \frac{\text{Re}(a_3)}{|a_3|}\right)^{1/2}\right\}, \\ b_3 &= |a_3|. \end{aligned} \tag{13.1.5}$$

$u = x$

$v = -y$

$|h(z)| = |z|$

$\arg h(z) = -\arg z$

If $f(z) = u + iv$

Let $h(z) = \overline{f(\bar{z})}$ ~~$\Rightarrow \frac{\partial h}{\partial \bar{z}} = \frac{\partial h}{\partial \bar{z}}$~~

Let ~~$h(z) = f(\bar{z})$~~

$$\frac{\partial f}{\partial \bar{z}} = f'(\bar{z})$$

If $z = x + iy$

$\bar{z} =$

Let $g(z) = e^{-i\theta} f(e^{i\theta} z)$

$|a_2| \leq 2$

$a_2 = \operatorname{Re} a_2 + i \operatorname{Im} a_2$

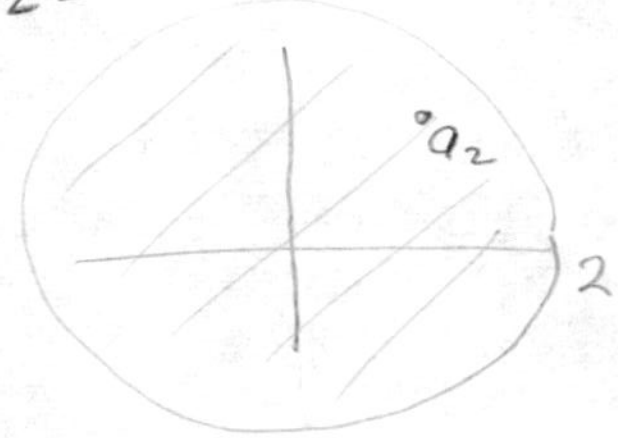

If $w = \overline{f(\bar{z})} = \bar{\zeta}$, where $\zeta = f(\bar{z})$

$\overline{f(\bar{z})} = h(z)$

$\frac{\partial h}{\partial \bar{z}} = \frac{\partial h}{\partial \bar{\zeta}} \cdot \frac{\partial \bar{\zeta}}{\partial \bar{z}} =$

$z \longrightarrow \bar{z}$

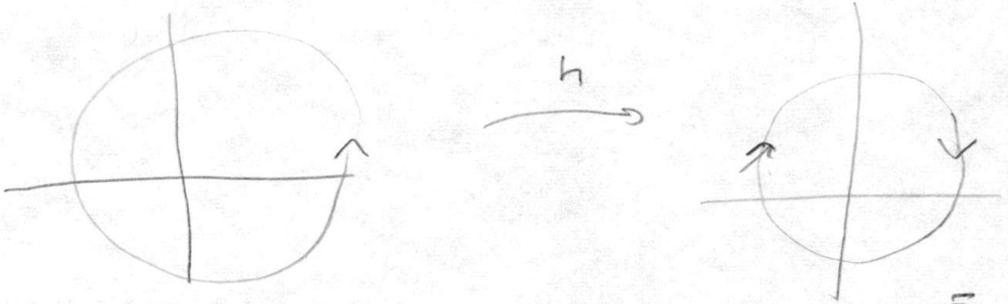

z $\bar{z}$

i.e. $h(z) = \bar{z} = re^{-i\theta}$

h not analytic but

Plate II, pp. xiv–xv, shows the portion of V_3 for which Im $(a_3) = 0$, Re $(a_2) \geqq 0$. This portion is one-half of the cross-section Im $(a_3) = 0$. The surface Π_1 is shown in yellow, Π_2 in blue. Table II of the Appendix gives points on the boundary of the portion of V_3 for which

$$\text{Im }(a_3) = 0, \qquad \text{Re }(a_2) \geqq 0, \qquad \text{Im }(a_2) \geqq 0$$

(one-fourth of the cross-section Im $(a_3) = 0$).

Other points (a_2, a_3) lying on the boundary of V_3 may easily be computed from Tables I and II by using rotations and reflections.

Plates I and II reproduce colored photographs of actual models each of which was constructed from the data in the tables by the following method: brass rods of proper heights were inserted in a board and the spaces between the rods were filled with plaster, the surface being then smoothed until the pointed tips of the rods appeared; an aluminum casting was then made from the plaster pattern.

13.2. Any function $w = f(z)$ belonging to a point (a_2, a_3) on the boundary of V_3 satisfies an equation of the form

$$\left(\frac{z}{w}\frac{dw}{dz}\right)^2 P(w) = Q(z) \tag{13.2.1}$$

where

$$P(w) = \frac{A_1 w + A_2}{w^2}, \qquad Q(z) = \frac{\bar{B}_2 z^4 + \bar{B}_1 z^3 + B_0 z^2 + B_1 z + B_2}{z^2}. \tag{13.2.2}$$

Here $A_2 = B_2$, and in the case $n = 3$ we have seen (**8.2**) that

$$A_1 = B_1 .$$

Thus in the case $n = 3$ the vectors $\mathfrak{A} = (A_1, A_2)$, $\mathfrak{B} = (B_1, B_2)$ are equal.

Suppose first that $A_2 = B_2 = 0$. Then $A_1 = B_1 \neq 0$. By making a rotation (that is, replacing w by $we^{-i\theta}$, z by $ze^{i\theta}$) we may suppose that $A_1 = B_1 > 0$. Dividing both sides of (13.2.1) by A_1, we obtain the simple equation

$$\left(\frac{z}{w}\frac{dw}{dz}\right)^2 \frac{1}{w} = B_0' + \frac{1}{z} + z$$

where $B_0' > 0$. Since the right side of this equation must be non-negative on $|z| = 1$ with at least one zero there, we see that $B_0' = 2$, and the equation may be written

$$\left(\frac{z}{w}\frac{dw}{dz}\right)^2 \frac{1}{w} = \frac{(1 + z)^2}{z}.$$

Taking square roots of both sides and integrating, we obtain

$$\frac{1}{w^{1/2}} = \pm\frac{1 - z}{z^{1/2}} + C,$$

where C is a constant of integration. Thus

$$w = \frac{z}{(1 + Cz^{1/2} - z)^2}.$$

This function will be regular in a neighborhood of $z = 0$ only if $C = 0$, and so

$$w = \frac{z}{(1 - z)^2} = z + 2z^2 + 3z^3 + \cdots.$$

Thus suppose that $A_2 = B_2 \neq 0$. Making a rotation, we may then suppose that $A_2 = B_2 = 1$. Since $Q(z)$ has at least one double zero on $|z| = 1$, we have

$$Q(z) = \frac{1}{z^2} \cdot (z - e^{i\theta})^2 (z - \rho e^{-i\varphi}) \left(z - \frac{1}{\rho} e^{-i\varphi}\right),$$

where $0 < \rho \leqq 1$, and, since the constant term is unity, $e^{2i(\theta-\varphi)} = 1$; so $e^{i(\theta-\varphi)} = \delta$, where $\delta = \pm 1$, and $\theta = \varphi + m\pi$. Then, writing $p = (\rho^2 + 1)/(2\rho)$, we have

$$B_0 = 2(2p\delta + \cos 2\varphi), \qquad B_1 = -2(pe^{i\varphi} + \delta e^{-i\varphi}).$$

Since $B_0 > 0$, we must take $\delta = 1$ and then

$$Q(z) = \frac{1}{z^2} (z - e^{i\varphi})^2 (z - \rho e^{-i\varphi}) \left(z - \frac{1}{\rho} e^{-i\varphi}\right),$$

where

$$B_0 = 2(2p + \cos 2\varphi), \qquad B_1 = -2(pe^{i\varphi} + e^{-i\varphi}). \tag{13.2.3}$$

By equation (8.3.4) of **8.3** (with $n = 3$) we have

$$B_0 = a_2 F_2 + 2a_3 F_3. \tag{13.2.4}$$

Solving equations (8.3.3) for F_2 and F_3, we obtain

$$F_3 = B_2 = 1, \qquad F_2 = B_1 - 2a_2 B_2 = B_1 - 2a_2. \tag{13.2.5}$$

On substituting from (13.2.5) into (13.2.4):

$$B_0 = B_1 a_2 + 2(a_3 - a_2^2)$$

and so by (13.2.3)

$$a_3 = a_2^2 + \left\{\frac{(1 + \rho)^2}{2\rho} \cos \varphi + i \frac{(1 - \rho)^2}{2\rho} \sin \varphi\right\} a_2 + \rho + \frac{1}{\rho} + \cos 2\varphi. \tag{13.2.6}$$

Formula (13.2.6) establishes a relation between a_2 and a_3 which will be used later.

Replacing in the differential equation z by $-z$, w by $-w$, we see that we can replace B_1 by $-B_1$. Similarly, replacing B_1 by $\bar{B}_1$ and the coefficients of $w = f(z)$ by their conjugates, we may suppose that B_1 lies in any one of the four quadrants of the plane. Let us take B_1 in the third quadrant, in which case

$$Q(z) = \frac{1}{z^2} (z - e^{i\varphi})^2 (z - \rho e^{-i\varphi}) \left(z - \frac{1}{\rho} e^{-i\varphi}\right),$$

$$(13.2.7)\qquad \begin{aligned} B_0 &= 2\left(\rho + \frac{1}{\rho} + \cos 2\varphi\right), \\ B_1 &= -\left\{\frac{(1+\rho)^2}{\rho}\cos\varphi + i\,\frac{(1-\rho)^2}{\rho}\sin\varphi\right\} \end{aligned}$$

where $0 < \rho \leqq 1$, $0 \leqq \varphi \leqq \pi/2$.

Writing

$$W = \int (P(w))^{1/2}\,\frac{dw}{w}, \qquad Z = \int (Q(z))^{1/2}\,\frac{dz}{z},$$

let us integrate formally.

Suppose first that $B_1 = 0$. It is then clear that $B_0 = 2$ and we have (omitting constants of integration)

$$W = \int \frac{dw}{w^2} = -\frac{1}{w}, \qquad Z = \int \frac{z^2+1}{z^2}\,dz = z - \frac{1}{z}.$$

Since $w = z + \cdots$ near $z = 0$, we have

$$-\frac{1}{w} = -\frac{1}{z} + z - C.$$

On the boundary in the w-plane

$$\operatorname{Re} \int (P(w))^{1/2}\,\frac{dw}{w} = \operatorname{Re}\left(\frac{-1}{w}\right) = \text{constant},$$

and so the slit is defined by $\operatorname{Re}(1/w) = 0$. Hence C is imaginary, $C = ic$, and we obtain

$$w = \frac{z}{1 + icz - z^2} = z - icz^2 + (1 - c^2)z^3 + \cdots.$$

Here it is necessary that $|c| \leqq 2$ in order that $w = f(z)$ should be regular in $|z| < 1$. Thus

$$(13.2.8)\qquad \begin{aligned} a_2 &= -ic, \\ a_3 &= 1 - c^2, \end{aligned} \qquad -2 \leqq c \leqq 2.$$

Assume now that $B_1 \neq 0$, and write

$$(13.2.9)\qquad v = 1 + B_1 w, \qquad \eta = \frac{1 - v^{1/2}}{1 + v^{1/2}}, \qquad v^{1/2} = \frac{1-\eta}{1+\eta}.$$

If $0 < \rho < 1$, we also write

$$(13.2.10)\qquad \zeta = \frac{z - \rho e^{-i\varphi}}{z - (1/\rho)e^{-i\varphi}} = -\rho e^{i\varphi}\,\frac{z - \rho e^{-i\varphi}}{1 - \rho e^{i\varphi} z}, \qquad z = \frac{e^{-i\varphi}}{\rho}\,\frac{\zeta - \rho^2}{\zeta - 1}.$$

Formal integration gives (omitting constants of integration)

$$W = \frac{1}{2} B_1 \left\{ \frac{2v^{1/2}}{1 - v} + \log \frac{1 - v^{1/2}}{1 + v^{1/2}} \right\} = \frac{1}{2} B_1 \left\{ \frac{1}{2}\left(\frac{1}{\eta} - \eta\right) + \log \eta \right\},$$

$$Z = z - \frac{1}{z} - 2 (\cos \varphi) \log z, \qquad \rho = 1,$$

(13.2.11)

$$Z = \frac{1}{2} B_1 \log \frac{\rho - \zeta^{1/2}}{\rho + \zeta^{1/2}} - \frac{1}{2} \bar{B}_1 \log \frac{1 - \zeta^{1/2}}{1 + \zeta^{1/2}}$$

$$+ \frac{1 - \rho^2}{\rho} e^{-i\varphi} \frac{\zeta^{1/2}}{\zeta - 1} + (1 - \rho^2) e^{i\varphi} \frac{\zeta^{1/2}}{\zeta - \rho^2}, \qquad 0 < \rho < 1.$$

In the integration of Z when $\rho < 1$, it is convenient to make the substitution

$$z = \frac{e^{-i\varphi}}{2\rho} \{(1 + \rho^2) - (1 - \rho^2)x\},$$

$$z - \rho e^{-i\varphi} = -\frac{e^{-i\varphi}}{2\rho}(1 - \rho^2)(x - 1), \qquad z - \frac{1}{\rho} e^{-i\varphi} = -\frac{e^{-i\varphi}}{2\rho}(1 - \rho^2)(x + 1).$$

If $0 < \rho < 1$, $Q(z)$ has a zero in $|z| < 1$ at $z = \rho e^{-i\varphi}$ and it follows that $P(w)$

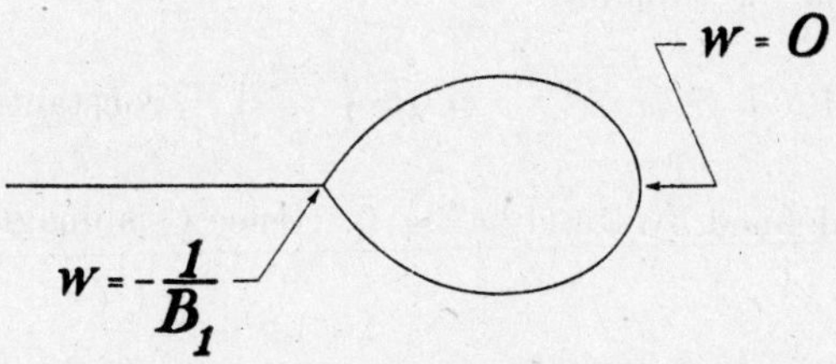

FIG. 9.

must vanish at the point $w = f(\rho e^{-i\varphi})$. For otherwise $f(z)$ would not be regular at $z = \rho e^{-i\varphi}$. Hence when $0 < \rho < 1$, the zero of $Q(z)$ is mapped by $w = f(z)$ into the zero of $P(w)$. Since $Q(z)$ has exactly one zero on $|z| = 1$, there can be only one terminal tip in the w-plane (at which dw/dz vanishes) and so the boundary in the w-plane is a single analytic slit without critical points. The function

$$\zeta = -\rho e^{i\varphi} \frac{z - \rho e^{-i\varphi}}{1 - \rho e^{i\varphi} z}$$

maps $|z| < 1$ into $|\zeta| < \rho$ with $z = 0$ going into $\zeta = \rho^2$ and $\zeta = 0$ going into $z = \rho e^{-i\varphi}$. Thus the mapping $\eta = \eta(\zeta)$, where η is defined by (13.2.9), carries $\zeta = 0$ into $\eta = 1$ and $\zeta = \rho^2$ into $\eta = 0$ or $\eta = \infty$.

If $\rho = 1$, then $P(w)$ does not vanish at any point w corresponding to a point z in $|z| < 1$. Since $B_1 \neq 0$, $P(w)$ must then vanish at some finite point of the boundary in the w-plane.

13.3. If $B_1 = A_1 \neq 0$, $P(w)$ has a finite zero from which three loci Re $(W) =$ constant issue. One locus issues from $w = \infty$. The loci issuing from the finite zero of $P(w)$ and from $w = \infty$ form the set $\mathbf{\Gamma}_w$, which has one of two possible configurations. If one of the three loci issuing from the finite zero of $P(w)$ tends to $w = \infty$, then $\mathbf{\Gamma}_w$ is as shown in Figure 9. If all three loci issuing from the zero of $P(w)$ tend to $w = 0$, $\mathbf{\Gamma}_w$ is as shown in Figure 10 and Figure 11. In the first

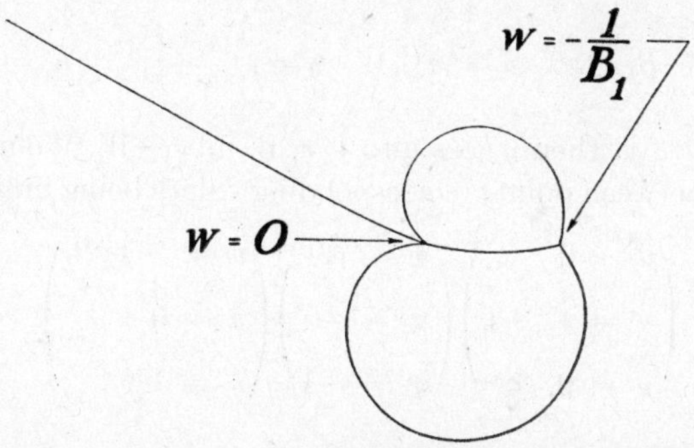

FIG. 10.

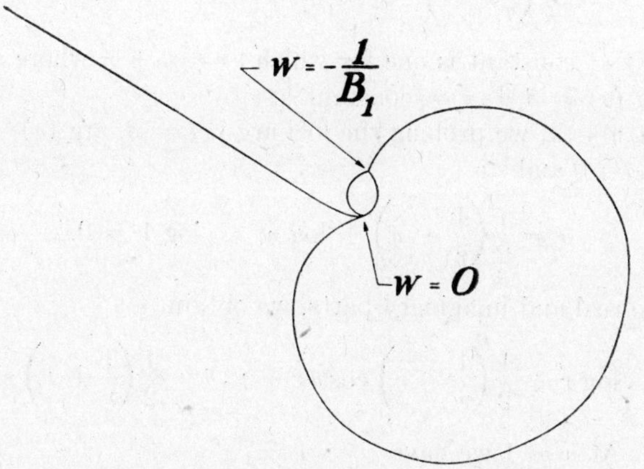

FIG. 11.

case (Figure 9) the complement of $\mathbf{\Gamma}_w$ consists of two end domains and in the second (Figure 10 and Figure 11) the complement consists of two end domains and one strip domain. The strip domain has $w = \infty$ as a boundary point.

By Lemma XXII, $\mathbf{\Gamma}_w = \mathbf{\Gamma}_w(B_1)$ depends continuously on B_1. The subsequent argument will show that if B_1 is real, $\mathbf{\Gamma}_w$ corresponds to Figure 9; otherwise, $\mathbf{\Gamma}_w$ resembles Figure 10. As Im (B_1) tends to zero, one of the two end domains expands as shown in Figure 11 and the boundary of this end domain approaches the locus extending to $w = \infty$. In the limit we obtain the configuration of Figure 9. In Figure 11 it appears as though the smaller end domain were contracting, but this is only because of the choice of scale.

If $B_1 = 0$, $\mathbf{\Gamma}_w$ consists of the imaginary axes plus the point at $w = \infty$. In this case the complement of $\mathbf{\Gamma}_w$ consists of two end domains. We suppose henceforth that $B_1 \neq 0$, and we shall now find the equations defining $\mathbf{\Gamma}_w$.

By (13.2.11) we have

$$W = \frac{1}{2} B_1 \left\{ \frac{1}{2}\left(\frac{1}{\eta} - \eta\right) + \log \eta \right\}$$

where

$$(13.3.1) \quad v = 1 + B_1 w, \qquad u = v^{1/2}, \qquad \eta = \frac{1-u}{1+u}, \qquad u = \frac{1-\eta}{1+\eta}.$$

If we replace u by $-u$, then η goes into $1/\eta$, W into $-W$. We note the following correspondences between points (corresponding values being bracketed together):

$$\begin{pmatrix} w = 0 \\ u = 1, -1 \\ \eta = 0, \infty \end{pmatrix} \begin{pmatrix} w = \infty \\ u = \infty \\ \eta = -1 \end{pmatrix} \begin{pmatrix} w = -1/B_1 \\ u = 0 \\ \eta = 1 \end{pmatrix}.$$

Let

$$s = \frac{1}{2}\left(\frac{1}{\eta} - \eta\right) + \log \eta, \qquad B_1 = i \mid B_1 \mid e^{-i\beta}.$$

A locus Re (W) = constant is one for which $s = s_0 + \sigma$ where arg $(\sigma) = \beta =$ constant or arg $(\sigma) = \beta + \pi =$ constant.

Beginning at $\eta = 1$, we prolong the loci arg $(\sigma) = \beta$, arg $(\sigma) = \beta + \pi$. Here we may take $s_0 = 0$ and

$$\sigma = \frac{1}{2}\left(\frac{1}{\eta} - \eta\right) + \log \eta, \qquad \log 1 = 0.$$

Separating into real and imaginary parts, we obtain

$$\sigma = \left\{ \log r + \frac{1}{2}\left(\frac{1}{r} - r\right) \cos \theta \right\} + i \left\{ \theta - \frac{1}{2}\left(\frac{1}{r} + r\right) \sin \theta \right\}$$

where $\eta = re^{i\theta}$. At $\eta = 1$ we have

$$\frac{d\sigma}{d\eta} = -\frac{1}{2}\left(\frac{1}{\eta^2} + 1\right) + \frac{1}{\eta} = 0,$$

$$\frac{d^2\sigma}{d\eta^2} = \frac{1}{\eta^3} - \frac{1}{\eta^2} = 0,$$

$$\frac{d^3\sigma}{d\eta^3} = -\frac{3}{\eta^4} + \frac{2}{\eta^3} = -1.$$

Hence

$$\sigma = \frac{-1}{6}(\eta - 1)^3\{1 + b_1(\eta - 1) + \cdots\}$$

near $\eta = 1$. Writing $\sigma = t^3$, we can invert, obtaining

$$\eta - 1 = (-6)^{1/3} \cdot t \{1 + c_1 t + \cdots\}.$$

Thus for arg $(\sigma) = \beta, \beta + \pi$, there are six possible directions for the loci at $\eta = 1$, and two adjacent directions make angle $2\pi/6$ with each other.

The loci issuing from $\eta = 1$ may be calculated by setting

$$\frac{\theta - (1/2)(1/r + r) \sin \theta}{\log r + (1/2)(1/r - r) \cos \theta} = \tan \beta, \quad \beta \neq \pm\frac{\pi}{2},$$

$$\log r + \frac{1}{2}\left(\frac{1}{r} - r\right) \cos \theta = 0, \qquad \beta = \pm\frac{\pi}{2}.$$

If $\beta \neq \pm \pi/2$, the loci cannot again intersect the unit circumference, for the only possible point of intersection is $\theta = 0, r = 1$; and the remark at the end of **4.1** shows that any Jordan curve Re (W) = constant must enclose the origin, $\eta = 0$. In this case three of the arcs issuing from $\eta = 1$ remain in $|\eta| > 1$ and tend to $\eta = \infty$ while the other three remain in $|\eta| < 1$ and tend to $\eta = 0$. The arcs interior to $|\eta| = 1$ are obtained from those exterior by the transformation $1/\eta$. Since points η and $1/\eta$ correspond to the same point in the w-plane, it is sufficient to determine the three arcs in $|\eta| < 1$ (or in $|\eta| > 1$).

If $\beta = \pm\pi/2$, two of the six arcs issuing from $\eta = 1$ combine to form the unit circumference $|\eta| = 1$. In this case $\eta = 1$ and $\eta = -1$ are connected by a locus not passing through $\eta = 0$ or $\eta = \infty$ and it follows that $w = -1/B_1$ and $w = \infty$ are connected by an arc of $\mathbf{\Gamma}_w$ not containing $w = 0$.

If $\beta \neq \pi/2$, there is a locus passing through $\eta = -1$ which meets the loci issuing from $\eta = 1$ only at $\eta = 0$ or at $\eta = \infty$. Defining $\log(-1) = i\pi$, we have

$$\sigma = \frac{1}{2}\left(\frac{1}{\eta} - \eta\right) + \log \eta - i\pi.$$

Writing $\eta = -\tau$, we obtain

$$\sigma = \frac{1}{2}\left(\tau - \frac{1}{\tau}\right) + \log \tau$$

where

$$\frac{d\sigma}{d\tau} = \frac{1}{2}\left(1 + \frac{1}{\tau^2}\right) + \frac{1}{\tau} = 2$$

at $\tau = 1$. Hence two loci issue from $\tau = 1$; these loci make angle π at $\tau = 1$ and form a single locus passing through $\tau = 1$. Let $\tau = re^{i\theta}$ and we have

$$\sigma = \left\{\log r - \frac{1}{2}\left(\frac{1}{r} - r\right) \cos \theta\right\} + i\left\{\theta + \frac{1}{2}\left(\frac{1}{r} + r\right) \sin \theta\right\}.$$

Since $\beta \neq \pm\pi/2$, the locus through $\tau = 1$ is computed by setting

$$\frac{\theta + (1/2)(1/r + r)\sin\theta}{\log r - (1/2)(1/r - r)\cos\theta} = \tan\beta.$$

Figures 9, 10, and 11 were made by computing points on the curves in the η-plane. The above formulas were used and afterwards the curves were transformed into the w-plane.

13.4. We now consider the two cases $\rho = 1$ and $0 < \rho < 1$ in detail and we begin with the case $\rho = 1$. The integrated equation then is

$$\pm\frac{1}{2} B_1 \left\{\frac{1}{2}\left(\frac{1}{\eta} - \eta\right) + \log\eta\right\} = z - \frac{1}{z} - 2(\cos\varphi)\cdot\log z + 2iC$$

where C is a constant of integration. We observe that either branch of $u = v^{1/2} = (1 + B_1 w)^{1/2} = [1 + B_1 f(z)]^{1/2}$ is single-valued for $|z| < 1$ since $w = -1/B_1$ is a boundary point. Since the two points η, $1/\eta$ correspond to the same point w, it is immaterial which branch we take. Let us take the branch for which $u = v^{1/2} = 1$ when $w = 0$. Since $w = z + \cdots$ near $z = 0$, we have

$$u = v^{1/2} = 1 + \frac{1}{2}B_1 z + \cdots,$$

$$\eta = -\frac{1}{4}B_1 z + \cdots,$$

$$\frac{1}{\eta} = -\frac{4}{B_1}\frac{1}{z} + \cdots.$$

The function $\eta(z)$ is regular and schlicht in $|z| < 1$ and maps $|z| < 1$ onto $|\eta| < 1$ cut along portions of two Jordan arcs issuing from $\eta = 1$. The indeterminancy in the integrated equation may thus be eliminated and we have

$$\frac{1}{2} B_1 \left\{\frac{1}{2}\left(\frac{1}{\eta} - \eta\right) + \log\eta\right\} = z - \frac{1}{z} - 2(\cos\varphi)\log z + 2iC. \tag{13.4.1}$$

By (13.2.7) with $\rho = 1$,

$$B_1 = -4\cos\varphi$$

and equation (13.4.1) becomes

$$\cos\varphi\left\{\left(\eta - \frac{1}{\eta}\right) - 2\log\eta\right\} = z - \frac{1}{z} - 2(\cos\varphi)\log z + 2iC \tag{13.4.2}$$

where we recall from **13.2** that $0 \leqq \varphi \leqq \pi/2$. Some point on $|z| = 1$ corresponds to the point $\eta = 1$ and this shows that C is real. By suitable choice of C we may suppose that $\log\eta = 0$ when $\eta = 1$, $\log z = 0$ when $z = 1$.

On taking $z = e^{i\theta}$ the equation (13.4.2) is

$$\cos \varphi \left\{ \eta - \frac{1}{\eta} - 2 \log \eta \right\} = 2i \{ \sin \theta - \theta \cos \varphi + C \}.$$

Write

$$g(\theta) = \sin \theta - \theta \cos \varphi, \qquad 0 \leqq \varphi < \pi/2,$$

$$g'(\theta) = \cos \theta - \cos \varphi.$$

The graph of $g(\theta)$ is shown in Figure 12.

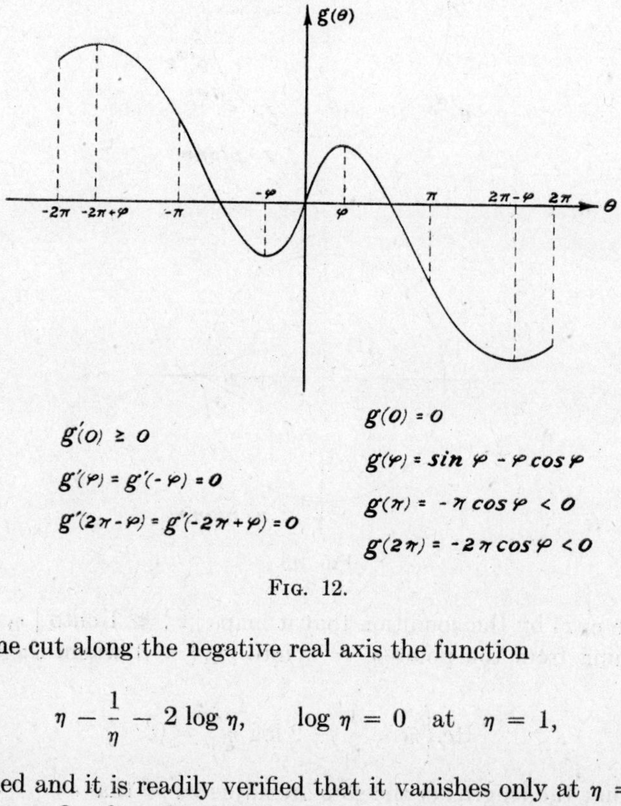

FIG. 12.

In the plane cut along the negative real axis the function

$$\eta - \frac{1}{\eta} - 2 \log \eta, \qquad \log \eta = 0 \quad \text{at} \quad \eta = 1,$$

is single-valued and it is readily verified that it vanishes only at $\eta = 1$ where it has a triple zero. On the other hand, in the z-plane cut along the negative real axis the function

$$z - \frac{1}{z} - 2(\cos \varphi) \log z, \qquad \log z = 0 \quad \text{at} \quad z = 1, \qquad 0 < \varphi < \pi/2,$$

has three simple zeros on $|z| = 1$ and no other zeros there; its derivative has simple zeros at $z = e^{i\varphi}$ and $z = e^{-i\varphi}$.

Differentiating both sides of the equation (13.4.2), we obtain

$$\cos \varphi \left(1 - \frac{1}{\eta}\right)^2 \frac{d\eta}{dz} = \frac{1}{z^2} (z - e^{i\varphi})(z - e^{-i\varphi}).$$

Thus $d\eta/dz$ can vanish only at the two points $z = e^{i\varphi}$, $z = e^{-i\varphi}$.

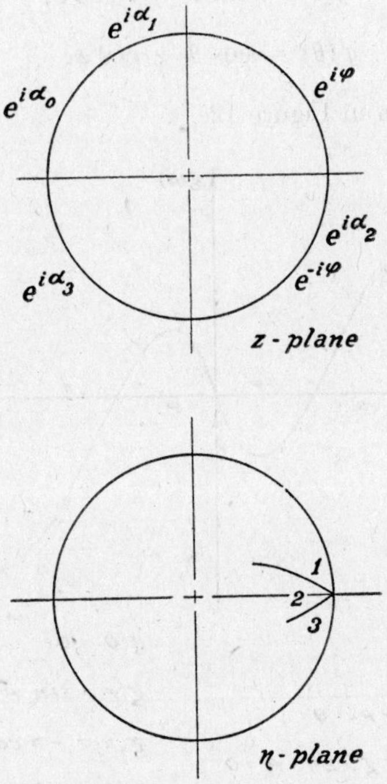

FIG. 13.

Now define $\eta_1(z)$ by the condition that it maps $|z| < 1$ onto $|\eta_1| < 1$ minus two slits issuing from the point $\eta_1 = 1$ into $|\eta_1| < 1$ where the slits are loci satisfying

$$\text{Re}\left\{\eta_1 - \frac{1}{\eta_1} - 2 \log \eta_1\right\} = 0.$$

We assume that the lengths of the slits are adjusted so that

(13.4.3) $$\eta_1 = (\cos \varphi)\, z + \cdots .$$

By lengthening one slit and shortening the other we obtain a one-parameter family of functions of the form (13.4.3) near $z = 0$. For any one of these functions the expression

$$\cos \varphi \left\{\eta_1 - \frac{1}{\eta_1} - 2 \log \eta_1\right\} - z + \frac{1}{z} + 2(\cos \varphi) \log z$$

has a vanishing real part everywhere on $|z| = 1$ and it is regular and single-valued in $|z| < 1$. Hence it is identically equal to some imaginary constant, that is,

$$\cos\varphi\left\{\eta_1 - \frac{1}{\eta_1} - 2\log\eta_1\right\} = z - \frac{1}{z} - 2(\cos\varphi)\log z + 2iC$$

where C is real. Thus $\eta_1(z)$ is a function $\eta(z)$ of the form discussed above.

When $\varphi = 0$, there is only one function of the form (13.4.3), namely $\eta = z$, and in this case there can be no slits. If $0 < \varphi < \pi/2$, the derivative $d\eta/dz$ must vanish at the two points on $|z| = 1$ which correspond to the tips of the slits and so these points must be $e^{i\varphi}$, $e^{-i\varphi}$. Here we assume that both slits are present. Let $e^{i\alpha_1}$, $e^{i\alpha_2}$, $e^{i\alpha_3}$ be the three points on $|z| = 1$ which map into $\eta = 1$ (see Figure 13) and let $e^{i\alpha_0}$ map into $\eta = -1$. At these three points the expression

$$z - \frac{1}{z} + 2(\cos\varphi)\log z + 2iC$$

must vanish since the left side of (13.4.2) does, and this is only possible if

$$-\sin\varphi + \varphi\cos\varphi < C < \sin\varphi - \varphi\cos\varphi.$$

If only one of the two slits is present, one of the points, say $e^{i\varphi}$, corresponds to the tip while the other, $e^{-i\varphi}$, corresponds to the larger of the two angles formed at $\eta = 1$. The function

$$z - \frac{1}{z} - 2(\cos\varphi)\log z + 2iC$$

can vanish at only two distinct points of $|z| = 1$ and so

$$C = \pm(\sin\varphi - \varphi\cos\varphi).$$

The domain $|\eta| < 1$ minus the two slits may be mapped into the w-plane by (13.3.1), the point $\eta = 1$ going into $w = -1/B_1 = 1/(4\cos\varphi)$. The circumference $|\eta| = 1$ is mapped into the segment of the positive real axis extending from $w = 1/(4\cos\varphi)$ to $w = \infty$, and the two slits at $\eta = 1$ become slits at $w = 1/(4\cos\varphi)$ making angles $2\pi/3$ with the segment of the real axis from $1/(4\cos\varphi)$ to infinity. Thus $w = f(z)$ maps $|z| < 1$ onto the w-plane minus a forked slit.

When $\varphi = 0$, $\eta = z$ and so

$$w = \frac{z}{(1+z)^2}.$$

In this case the function $w = f(z)$ maps $|z| < 1$ onto the w-plane minus the segment of the positive real axis extending from $w = 1/4$ to $w = \infty$.

As φ tends to $\pi/2$, the point $w = 1/(4 \cos \varphi)$ moves out along the positive axis to infinity and

$$\cos \varphi \left\{ \eta - \frac{1}{\eta} - 2 \log \eta \right\} \to -\frac{1}{w},$$

$$z - \frac{1}{z} - 2 \cos \varphi \log z + 2iC \to z - \frac{1}{z} + 2iC,$$

$$\sin \varphi - \varphi \cos \varphi \to 1.$$

Hence

$$w = f(z) \to \frac{z}{1 - 2iCz - z^2}, \qquad -1 \leqq C \leqq 1,$$

the function obtained in **13.2** by direct integration when $B_1 = 0$.

We have

$$\eta = \cos \varphi \, \{z + (2 \cos \varphi + a_2)z^2 + \cdots\},$$

$$\frac{1}{\eta} = \frac{1}{(\cos \varphi) \, z} \{1 - (2 \cos \varphi + a_2)z + \cdots\},$$

$$\log \eta = \log z + \log (\cos \varphi) + (2 \cos \varphi + a_2)z \pm \cdots .$$

Thus, as z tends to zero,

$$\cos \varphi \left\{ \eta - \frac{1}{\eta} - 2 \log \eta \right\} = -\frac{1}{z} - 2 (\cos \varphi) \log z + 2 \cos \varphi + a_2 - 2(\cos \varphi) \log(\cos \varphi) + o(1),$$

and so

$$a_2 = 2(\cos \varphi) \log (\cos \varphi) - 2 \cos \varphi + 2iC.$$

By (13.2.6) (with $\rho = 1$) we obtain

$$a_3 = a_2^2 + 2(\cos \varphi)a_2 + 2 \cos^2 \varphi + 1.$$

Therefore the parametric equations of this part of the boundary are

$$\text{(13.4.4)} \qquad \begin{aligned} a_2 &= 2(\cos \varphi) \log (\cos \varphi) - 2 \cos \varphi + 2iC, \\ a_3 &= a_2^2 + 2(\cos \varphi) \, a_2 + 2 \cos^2 \varphi + 1, \end{aligned}$$

where

$$- \sin \varphi + \varphi \cos \varphi \leqq C \leqq \sin \varphi - \varphi \cos \varphi, \qquad 0 \leqq \varphi \leqq \pi/2.$$

We observe that the parametric equations do not involve $B_1 = -4 \cos \varphi$ algebraically. This is owing to the fact that $n = 3$ is an odd number and so by Lemma XXIV we cannot expect to define a_2, a_3 algebraically in terms of the coefficients of $P(w)$ and $Q(z)$.

The equations (13.1.2) are obtained from (13.4.4) by writing $\mu = 2C$ and performing a rotation to make $a_2 \geqq 0$.

13.5. Suppose finally that $0 < \rho < 1$. We have

$$W = \frac{1}{2} B_1 \left\{ \frac{1}{2}\left(\frac{1}{\eta} - \eta\right) + \log \eta \right\},$$

$$(13.5.1) \quad Z = \frac{1}{2} B_1 \log \frac{\rho - \zeta^{1/2}}{\rho + \zeta^{1/2}} - \frac{1}{2} \bar{B}_1 \log \frac{1 - \zeta^{1/2}}{1 + \zeta^{1/2}} + \frac{1 - \rho^2}{\rho} e^{-i\varphi} \frac{\zeta^{1/2}}{\zeta - 1} + (1 - \rho^2) e^{i\varphi} \frac{\zeta^{1/2}}{\zeta - \rho^2},$$

where w and η, z and ζ are connected by formulas (13.2.9) and (13.2.10). The point $\zeta = 0$ corresponds to $\eta = 1$.

As z tends to zero,

$$\pm W = -\frac{1}{z} + \frac{1}{2} B_1 \log\left(-\frac{B_1}{4} z\right) + a_2 - \frac{1}{2} B_1 + o(1),$$

$$(13.5.2) \quad \pm Z = -\frac{1}{z} + \frac{1}{2} B_1 \log \frac{(1 - \rho^2) e^{i\varphi} z}{4\rho} - \frac{1}{2} \bar{B}_1 \log \frac{1 - \rho}{1 + \rho} - e^{-i\varphi} + \frac{1 + \rho^2}{2\rho} e^{i\varphi} + o(1).$$

Writing

$$(13.5.3) \quad 1 + 2\rho e^{-2i\varphi} + \rho^2 = r e^{-i\alpha}, \qquad 0 \leqq \alpha < \pi/2,$$

we have from (13.2.7)

$$-\frac{B_1 \rho}{(1 - \rho^2) e^{i\varphi}} = \frac{1 + 2\rho e^{-2i\varphi} + \rho^2}{1 - \rho^2} = \frac{r e^{-i\alpha}}{1 - \rho^2}.$$

Starting at a point on the positive real axis in the ζ-plane with $\zeta^{1/2} > 0$ and with each of the logarithms in (13.5.1) having their principal values, let ζ approach ρ^2 through real values. We obtain from (13.5.2)

$$a_2 = \frac{(1 + \rho)^2}{2\rho} (\cos \varphi) \log \frac{r}{(1 + \rho)^2} + \frac{(1 - \rho)^2}{2\rho} \alpha \sin \varphi - 2 \cos \varphi$$

$$(13.5.4) \quad + i \left\{ \frac{(1 - \rho)^2}{2\rho} (\sin \varphi) \log \frac{r}{(1 - \rho)^2} - \frac{(1 + \rho)^2}{2\rho} \alpha \cos \varphi + 2 \sin \varphi \right\} + k\pi i B_1$$

where the logarithms have their principal values and k is an integer.

We show now that $k = 0$ and we go back to the differential equation

$$(13.5.5) \quad \left(\frac{z}{w} \frac{dw}{dz}\right)^2 P(w) = Q(z)$$

where

$$(13.5.6) \qquad P(w) = \frac{1 + B_1 w}{w^2}, \qquad Q(z) = \frac{1}{z^2}\{z^4 + \bar{B}_1 z^3 + B_0 z^2 + B_1 z + 1\},$$

and

$$(13.5.7) \qquad \begin{aligned} B_0 &= 2\left(\rho + \frac{1}{\rho} + \cos 2\varphi\right), \\ B_1 &= -\frac{(1+\rho)^2}{\rho}\cos\varphi - i\frac{(1-\rho)^2}{\rho}\sin\varphi. \end{aligned}$$

Given any ρ, φ, $0 < \rho < 1$, $0 \leqq \varphi \leqq \pi/2$, let B_1 be defined by (13.5.7). Then

$$P(w) = P_{B_1}(w) = \frac{1 + B_1 w}{w^2}$$

is defined and therefore also the set of loci $\mathbf{\Gamma}_w(B_1)$. Since $B_1 \neq 0$, there is exactly one locus

$$\operatorname{Re} \int (P(w))^{1/2} \frac{dw}{w}$$

issuing from $w = \infty$. Let

$$w = f(z) = z + a_2 z^2 + a_3 z^3 + \cdots$$

be any function mapping $|z| < 1$ onto the w-plane minus a subcontinuum $\mathbf{C}_w$ of $\mathbf{\Gamma}_w$ which contains $w = \infty$ and has mapping radius unity. If $1 + B_1 w = 0$ anywhere on $\mathbf{C}_w$, we have seen in **13.3** that B_1 is real. Then by (13.5.7)

$$B_1 = -\frac{(1+\rho)^2}{\rho}, \qquad |B_1| = \frac{(1+\rho)^2}{\rho} > 4,$$

and so $P(w) = 0$ at a point of $\mathbf{C}_w$ lying in $|w| < 1/4$. But since $|f(z)| \geqq 1/4$ on $|z| = 1$ (as is well known), $\mathbf{C}_w$ lies in $|w| \geqq 1/4$.

It follows that for $0 < \rho < 1$ the function $w = f(z)$ is uniquely defined and it maps $|z| < 1$ onto the w-plane minus a single analytic slit without critical points (that is to say, the slit is unforked). From the remarks made in **8.2** we see that $w = f(z)$ satisfies the differential equation (13.5.5) where $Q(z)$ as well as $P(w)$ is given by (13.5.6). By Lemma XXII, $N_w(B_1)$ depends on B_1 and therefore on ρ, φ in a continuous way, and so $w = f(z)$ depends on ρ, φ continuously. In particular, the coefficient a_2 of $f(z)$ varies continuously with ρ and φ. Taking $\varphi = 0$ in (13.5.4), we obtain

$$a_2 = -2 + k\pi i B_1 = -2 - \frac{(1+\rho)^2}{\rho} k\pi i.$$

Since $|a_2| \leqq 2$, this is impossible unless $k = 0$. Thus a_2 is given by (13.5.4) with $k = 0$ and we obtain the formula for a_3 from (13.2.6). The parametric formulas for this part of the boundary therefore are

$$\begin{aligned} a_2 = {} & \frac{(1+\rho)^2}{2\rho} (\cos\varphi) \log \frac{r}{(1+\rho)^2} + \frac{(1-\rho)^2}{2\rho} \alpha \sin\varphi - 2\cos\varphi \\ & + i\left\{ \frac{(1-\rho)^2}{2\rho} (\sin\varphi) \log \frac{r}{(1-\rho)^2} - \frac{(1+\rho)^2}{2\rho} \alpha \cos\varphi + 2\sin\varphi \right\}, \\ a_3 = {} & a_2^2 + \left\{ \frac{(1+\rho)^2}{2\rho} \cos\varphi + i \frac{(1-\rho)^2}{2\rho} \sin\varphi \right\} a_2 + \rho + \frac{1}{\rho} + \cos 2\varphi. \end{aligned} \tag{13.5.8}$$

The equations (13.1.1) are obtained from (13.5.8) by performing a rotation to make $a_2 \geqq 0$. The numbers r and α are defined by (13.5.3).

CHAPTER XIV

A METHOD FOR INVESTIGATING THE CONJECTURE $|a_4| \leqq 4$

14.1. Some interest has been focused on the family $\mathfrak{S}$ of schlicht functions because of the conjecture that

$$(14.1.1) \qquad |a_n| \leqq n$$

with equality for any n, $n \geqq 2$, only if

$$(14.1.2) \qquad w = f(z) = \frac{z}{(1 - e^{i\theta}z)^2} = z + 2e^{i\theta}z^2 + 3e^{2i\theta}z^3 + 4e^{3i\theta}z^4 + \cdots$$

where θ is real. In terms of the coefficient region V_n, the inequality (14.1.1) is a precise statement concerning the maximum distance of the boundary of V_n from the hyperplane $a_n = 0$.

In this chapter we derive a method for proving or disproving the conjecture in the case $n = 4$. The method might have to be modified somewhat if $|a_4| \leqq 4$ with equality occurring for some function which is different from the functions (14.1.2), but the probability of such an occurrence seems very small.

Considerable numerical work is involved in the method. Computations begun in the winter of 1946–1947 are being carried out as part of a project sponsored by the Office of Naval Research but the work is not yet complete and final results cannot be announced.

Although we have chosen this problem as one specific example of the methods developed in the preceding chapters, it should be emphasized that certain of the techniques employed are applicable (at least in principle) to a wider class of problems.

Points on the boundary of V_4 can be calculated by integrating the system of differential equations discussed in Chapter XI. However, these calculations cannot yield the inequality $|a_4| \leqq 4$ unless we know that the inequality $|a_4| \leqq 4$ is true for all points of the boundary of V_4 lying in a neighborhood of the curve $(2e^{i\theta}, 3e^{2i\theta}, 4e^{3i\theta})$, θ real. Also it is obviously desirable to exclude from consideration as much of the boundary of V_4 as possible before calculating boundary points from the system of differential equations. Our program thus consists of three parts: (a) proof that points of the boundary corresponding to functions (14.1.2) have the local maximum property with respect to neighboring points; (b) proof that over a large part of the boundary of V_4 an inequality $|a_4| \leqq 4 - \delta$ is valid where δ is some fixed positive number; (c) calculation of points on the remaining portion of the boundary for which $|a_4| > 4 - \delta$ by integrating a system of differential equations.

14.2. We begin with some purely formal remarks concerning $\mathfrak{D}_4$-equations. Solving the equations (8.3.2) and (8.3.3) for the F_ν, we obtain

(14.2.1)
$$\begin{aligned} F_2 &= A_1 - 2a_2 A_2 + (5a_2^2 - 2a_3)A_3, \\ F_3 &= A_2 - 3a_2 A_3, \\ F_4 &= A_3 \end{aligned}$$

and

(14.2.2)
$$\begin{aligned} F_2 &= B_1 - 2a_2 B_2 + (4a_2^2 - 3a_3)B_3, \\ F_3 &= B_2 - 2a_2 B_3, \\ F_4 &= B_3. \end{aligned}$$

Solving equations (8.3.3) for a_2 and a_3, we obtain

(14.2.3)
$$\begin{aligned} 2a_2 F_4 &= B_2 - F_3, \\ 3a_3 F_4^2 &= B_1 F_4 - F_2 F_4 - B_2 F_3 + F_3^2. \end{aligned}$$

Finally, equation (8.3.4) is

(14.2.4)
$$B_0 = a_2 F_2 + 2a_3 F_3 + 3a_4 F_4.$$

Assume that $F_4 = A_3 = B_3 \neq 0$. We have

(14.2.5)
$$\begin{aligned} B_3 a_2 &= A_2 - B_2, \\ B_3^2 a_3 &= (A_2 - B_2)^2 + B_3(B_1 - A_1), \\ 3B_3^3 a_4 &= B_0 B_3^2 - B_1 B_3(A_2 - B_2) \\ &\qquad + 3(A_2 - B_2)^3 - B_3(7A_2 - 9B_2)(A_1 - B_1). \end{aligned}$$

The first two equations are obtained by equating the expressions for F_2 and F_3 in formulas (14.2.1) and (14.2.2). The third equation is obtained by solving (14.2.4) for a_4 and expressing a_2, a_3, F_2, F_3 in terms of the A_ν, B_ν. Equations (14.2.5) express a_2, a_3, a_4 algebraically in terms of the numbers A_ν, B_ν; this is possible because $n = 4$ is even.

When $F_2 = F_3 = 0$, $F_4 = 1$, we obtain the $\mathfrak{D}_4$-equation which is satisfied by functions

$$w = f(z) = z + a_2 z^2 + a_3 z^3 + a_4 z^4 + \cdots$$

of class $\mathfrak{S}$ which maximize Re (a_4). In this case the formulas (8.3.2), (8.3.3), and (8.3.4) give

(14.2.6)
$$\begin{aligned} &A_1 = a_4^{(2)} = a_2^2 + 2a_3, \quad A_2 = a_4^{(3)} = 3a_2, \quad A_3 = 1, \\ &B_0 = 3a_4, \quad B_1 = 3a_3, \quad B_2 = 2a_2, \quad B_3 = 1. \end{aligned}$$

Thus a function $w = f(z)$ which maximizes Re (a_4) satisfies an equation

$$\left(\frac{z}{w}\frac{dw}{dz}\right)^2 P(w) = Q(z) \tag{14.2.7}$$

where

$$P(w) = \sum_{\nu=2}^{4} \frac{a_4^{(\nu)}}{w^{\nu-1}} = \frac{a_2^2 + 2a_3}{w} + \frac{3a_2}{w^2} + \frac{1}{w^3}, \tag{14.2.8}$$

$$Q(z) = 3a_4 + \frac{1}{z^3} + \frac{2a_2}{z^2} + \frac{3a_3}{z} + 3\bar{a}_3 z + 2\bar{a}_2 z^2 + z^3. \tag{14.2.9}$$

14.3. By Lemmas XXVIII and XXX we know that a function $w = f(z)$ maximizing Re (a_4) either maps $|z| < 1$ onto the w-plane minus a single analytic arc not passing through any zero of $P(w)$ or it is the function

$$w = f(z) = \frac{z}{(1 - e^{i\theta}z)^2}$$

where $\theta \equiv 0, 2\pi/3, 4\pi/3, \pmod{2\pi}$. The authors have shown that $A_1 \neq 0$ (see [22a]), and Golusin [6d] has extended the method to the case $n = 5$.

To prove that $a_4^{(2)} \neq 0$ for functions maximizing Re (a_4) we use the area principle. By an inequality of Prawitz [19], we have

$$\sum_{\nu=1}^{\infty} \frac{2\nu - \alpha}{\alpha} |b_\nu|^2 \leqq 1 \tag{14.3.1}$$

where α is any positive number and

$$\left[\frac{f(z)}{z}\right]^{-\alpha/2} = \sum_{\nu=0}^{\infty} b_\nu z^\nu, \qquad |z| < 1.$$

Expressing the b_ν in terms of the coefficients of $f(z)$, we obtain the inequality

$$\begin{aligned} &\frac{1}{4}\alpha(2 - \alpha)|a_2|^2 + \frac{1}{64}\alpha(4 - \alpha)|(\alpha + 2)a_2^2 - 4a_3|^2 \\ &\qquad + \alpha(6 - \alpha)\left|\frac{\alpha^2 + 6\alpha + 8}{48} a_2^3 - \frac{\alpha + 2}{4} a_2 a_3 + \frac{a_4}{2}\right|^2 \leqq 1. \end{aligned} \tag{14.3.2}$$

Assume now that $a_4^{(2)} = a_2^2 + 2a_3 = 0$. If we make the substitution $a_3 = -a_2^2/2$,

$$\begin{aligned} &\frac{1}{4}\alpha(2 - \alpha)|a_2|^2 + \frac{1}{64}\alpha(4 - \alpha)(4 + \alpha)^2|a_2|^4 \\ &\qquad + \alpha(6 - \alpha)\left|\frac{\alpha^2 + 12\alpha + 20}{48} a_2^3 + \frac{a_4}{2}\right|^2 \leqq 1. \end{aligned}$$

If we take $\alpha = 2$, the second term on the left shows that $|a_2| < 1$. Then if we take $\alpha = 1$, the third term on the left gives

$$\left|\frac{11}{16} a_2^3 + \frac{a_4}{2}\right| \leqq \frac{1}{5^{1/2}} < 1,$$

and so

$$|a_4| < 2\left(1 + \frac{11}{16}\right) < 4.$$

Since we know that max Re $(a_4) \geqq 4$, this shows that any function maximizing Re (a_4) must have $a_4^{(2)} \neq 0$. The function $Q(z)$ can thus vanish at only one point of $|z| = 1$, where it has a zero of precise order 2.

We shall further restrict the portion of the boundary of V_4 which it is necessary to consider in a later part of this chapter. Now we turn to part (a) of our program and show that the functions (14.1.2) have the desired local maximum property. For this we refine the estimates given by Krzywoblocki [11].

14.4. Since we need only consider functions

$$w = f(z) = z + a_2 z^2 + a_3 z^3 + a_4 z^4 + \cdots \tag{14.4.1}$$

which map $|z| < 1$ onto the plane minus a single analytic slit without critical points, the coefficient a_4 may be represented by Löwner's formula, namely

$$\begin{aligned} a_4 = 2 \int_0^1 u^2 e^{3i\theta(u)}\, du - 12 \int_0^1 u e^{2i\theta(u)}\, du \cdot \int_0^1 e^{i\theta(u)}\, du \\ + 4 \int_{v=0}^1 \int_{u=0}^v v e^{i[\theta(u)+2\theta(v)]}\, du\, dv + 8 \left(\int_0^1 e^{i\theta(u)}\, du \right)^3, \end{aligned} \tag{14.4.2}$$

where $\theta(u)$ is a real-valued continuous function of u, $0 \leqq u \leqq 1$. This formula is obtained from (9.2.1) on replacing $\exp(i\alpha)$ by $-\exp(i\theta)$. Taking real and imaginary parts, we obtain

$$\begin{aligned} a_4 = {} & 2 \int_0^1 u^2 \cos 3\theta(u)\, du - 12 \int_0^1 u \cos 2\theta(u)\, du \cdot \int_0^1 \cos \theta(u)\, du \\ & + 12 \int_0^1 u \sin 2\theta(u)\, du \cdot \int_0^1 \sin \theta(u)\, du + 4 \int_{v=0}^1 \int_{u=0}^v v \cos [\theta(u) + 2\theta(v)]\, du\, dv \\ & + 8 \left(\int_0^1 \cos \theta(u)\, du \right)^3 - 24 \int_0^1 \cos \theta(u)\, du \cdot \left(\int_0^1 \sin \theta(u)\, du \right)^2 \\ & + i \Big\{ 2 \int_0^1 u^2 \sin 3\theta(u)\, du - 12 \int_0^1 u \sin 2\theta(u)\, du \cdot \int_0^1 \cos \theta(u)\, du \\ & - 12 \int_0^1 u \cos 2\theta(u) du \cdot \int_0^1 \sin \theta(u)\, du + 4 \int_{v=0}^1 \int_{u=0}^v v \sin [\theta(u) + 2\theta(v)]\, du\, dv \\ & - 8 \left(\int_0^1 \sin \theta(u)\, du \right)^3 + 24 \left(\int_0^1 \cos \theta(u)\, du \right)^2 \cdot \int_0^1 \sin \theta(u)\, du \Big\}. \end{aligned}$$

Let

$$\sin\theta(u) = \frac{2x}{1+x^2}, \qquad \cos\theta(u) = 1 - \frac{2x^2}{1+x^2},$$

$$\sin 2\theta(u) = \frac{4x}{1+x^2} - \frac{8x^3}{(1+x^2)^2},$$

$$\cos 2\theta(u) = 1 - \frac{8x^2}{(1+x^2)^2}, \qquad \sin 3\theta(u) = \frac{6x}{1+x^2} - \frac{32x^3}{(1+x^2)^3},$$

$$\cos 3\theta(u) = 1 - 18\frac{x^2}{1+x^2} + 48\frac{x^4}{(1+x^2)^2} - 32\frac{x^6}{(1+x^2)^3},$$

where $x = x(u)$, $-\infty < x < \infty$, and let Δ denote the triangular region $0 \leqq u \leqq v \leqq 1$. Then

(14.4.3) $$a_4 = 4 - J_2 - J_4 - J_6 + i(I_1 + I_3 + I_5)$$

where (on writing $x = x(u)$, $y = x(v)$)

$$J_2 = 8\int_0^1 (8u^2 - 12u + 5)\frac{x^2}{(1+x^2)^2}\,du$$

(14.4.4) $$+ 32\iint_\Delta v\frac{xy}{(1+x^2)(1+y^2)}\,du\,dv - 24\left(\int_0^1 u\frac{x}{1+x^2}\,du\right)^2$$

$$+ 24\left\{2\int_0^1 \frac{x}{1+x^2}\,du - \int_0^1 u\frac{x}{1+x^2}\,du\right\}^2,$$

$$J_4 = -8\int_0^1 (8u^2 - 5)\frac{x^4}{(1+x^2)^2}\,du + 192\int_0^1 u\frac{x^2}{(1+x^2)^2}\,du\int_0^1 \frac{x^2}{1+x^2}\,du$$

(14.4.5) $$+ 192\int_0^1 u\frac{x^3}{(1+x^2)^2}\,du \cdot \int_0^1 \frac{x}{1+x^2}\,du - 192\int_0^1 \frac{x^2}{1+x^2}\,du \cdot \left(\int_0^1 \frac{x}{1+x^2}\,du\right)^2 - 96\left(\int_0^1 \frac{x^2}{1+x^2}\,du\right)^2$$

$$- 64\iint_\Delta v\frac{x^2}{1+x^2}\cdot\frac{y^2}{(1+y^2)^2}\,du\,dv - 64\iint_\Delta v\frac{xy^3}{(1+x^2)(1+y^2)^2}\,du\,dv,$$

(14.4.6) $$J_6 = 64\left\{\int_0^1 u^2\frac{x^6}{(1+x^2)^3}\,du + \left(\int_0^1 \frac{x^2}{1+x^2}\,du\right)^3\right\};$$

and

(14.4.7) $$I_1 = 8\int_0^1 (3u^2 - 6u + 5)\frac{x}{1+x^2}\,du,$$

$$
\begin{aligned}
I_3 = {} & -96 \int_0^1 u^2 \frac{x^3}{(1+x^2)^3}\, du + 96 \int_0^1 u \frac{x^3}{(1+x^2)^3}\, du \\
& + 96 \int_0^1 u \frac{x}{1+x^2}\, du \cdot \int_0^1 \frac{x^2}{(1+x^2)^2}\, du + 192 \int_0^1 \frac{x}{1+x^2}\, du \\
& \cdot \int_0^1 (u-1) \frac{x^2}{(1+x^2)^2}\, du \\
& - 32 \iint_\Delta v \left\{ \frac{x^2}{(1+x^2)^2} \frac{y}{1+y^2} + 2 \frac{x}{1+x^2} \frac{y^2}{(1+y^2)^2} \right\} du\, dv \\
& - 64 \left(\int_0^1 \frac{x}{1+x^2}\, du \right)^3,
\end{aligned} \tag{14.4.8}
$$

$$
\begin{aligned}
I_5 = {} & -32 \int_0^1 u(u-3) \frac{x^5}{(1+x^2)^3}\, du + 96 \int_0^1 (u-2) \frac{x}{1+x^2}\, du \\
& \cdot \int_0^1 \frac{x^4}{(1+x^2)^2}\, du - 192 \int_0^1 u \frac{x^3}{(1+x^2)^2}\, du \cdot \int_0^1 \frac{x^2}{1+x^2}\, du \\
& + 192 \left(\int_0^1 \frac{x^2}{1+x^2}\, du \right)^2 \cdot \left(\int_0^1 \frac{x}{1+x^2}\, du \right) \\
& + 64 \iint_\Delta v \frac{x^2}{1+x^2} \cdot \frac{y^3}{(1+y^2)^2}\, du\, dv - 32 \iint_\Delta v \frac{x^4}{(1+x^2)^2} \cdot \frac{y}{1+y^2}\, du\, dv.
\end{aligned} \tag{14.4.9}
$$

Now let

$$
X = X(u) = \frac{x}{1+x^2}, \qquad Y = Y(v) = \frac{y}{1+y^2}. \tag{14.4.10}
$$

Then

$$
\begin{aligned}
\frac{1}{8} J_2 = {} & \iint_\Delta \{(8u^2 - 12u + 5)X^2 + 2v(2-3u)XY \\
& + (8v^2 - 12v + 5)Y^2\}\, du\, dv + 3 \left(\int_0^1 (u-2)X\, du \right)^2
\end{aligned} \tag{14.4.11}
$$

and

$$
\frac{1}{8} I_1 = \int_0^1 (3u^2 - 6u + 5)\, X\, du. \tag{14.4.12}
$$

14.5. We have

$$
\frac{1}{8} J_2 \geqq \iint_\Delta G\, du\, dv \tag{14.5.1}
$$

where

$$G = (8u^2 - 12u + 5)X^2 + 2v(2 - 3u)XY + (8v^2 - 12v + 5)Y^2. \tag{14.5.2}$$

Let $D = D(u, v)$ denote the discriminant of the quadratic form $G(X, Y)$. Then, since $8t^2 + 12t + 5 > 0$, G will be positive semi-definite provided that $D \leqq 0$. Let

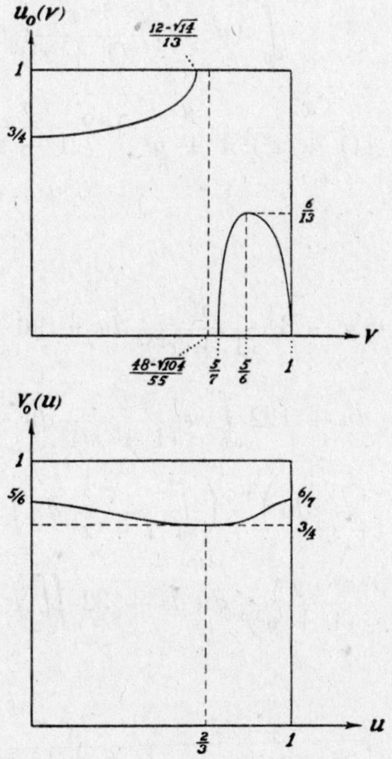

FIG. 14

$$g(u, v) = -\frac{1}{4} D(u, v). \tag{14.5.3}$$

It is to be shown that $g(u, v) \geqq 0$ in Δ, and to show this, we investigate local minima of $g(u, v)$ in Δ. Now

$$g(u, v) = \begin{cases} (55v^2 - 96v + 40)u^2 - 12(7v^2 - 12v + 5)u \\ \qquad\qquad + (36v^2 - 60v + 25), \\ (55u^2 - 84u + 36)v^2 - 12(8u^2 - 12u + 5)v \\ \qquad\qquad + 5(8u^2 - 12u + 5) \end{cases} \tag{14.5.4}$$

and so

$$\begin{aligned} \frac{\partial g(u, v)}{\partial u} &= 2\{(55v^2 - 96v + 40)u - 6(7v^2 - 12v + 5)\}, \\ \frac{\partial g(u, v)}{\partial v} &= 2\{(55u^2 - 84u + 36)v - 6(8u^2 - 12u + 5)\}. \end{aligned} \tag{14.5.5}$$

Thus

$$(14.5.6)\qquad \begin{aligned} \frac{\partial g(u, v)}{\partial u} &= 0 \quad \text{for} \quad u = u_0(v) = 6\,\frac{7v^2 - 12v + 5}{55v^2 - 96v + 40}, \\ \frac{\partial g(u, v)}{\partial v} &= 0 \quad \text{for} \quad v = v_0(u) = 6\,\frac{8u^2 - 12u + 5}{55u^2 - 84u + 36}. \end{aligned}$$

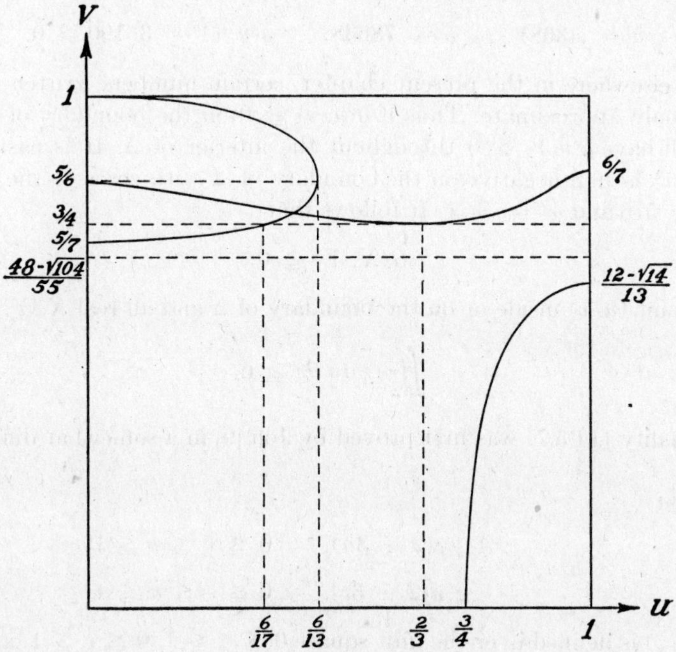

FIG. 15

We observe that

$$\begin{cases} 55v^2 - 96v + 40 = 0 \quad \text{for} \quad v = \dfrac{48 \pm (104)^{1/2}}{55} \cong 0.687,\ 1.058, \\ 7v^2 - 12v + \ 5 = 0 \quad \text{for} \quad v = \dfrac{5}{7}, 1; \end{cases}$$

$$\begin{cases} 55u^2 - 84u + 36 \geqq \dfrac{216}{55}, \qquad \text{equality at } u = \dfrac{42}{55}, \\ 8u^2 - 12u + \ 5 \geqq \dfrac{1}{2}, \qquad \text{equality at } u = \dfrac{3}{4}; \end{cases}$$

$$\begin{cases} \dfrac{du_0}{dv} = 12v\,\dfrac{5 - 6v}{(55v^2 - 96v + 40)^2} = 0 \quad \text{for} \quad v = 0, \dfrac{5}{6}, \\ \dfrac{dv_0}{du} = -12\,\dfrac{6u^2 - 13u + 6}{(55u^2 - 84u + 36)^2} = 0 \quad \text{for} \quad u = \dfrac{2}{3}, \dfrac{3}{2}. \end{cases}$$

The graphs of $u_0(v)$ and $v_0(u)$ are shown in Figure 14 and the two graphs are superimposed in Figure 15. It is clear that

$$\left(\frac{\partial g}{\partial u}\right)^2 + \left(\frac{\partial g}{\partial v}\right)^2 > 0$$

throughout the unit square $0 \leqq u \leqq 1, 0 \leqq v \leqq 1$ except at the one point $(\tilde{u}, \tilde{v})$, $6/17 < \tilde{u} < 6/13, 3/4 < \tilde{v} < 5/6$, where we have

$$\tilde{u} = .43681, \qquad \tilde{v} = .78638, \qquad g(\tilde{u}, \tilde{v}) = .36190 > 0.$$

Here and elsewhere in the present chapter, certain numbers written in decimal form are only approximate. Thus if $g(u, v) \geqq 0$ on the boundary of the region Δ, we shall have $g(u, v) > 0$ throughout the interior of Δ. It is easily verified that $g(u, v)$ is non-negative on the boundary of Δ with zeros at the two points $u = 0, v = 5/6$ and $u = v = 1$. It follows that

$$G(X, Y) \geqq 0 \tag{14.5.7}$$

for any point (u, v) inside or on the boundary of Δ and all real X, Y, and hence

$$\iint_\Delta G\, du\, dv \geqq 0. \tag{14.5.8}$$

The inequality (14.5.7) was first proved by Joh [9] in a somewhat different way.

14.6. Let

$$\beta(u, v) = \begin{cases} v(2 - 3u), & 0 \leqq u \leqq v \leqq 1, \\ u(2 - 3v), & 0 \leqq v \leqq u \leqq 1. \end{cases} \tag{14.6.1}$$

Then $\beta(u, v)$ is defined over the unit square $0 \leqq u \leqq 1, 0 \leqq v \leqq 1$, and it takes equal values at points symmetrical with respect to the diagonal $u = v$. Write

$$\alpha(u) = 8u^2 - 12u + 5. \tag{14.6.2}$$

Then $\alpha(u) \geqq 1/2$ with equality at $u = 3/4$, and

$$F(h) = \int_0^1 \int_0^1 \{\alpha(u)h(u)^2 + 2\beta(u, v)h(u)h(v) + \alpha(v)h(v)^2\}\, du\, dv. \tag{14.6.3}$$

Then

$$F(h) = 2 \iint_\Delta G(h(u), h(v))\, du\, dv.$$

Let c be the greatest lower bound of $F(h)$ for all real functions h satisfying

$$\int_0^1 h(u)^2\, du = 1. \tag{14.6.4}$$

By (14.5.8) we see that $c \geqq 0$.

To show that $c < 1$, consider the special function

$$h(u) = \begin{cases} \dfrac{1}{\delta^{1/2}}, & \dfrac{3}{4} - \dfrac{\delta}{2} \leqq u \leqq \dfrac{3}{4} + \dfrac{\delta}{2}, \\ 0, & \text{otherwise.} \end{cases}$$

This function satisfies the condition (14.6.4) and we have

$$\int_0^1 \alpha(u)h(u)^2\,du = \frac{1}{\delta}\int_{3/4-\delta/2}^{3/4+\delta/2} \left\{\frac{1}{2} + 8\left(u - \frac{3}{4}\right)^2\right\} du = \frac{1}{2} + \frac{2}{3}\delta^2.$$

Writing

$$v(2 - 3u) = -\frac{3}{16} - \frac{9}{4}\left(u - \frac{3}{4}\right) - \frac{1}{4}\left(v - \frac{3}{4}\right) - 3\left(u - \frac{3}{4}\right)\left(v - \frac{3}{4}\right),$$

we obtain

$$\int_{u=3/4-\delta/2}^{v} v(2 - 3u)\,du = -\frac{3}{16}\left(v - \frac{3}{4}\right) - \frac{11}{8}\left(v - \frac{3}{4}\right)^2 - \frac{3}{2}\left(v - \frac{3}{4}\right)^3$$
$$- \frac{3}{32}\delta + \frac{9}{32}\delta^2 - \frac{\delta}{8}\left(v - \frac{3}{4}\right) + \frac{3}{8}\delta^2\left(v - \frac{3}{4}\right),$$

and so

$$\int_{v=3/4-\delta/2}^{3/4+\delta/2} \int_{u=3/4-\delta/2}^{v} 2v(2 - 3u)h(u)h(v)\,du\,dv = -\frac{3}{16}\delta + \frac{1}{3}\delta^2.$$

Hence

$$F(h) = I - \frac{3}{8}\delta + 2\delta^2 = 1 - \frac{9}{512} \quad \text{for} \quad \delta = \frac{3}{32}.$$

In order to show that there is a function $h(u)$ which minimizes $F(h)$ under condition (14.6.4) and to investigate some of its properties we shall study a related quadratic form. Given a positive integer m, let u_μ be such that

(14.6.5)
$$\frac{\mu - 1}{m} \leqq u_\mu < \frac{\mu}{m}, \qquad \mu = 1, 2, \cdots, m,$$

and let

(14.6.6)
$$\alpha_\mu = \alpha(u_\mu),$$
$$\beta_{\mu\nu} = \beta(u_\mu, v_\nu).$$

Write

$$F_m = \frac{1}{m^2}\sum_{\mu=1}^{m}\sum_{\nu=1}^{m} \{\alpha_\mu x_\mu^2 + 2\beta_{\mu\nu}x_\mu x_\nu + \alpha_\nu x_\nu^2\}.$$

We see that $\alpha_\mu > 0$ and

$$\alpha_\mu\alpha_\nu - \beta_{\mu\nu}^2 = g(u_\mu, v_\nu) \geqq 0;$$

so F_m is a positive definite or positive semi-definite quadratic form depending on the choice of the sequence (14.6.5). Let c_m be the minimum of F_m subject to the condition that

$$\frac{1}{m}\sum_{\mu=1}^{m} x_\mu^2 = 1. \tag{14.6.7}$$

Points $x = (x_1, x_2, \cdots, x_m)$ satisfying (14.6.7) form a closed set and F_m is a continuous function of $x_1, x_2, \cdots, x_m$ over this set; so F_m attains its minimum c_m. If

$$\alpha^*(u) = \alpha_\mu, \qquad \frac{\mu-1}{m} \leqq u < \frac{\mu}{m},$$

$$\beta^*(u, v) = \beta_{\mu\nu}, \qquad \frac{\mu-1}{m} \leqq u < \frac{\mu}{m}, \qquad \frac{\nu-1}{m} \leqq v < \frac{\nu}{m},$$

then F_m can be written in the form

$$F_m = \int_0^1\int_0^1 \{\alpha^*(u)x(u)^2 + 2\beta^*(u, v)x(u)x(v) + \alpha^*(v)x(v)^2\}\, du\, dv$$

where

$$x(u) = x_\mu, \qquad \frac{\mu-1}{m} \leqq u < \frac{\mu}{m}.$$

There is a constant C_1 such that

$$\begin{aligned} &|\alpha(u') - \alpha(u'')| \leqq C_1 |u' - u''|, \\ &|\beta(u', v') - \beta(u'', v'')| \leqq C_1 \{|u' - u''| + |v' - v''|\}, \end{aligned} \tag{14.6.8}$$

for any two points in the unit square $0 \leqq u \leqq 1, 0 \leqq v \leqq 1$. In particular,

$$\begin{aligned} &|\alpha^*(u) - \alpha(u)| \leqq \frac{C_1}{m}, \\ &|\beta^*(u, v) - \beta(u, v)| \leqq \frac{2C_1}{m}. \end{aligned} \tag{14.6.9}$$

It is clear that

$$|\alpha(u)| \leqq 5, \qquad |\beta(u)| \leqq 5, \tag{14.6.10}$$

for $0 \leqq u \leqq 1, 0 \leqq v \leqq 1$.

To show that as m becomes infinite, the limit of c_m exists and is equal to c, we note first that

$$c \leqq \liminf_{m\to\infty} c_m. \tag{14.6.11}$$

For if $(x_1, x_2, \cdots, x_m)$ minimizes F_m under condition (14.6.7), let $h(u) = x_\mu$,

$(\mu - 1)/m \leqq u < \mu/m$. Then by using (14.6.9) it follows that $F(h) \leqq c_m + 6C_1/m$, which implies (14.6.11).

To show that

$$\limsup_{m\to\infty} c_m \leqq c, \tag{14.6.12}$$

let $h(u)$ be a function which satisfies (14.6.4) and for which

$$F(h) < c + \epsilon$$

where ϵ is some positive number, $0 < \epsilon < 1/4$. If $h^*(u)$ is the sum of the first k terms of the Fourier series of $h(u)$ and k is sufficiently large, then according to the Riesz-Fischer theorem,

$$\int_0^1 [h(u) - h^*(u)]^2\, du < \epsilon^2.$$

Let

$$x_\mu^* = m \int_{(\mu-1)/m}^{\mu/m} h^*(u)\, du,$$

and let

$$x^*(u) = x_\mu^*, \qquad \frac{\mu - 1}{m} \leqq u < \frac{\mu}{m}.$$

Since $h^*(u)$ is continuous, it follows that if m is sufficiently large,

$$\int_0^1 [x^*(u) - h^*(u)]^2\, du < \epsilon^2.$$

Then

$$\int_0^1 [x^*(u) - h(u)]^2\, du < 4\epsilon^2. \tag{14.6.13}$$

Expanding the left side of this inequality and making use of (14.6.4), we have

$$\int_0^1 (x^*)^2\, du + 1 - 4\epsilon^2 < 2\int_0^1 x^* h\, du < 2\left\{\int_0^1 (x^*)^2\, du\right\}^{1/2}.$$

Hence

$$(1 - 2\epsilon)^2 < \int_0^1 (x^*(u))^2\, du < (1 + 2\epsilon)^2.$$

Writing

$$x(u) = \lambda x^*(u),$$

where λ is a suitably chosen constant, $1/(1 + 2\epsilon) < \lambda < 1/(1 - 2\epsilon)$, we obtain a function $x(u)$ which satisfies

$$\int_0^1 x(u)^2\,du = 1.$$

From (14.6.13) it follows that

$$\int_0^1 [x(u) - h(u)]^2\,du < 64\epsilon^2. \tag{14.6.14}$$

Then

$$\int_0^1 |\,x(u)^2 - h(u)^2\,|\,du \leqq \left\{\int_0^1 [x(u) - h(u)]^2\,du\right\}^{1/2}\left\{\int_0^1 [x(u) + h(u)]^2\,du\right\}^{1/2} \leqq 16\epsilon.$$

Making use of (14.6.9), we see that

$$\begin{aligned} c_m &\leqq \int_0^1\int_0^1 \{\alpha^*(u)x(u)^2 + 2\beta^*(u, v)x(u)x(v) + \alpha^*(v)x(v)^2\}\,du\,dv \\ &\leqq F(x) + 6C_1/m \leqq F(h) + 6C_1/m + 10\cdot 16\epsilon \\ &\qquad +10\int_0^1\int_0^1 |\,x(u)x(v) - h(u)h(v)\,|\,du\,dv \\ &\leqq c + 6C_1/m + 160\epsilon + 160\epsilon. \end{aligned}$$

Thus (14.6.12) is proved.

By (14.6.11) and (14.6.12)

$$\lim_{m\to\infty} c_m = c, \qquad 0 \leqq c \leqq 1 - \frac{9}{512}. \tag{14.6.15}$$

Therefore, if m is large enough,

$$c_m \leqq 1 - \frac{1}{100}. \tag{14.6.16}$$

Now let $x = (x_1, x_2, \cdots, x_m)$ minimize F_m, that is, $F_m(x) = c_m$. Then

$$F_m(x + \epsilon y) = F_m(x) + \frac{4\epsilon}{m}\sum_{\mu=1}^m \alpha_\mu x_\mu y_\mu + \frac{4\epsilon}{m^2}\sum_{\mu=1}^m\sum_{\nu=1}^m \beta_{\mu\nu}x_\mu y_\nu + \epsilon^2 F_m(y),$$

and

$$\begin{aligned} &\frac{F_m(x + \epsilon y)}{\dfrac{1}{m}\displaystyle\sum_{\mu=1}^m (x_\mu + \epsilon y_\mu)^2} - c_m \\ &= \frac{\dfrac{4\epsilon}{m}\displaystyle\sum_{\mu=1}^m \alpha_\mu x_\mu y_\mu + \dfrac{4\epsilon}{m^2}\sum_{\mu=1}^m\sum_{\nu=1}^m \beta_{\mu\nu}x_\mu y_\nu - \dfrac{2c_m\epsilon}{m}\sum_{\mu=1}^m x_\mu y_\mu - \dfrac{c_m\epsilon^2}{m}\sum_{\mu=1}^m y_\mu^2 + \epsilon^2 F_m(y)}{1 + \dfrac{2\epsilon}{m}\displaystyle\sum_{\mu=1}^m x_\mu y_\mu + \dfrac{\epsilon^2}{m}\sum_{\mu=1}^m y_\mu^2} \\ &\geqq 0. \end{aligned}$$

The coefficient of ϵ on the left of the inequality is

$$\frac{2}{m}\left\{2\sum_{\mu=1}^{m}\alpha_\mu x_\mu y_\mu + \frac{2}{m}\sum_{\mu=1}^{m}\sum_{\nu=1}^{m}\beta_{\mu\nu}x_\mu y_\nu - c_m\sum_{\mu=1}^{m}x_\mu y_\mu\right\}$$
$$= \frac{2}{m}\sum_{\nu=1}^{m}y_\nu\left[(2\alpha_\nu - c_m)x_\nu + \frac{2}{m}\sum_{\mu=1}^{m}\beta_{\mu\nu}x_\mu\right].$$

Since ϵ can be chosen positive or negative, we see that the coefficient of ϵ must vanish. Since the y_ν are arbitrary, we obtain

$$(14.6.17)\qquad (c_m - 2\alpha_\nu)x_\nu = \frac{2}{m}\sum_{\mu=1}^{m}\beta_{\mu\nu}x_\mu\,, \qquad \nu = 1, 2, \cdots, m.$$

We observe that

$$\left|\frac{1}{m}\sum_{\mu=1}^{m}\beta_{\mu\nu}x_\mu\right| \leqq \frac{1}{m}\left(\sum_{\mu=1}^{m}\beta_{\mu\nu}^2\right)^{1/2}\left(\sum_{\mu=1}^{m}x_\mu^2\right)^{1/2} = \left(\frac{1}{m}\sum_{\mu=1}^{m}\beta_{\mu\nu}^2\right)^{1/2}$$

where

$$\lim_{m\to\infty}\frac{1}{m}\sum_{\mu=1}^{m}\beta_{\mu\nu}^2 = \int_0^1 [\beta(u, v)]^2\,du \leqq 25$$

by (14.6.10). Thus for all sufficiently large m

$$|(c_m - 2\alpha_\nu)x_\nu| \leqq 15.$$

Since α_ν is a value taken by the function (14.6.2), we have $2\alpha_\nu \geqq 1$. Hence for all sufficiently large m we have by (14.6.16)

$$(14.6.18)\qquad |x_\nu| \leqq 1500.$$

Also

$$|x_{\nu+1} - x_\nu| = \frac{2}{m}\left|\frac{\sum_{\mu=1}^{m}\beta_{\mu,\nu+1}x_\mu}{c_m - 2\alpha_{\nu+1}} - \frac{\sum_{\mu=1}^{m}\beta_{\mu\nu}x_\mu}{c_m - 2\alpha_\nu}\right|$$
$$= \frac{2}{m}\left|\frac{\sum_{\mu=1}^{m}x_\mu\{c_m(\beta_{\mu,\nu+1} - \beta_{\mu\nu}) + 2[(\alpha_{\nu+1} - \alpha_\nu)\beta_{\mu\nu} + \alpha_\nu(\beta_{\mu\nu} - \beta_{\mu,\nu+1})]\}}{(c_m - 2\alpha_\nu)(c_m - 2\alpha_{\nu+1})}\right|.$$

Relation (14.6.8) shows that

$$|\alpha_{\nu+1} - \alpha_\nu| \leqq \frac{2C_1}{m}, \qquad |\beta_{\mu,\nu+1} - \beta_{\mu\nu}| \leqq \frac{3C_1}{m},$$

and it follows that there is a constant C_2 such that

$$(14.6.19)\qquad |x_{\nu+1} - x_\nu| \leqq C_2/m.$$

Thus if

$$h_m(u) = x_\mu, \qquad \frac{\mu - 1}{m} \leqq u < \frac{\mu}{m}$$

$(h_m(1) = x_m)$, it follows that the functions $h_m(u)$ are bounded by a constant which is independent of m, and they satisfy

$$| h_m(u') - h_m(u'') | \leqq C_2 | u' - u'' | + C_2/m$$

for any pair of points u', u'' in $0 \leqq u \leqq 1$. Then the method used to prove that a sequence of uniformly bounded equicontinuous functions defined over a closed finite interval has a subsequence which converges to a limit will show that there is a subsequence of the $h_m(u)$ which converges uniformly to a limit $h_0(u)$ in $0 \leqq u \leqq 1$. Moreover,

$$\int_0^1 h_0(u)^2\, du = 1,$$

$$\int_0^1 \int_0^1 \{\alpha(u)h_0(u)^2 + 2\beta(u, v)h_0(u)h_0(v) + \alpha(v)h_0(v)^2\}\, du\, dv = \lim_{m\to\infty} c_m = c.$$

From (14.6.17) it follows that $h_0(u)$ satisfies the integral equation

$$[c - 2\alpha(u)]h_0(u) = 2 \int_0^1 \beta(u, v)h_0(v)\, dv \tag{14.6.20}$$

for $0 \leqq u \leqq 1$.

Multiplying both sides of (14.6.20) by $h_0(u)\, du$ and integrating from 0 to 1, we obtain

$$c \int_0^1 h_0(u)^2\, du = 2 \int_0^1 \alpha(u)h_0(u)^2\, du + 2 \int_0^1 \int_0^1 \beta(u, v)h_0(u)h_0(v)\, du\, dv$$

$$= \int_0^1 \int_0^1 \{\alpha(u)h_0(u)^2 + 2\beta(u, v)h_0(u)h_0(v) + \alpha(v)h_0(v)^2\}\, du\, dv.$$

Hence we have only to determine the smallest eigen-value c_0 of the equation (14.6.20), and then we shall have

$$F(h) \geqq c_0 \int_0^1 h(u)^2\, du \tag{14.6.21}$$

for every function $h(u)$ belonging to $L^2(0, 1)$.

The equation (14.6.21) may be written

$$[c - 2\alpha(u)]h_0(u) = 4 \int_u^1 vh_0(v)\, dv + 4u \int_0^u h_0(v)\, dv - 6u \int_0^1 vh_0(v)\, dv.$$

Differentiating this equation twice with respect to u, we obtain

$$\{[c - 2\alpha(u)]h_0(u)\}' = 4 \int_0^u h_0(v)\, dv - 6 \int_0^1 vh_0(v)\, dv,$$

$$\{[c - 2\alpha(u)]h_0(u)\}'' = 4h_0(u).$$

Write

$$g(u) = [2\alpha(u) - c]h_0(u), \qquad p(u) = \frac{4}{2\alpha(u) - c}. \tag{14.6.22}$$

Then

$$g''(u) = -p(u)g(u),$$

$$g'(u) = -\int_0^u p(v)g(v)\,dv + \frac{3}{2}\int_0^1 vp(v)g(v)\,dv,$$

$$g(u) = -\int_u^1 vp(v)g(v)\,dv - u\int_0^u p(v)g(v)\,dv + \frac{3}{2}u\int_0^1 vp(v)g(v)\,dv.$$

In particular,

$$g(0) = -\int_0^1 vp(v)g(v)\,dv = \int_0^1 vg''(v)\,dv = vg'(v)\Big|_0^1 - \int_0^1 g'(v)\,dv$$

$$= g'(1) - g(1) + g(0)$$

and so

$$g(1) = g'(1).$$

Also

$$g'(0) = \frac{3}{2}\int_0^1 vp(v)g(v)\,dv = -\frac{3}{2}g(0).$$

Thus we obtain the Sturm-Liouville system

$$g''(u) + p(u)g(u) = 0,$$

$$3g(0) + 2g'(0) = 0, \tag{14.6.23}$$

$$g(1) - g'(1) = 0.$$

We know that $0 \leqq c_0 \leqq 1 - (9/512)$ and so we have only to investigate numbers c in the range $0 \leqq c \leqq 1 - (9/512)$. In this range we have

$$0 < p(u) \leqq 2048/9. \tag{14.6.24}$$

Also

$$\frac{\partial p}{\partial c} = \frac{1}{4}p^2 > 0. \tag{14.6.25}$$

14.7. Now let $p_0(u)$ and $p_1(u)$ be two real piecewise continuous functions of u in $0 \leqq u \leqq 1$ and suppose that

$$p_0(u) \leqq p_1(u), \qquad 0 \leqq u \leqq 1,$$

with equality only at a set of measure zero in the unit interval. Let $g_0(u)$ and $g_1(u)$ be non-trivial solutions respectively of the equations

$$g_0''(u) + p_0(u)g_0(u) = 0, \qquad g_1''(u) + p_1(u)g_1(u) = 0.$$

Multiplying the first of these equations by g_1, the second by g_0, and subtracting the second equation from the first, we obtain the Green's identity

$$\frac{d}{du}[g_0' g_1 - g_0 g_1'] = (p_1 - p_0)g_0 g_1 . \tag{14.7.1}$$

This may be used to prove Sturm's theorem that between any two zeros of g_0 there lies a zero of g_1. For let u_1 and u_2 denote consecutive zeros of g_0 and suppose that g_1 is not zero at any point of the open interval $u_1 < u < u_2$. Without loss of generality we may assume that both g_0 and g_1 are positive in this open interval. Integrating (14.7.1) between u_1 and u_2, we obtain

$$[g_0'g_1 - g_0g_1']_{u_1}^{u_2} = \int_{u_1}^{u_2} (p_1 - p_0)g_0 g_1 \, dv > 0. \tag{14.7.2}$$

Since $g_0' \geqq 0$ at u_1 and $g_0' \leqq 0$ at u_2, the left side of (14.7.2) is not positive, a contradiction.

A similar argument shows that if $g_0(0) = g_1(0)$, $g_0'(0) = g_1'(0)$; and if $g_0(u)$ is equal to zero at some point u_0, $0 < u_0 \leqq 1$, then $g_1(u)$ vanishes at some point u_1 satisfying $0 < u_1 < u_0$.

For future reference we now show that if

$$g_0(0) = g_1(0), \qquad g_0'(0) = g_1'(0),$$

and if both $g_0(u)$ and $g_1(u)$ have precisely one zero in $0 \leqq u \leqq 1$, these zeros lying in $0 < u < 1$, then

$$\frac{g_0'(1)}{g_0(1)} > \frac{g_1'(1)}{g_1(1)} . \tag{14.7.3}$$

In fact, for each t in the range $0 \leqq t \leqq 1$, let

$$p_t(u) = p_0(u) + t\{p_1(u) - p_0(u)\},$$

and let $g_t(u)$ satisfy the differential equation

$$g_t''(u) + p_t(u)g_t(u) = 0$$

with initial conditions

$$g_t(0) = a, \qquad g_t'(0) = b,$$

where $a = g_0(0) = g_1(0)$, $b = g_0'(0) = g_1'(0)$. Then if $0 \leqq t \leqq 1$, $g_t(u)$ has precisely one zero in $0 \leqq u \leqq 1$, and this zero lies in $0 < u < 1$. In particular, $g_t(1) \neq 0$. The equation

$$g_t(u) = a + bu - \int_0^u (u - v)p_t(v)g_t(v)\, dv$$

may be used to show that $g_t(u)$ and $g_t'(u)$ are differentiable with respect to t. From Green's identity we have, if $0 \leqq t \leqq 1$, $0 \leqq t + \delta \leqq 1$,

$$g_t'(1)g_{t+\delta}(1) - g_t(1)g_{t+\delta}'(1) = \int_0^1 (p_{t+\delta} - p_t)g_t g_{t+\delta}\, dv.$$

On dividing by δ and letting δ approach zero it follows that

$$g_t'(1)\frac{dg_t(1)}{dt} - g_t(1)\frac{dg_t'(1)}{dt} = \int_0^1 (p_1 - p_0)g_t^2\, dv > 0.$$

Thus

$$\frac{d}{dt}\left(\frac{g_t'(1)}{g_t(1)}\right) < 0$$

and (14.7.3) follows.

The functions $p(u)$ and $g(u)$ of the Sturm-Liouville system (14.6.23) depend on c. We write

(14.7.4) $$p(u, c) = \frac{4}{(4u - 3)^2 + 1 - c}$$

and let $g(u, c)$ satisfy the differential equation

(14.7.5) $$g''(u, c) + p(u, c)g(u, c) = 0$$

where the primes denote differentiation with respect to u. Since one of the boundary conditions is $3g(0, c) + 2g'(0, c) = 0$, we see that if $g(0, c) = 0$, then $g'(0, c) = 0$ and $g(u, c)$ is the trivial solution. Thus $g(0, c) \neq 0$, and, since the boundary conditions are homogeneous, we suppose without loss of generality that

$$g(0, c) = 1, \qquad g'(0, c) = -3/2.$$

It is to be shown by using calculations which are only outlined in the present text that

$$g'(1, c) - g(1, c) < 0$$

for $0 \leqq c \leqq .80$, and it will follow that the smallest eigen-value c_0 of the Sturm-Liouville system (14.6.23) satisfies $c_0 \geqq .80$. We shall therefore consider the range $0 \leqq c \leqq .85$.

If $u_1 = u_1(c)$ is the first positive zero of $g(u, c)$, then for $0 \leqq c \leqq .85$ we have

$$.4 < u_1(c) < 2/3.$$

For in the range $0 \leqq u \leqq u_1$, we have $g''(u) \leqq 0$; so

$$g(u) < 1 - 3/2u, \qquad 0 < u \leqq u_1.$$

This shows that $u_1(c) < 2/3$. To consider the lower bound we note that since $g(u, c)$ satisfies the differential equation (14.7.5),

$$g(u, c) = 1 - 3u/2 - \int_0^u (u - v)p(v, c)g(v, c)\, dv.$$

Now in $0 \leqq u \leqq .4$ we have

$$p(u, c) \leqq \frac{4}{(4u - 3)^2 + .15} < 2;$$

so if $0 \leqq u \leqq \min(u_1, .4)$, then

$$g(u, c) \geqq 1 - 3u/2 - \int_0^u (u - v)(2 - 3v)\, dv.$$

Thus

$$g(u, c) \geqq 1 - 3u/2 - u^2 + \frac{1}{2}u^3.$$

The right side of the last inequality is positive for $0 \leqq u \leqq .4$ and it follows that $u_1(c) > .4$.

If $0 \leqq c \leqq .85$, then $g(u, c)$ has only one zero in $0 \leqq u \leqq 1$. For in this case

$$p(u, c) \leqq 4/.15 = 80/3.$$

Hence, according to Sturm's theorem, between any two zeros of $g(u, c)$ there is a zero of the function $y(u)$ which satisfies $y'' + 80y/3 = 0$. The distance between consecutive zeros of $y(u)$ is $\pi/(80/3)^{1/2} > 3/5$; so by choosing a function $y(u)$ which vanishes at $u_1(c)$ it follows that the next zero $u_2(c)$ of $g(u, c)$ to the right of $u_1(c)$ must satisfy $u_2(c) > u_1(c) + 3/5 > 1$.

In the range $0 \leqq c \leqq .85$ we see that the function $g(u, c)$ has precisely one zero in $0 \leqq u \leqq 1$, this zero lies in $0 < u < 1$, and $p(u, c)$ is an increasing function of c,

$$\frac{\partial p}{\partial u} = [p(u, c)]^2/4 > 0.$$

Thus from (14.7.3),

$$\frac{g'(1, c)}{g(1, c)} > \frac{g'(1, c^*)}{g(1, c^*)} \tag{14.7.6}$$

if $0 \leqq c < c^* \leqq .85$. If we show that

$$\frac{g'(1, c)}{g(1, c)} > 1 \tag{14.7.7}$$

for a particular value c^* of c, then it will follow that this inequality is true for $0 \leqq c \leqq c^*$ and therefore that $c_0 > c^*$.

By a numerical calculation we have found that the smallest eigen-value c_0 of the system (14.6.23) exceeds 0.80. It is sufficient to show that (14.7.7) is true for $c = 0.80$, and this was done as follows. Take $c = 0.80$ and dissect the interval $0 \leqq u \leqq 1$ into subintervals. Let $p_1(u)$ be the step-function which is constant over each subinterval and equal to the maximum of $p(u) = p(u, .80)$ on the subinterval. Then $p_1(u) \geqq p(u)$, and the subintervals were chosen of various lengths depending on the magnitude of $p'(u)$ in order that the difference $p_1(u) - p(u)$ should be uniformly small. Over each subinterval, the solution of the equation

$$g_1'' + p_1 g_1 = 0$$

is trigonometric and so it is possible to compute g_1 with considerable precision. This was done and a precise bound was determined for the maximum possible error. Allowing for the error, it was still found that

$$\left(\frac{g_1'(1)}{g_1(1)}\right) - 1 > 0,$$

and this shows that $c_0 > 0.80$.

From (14.6.21) we thus have

$$F(h) \geqq 0.8 \int_0^1 h(u)^2\, du \tag{14.7.8}$$

for any function $h(u)$ belonging to $L^2(0, 1)$. Since

$$F(h) = 2 \iint_\Delta G(h(u), h(v))\, du\, dv,$$

we obtain by (14.5.1)

$$J_2 \geqq 3.2 \int_0^1 X(u)^2\, du. \tag{14.7.9}$$

14.8. Now we obtain an upper bound for the expression J_4. Writing

$$X = X(u) = \frac{x}{1 + x^2}, \qquad Y = Y(v) = \frac{y}{1 + y^2},$$

we have by (14.4.5), if Δ is the triangle $0 \leqq u \leqq v \leqq 1$ and Δ_1 is the triangle $0 \leqq v \leqq u \leqq 1$,

$$\begin{aligned} J_4 = {} & -8 \int_0^1 (8u^2 - 5)X^4\,(1 + x^2)^2\, du + 128 \iint_{\Delta_1} uX^2Y^2(1 + y^2)\, du\, dv \\ & + 192 \iint_\Delta uX^2Y^2(1 + y^2)\, du\, dv + 128 \iint_{\Delta_1} uX^3Y(1 + x^2)\, du\, dv \\ & + 192 \iint_\Delta uX^3Y(1 + x^2)\, du\, dv - 96 \left(\int_0^1 X^2(1 + x^2)\, du\right)^2 \\ & - 192 \int_0^1 X^2(1 + x^2)\, du \cdot \left(\int_0^1 X\, du\right)^2. \end{aligned}$$

Now suppose that

(14.8.1) $$| x(u) | \leqq \delta < 1.$$

Then

$$| X | \leqq \frac{\delta}{1 + \delta^2}$$

and

$$-J_4 \leqq 24(1 + \delta^2)^2 \int_0^1 X^4 \, du + 192(1 + \delta^2) \int_0^1 \int_0^1 u \, | X^3 Y | \, du \, dv$$
$$+ 96(1 + \delta^2)^2 \left(\int_0^1 X^2 \, du \right)^2 + 192(1 + \delta^2) \left(\int_0^1 X^2 \, du \right) \left(\int_0^1 X \, du \right)^2$$
$$= T_1 + T_2 + T_3 + T_4,$$

say. We have

$$T_1 \leqq 24 \, (1 + \delta^2)^2 \cdot \frac{\delta^2}{(1 + \delta^2)^2} \int_0^1 X^2 \, du = 24 \, \delta^2 \int_0^1 X^2 \, du,$$

$$T_2 = 192(1 + \delta^2) \left(\int_0^1 u \, | X |^3 \, du \right) \left(\int_0^1 | X | \, du \right)$$
$$\leqq 192(1 + \delta^2) \left(\int_0^1 u^2 X^4 \, du \right)^{1/2} \left(\int_0^1 X^2 \, du \right)^{1/2} \left(\int_0^1 X^2 \, du \right)^{1/2}$$
$$\leqq 192(1 + \delta^2) \frac{\delta^2}{(1 + \delta^2)^2} \left(\int_0^1 u^2 \, du \right)^{1/2} \left(\int_0^1 X^2 \, du \right)$$
$$= 192 \frac{\delta^2}{1 + \delta^2} \cdot \left(\frac{1}{3} \right)^{1/2} \int_0^1 X^2 \, du \leqq 111 \frac{\delta^2}{1 + \delta^2} \int_0^1 X^2 \, du,$$

$$T_3 \leqq 96(1 + \delta^2)^2 \cdot \frac{\delta^2}{(1 + \delta^2)^2} \int_0^1 X^2 \, du = 96\delta^2 \int_0^1 X^2 \, du,$$

$$T_4 \leqq 192(1 + \delta^2) \left(\int_0^1 X^2 \, du \right)^2 \leqq 192 \frac{\delta^2}{1 + \delta^2} \int_0^1 X^2 \, du.$$

Thus

(14.8.2)
$$-J_4 \leqq \left(24 + 111 \frac{1}{1 + \delta^2} + 96 + 192 \frac{1}{1 + \delta^2} \right) \delta^2 \int_0^1 X^2 \, du$$
$$= \left(120 + \frac{303}{1 + \delta^2} \right) \delta^2 \int_0^1 X^2 \, du.$$

Since $J_6 \geqq 0$, we obtain from (14.4.3), (14.7.9), and (14.8.2)

$$\operatorname{Re}(a_4) \leqq 4 - 3.2 \int_0^1 X^2\, du + \left(120 + \frac{303}{1+\delta^2}\right) \delta^2 \int_0^1 X^2\, du$$

$$= 4 - \left[3.2 - \left(120 + \frac{303}{1+\delta^2}\right)\delta^2\right] \int_0^1 X^2\, du$$

$$\leqq 4 - \frac{9}{10^4} \int_0^1 X^2\, du$$

if

$$\delta = .0872.$$

But $x = x(u) = \tan[\theta(u)/2]$, and so we have

(14.8.3) $\operatorname{Re}(a_4) \leqq 4 \quad \text{if} \quad |\theta(u)| \leqq .17395.$

Here there is equality if and only if $\theta(u)$ is zero almost everywhere in $0 \leqq u \leqq 1$.

14.9 The result to be proved may be stated as a lemma. In order to bring the notation into conformity with that of Chapter XI, we replace $\theta(u)$ by $\alpha(u)$ (which will no longer be confused with the function of (14.6.2)).

LEMMA XXXII. *Let a_4 be given by the formula*

(14.9.1)
$$a_4 = 2\int_0^1 u^2 e^{3i\alpha(u)}\, du - 12\left\{\int_0^1 u e^{2i\alpha(u)}\, du\right\}\left\{\int_0^1 e^{i\alpha(u)}\, du\right\} + 4\int_{v=0}^1 \int_{u=0}^1 v e^{i[\alpha(u)+2\alpha(v)]}\, du\, dv + 8\left(\int_0^1 e^{i\alpha(u)}\, du\right)^3$$

where $\alpha(u)$ is a real function. If there is a constant c such that

(14.9.2) $|\alpha(u) - c| \leqq 0.08697, \qquad 0 \leqq u \leqq 1,$

then $|a_4| \leqq 4$ with equality if and only if $\alpha(u)$ is almost everywhere equal to a constant.

If

(14.9.3) $|\alpha(u)| \leqq 0.17395, \qquad 0 \leqq u \leqq 1,$

then $\operatorname{Re}(a_4) \leqq 4$ *with equality if and only if $\alpha(u)$ is zero almost everywhere.*

PROOF: We have shown that if (14.9.3) is true, then $\operatorname{Re}(a_4) \leqq 4$. Now suppose that (14.9.2) is true and let the real constant θ be such that

$$a_4 e^{3i\theta} \geqq 0.$$

Then if we replace $\alpha(u)$ by $\theta + \alpha(u)$ in the right side of (14.9.1), the left side becomes $a_4 e^{3i\theta}$. The function $\theta + \alpha(u)$ differs from some constant by no more than

0.08697 according to (14.9.2); so there is no loss of generality in supposing that $a_4 \geqq 0$.

Thus suppose that $a_4 \geqq 0$ and let

$$\alpha(u) = \tau + \varphi(u), \qquad \varphi(u) = \varphi_\tau(u), \tag{14.9.4}$$

where τ is a real number independent of u. Then by **14.4**,

$$a_4 = e^{3i\tau}(p_\tau + iq_\tau) \tag{14.9.5}$$

where

$$\begin{aligned} p_\tau = {} & 2\int_0^1 u^2 \cos 3\varphi(u)\,du - 8\int_0^1\int_0^1 u\cos[2\varphi(u) + \varphi(v)]\,du\,dv \\ & - 4\iint_\Delta u\cos[2\varphi(u)+\varphi(v)]\,du\,dv \\ & - 8\left(\int_0^1 \cos\varphi(u)\,du\right)\left\{3\left(\int_0^1 \sin\varphi(u)\,du\right)^2 - \left(\int_0^1 \cos\varphi(u)\,du\right)^2\right\}, \\ q_\tau = {} & 2\int_0^1 u^2 \sin 3\varphi(u)\,du - 8\int_0^1\int_0^1 u\sin[2\varphi(u) + \varphi(v)]\,du\,dv \\ & - 4\iint_\Delta u\sin[2\varphi(u)+\varphi(v)]\,du\,dv \\ & + 8\left(\int_0^1 \sin\varphi(u)\,du\right)\left\{3\left(\int_0^1 \cos\varphi(u)\,du\right)^2 - \left(\int_0^1 \sin\varphi(u)\,du\right)^2\right\}. \end{aligned}$$

Write

$$q_\tau = \int_0^1 Q(u)\sin\varphi(u)\,du$$

where

$$\begin{aligned} Q(u) = {} & 6u^2 - 8u^2\sin^2\varphi(u) - 2au\cos\varphi(u) - b - 2j_1(u)u\cos\varphi(u) \\ & - j_2(u) + 8(c-d). \end{aligned}$$

Here

$$a = 8\int_0^1 \cos\varphi(v)\,dv, \qquad b = 8\int_0^1 v\cos 2\varphi(v)\,dv,$$

$$c = 3\left(\int_0^1 \cos\varphi(v)\,dv\right)^2, \qquad d = \left(\int_0^1 \sin\varphi(v)\,dv\right)^2,$$

$$j_1(u) = 4\int_{v=u}^1 \cos\varphi(v)\,dv, \qquad j_2(u) = 4\int_{v=0}^u v\cos 2\varphi(v)\,dv.$$

Suppose now that $|\varphi(v)| \leqq 1/2$. Then

$$|\sin \varphi(v)| \leqq 0.48, \qquad \cos \varphi(v) \geqq 0.87, \qquad \cos 2\varphi(v) \geqq 0.54,$$

$$a \leqq 8, \qquad b \leqq 4, \qquad c \geqq 3(0.87)^2 > 2.27, \qquad d \leqq (0.48)^2 = 0.2304,$$

$$j_1(u) \leqq 4(1 - u), \qquad j_2(u) \leqq 2u^2.$$

Hence

$$Q(u) \geqq u^2[6 - 8 \sin^2 \varphi(u)] - 16u \cos \varphi(u) - 4 - 8(1 - u)u \cos \varphi(u) - 2u^2 + 8(2.27 - 0.2304)$$

$$= 8u^2 \cos^2 \varphi(u) - 8u(3 - u) \cos \varphi(u) + (12.3168 - 4u^2)$$
$$> 8u^2 \cos^2 \varphi(u) - 8u(3 - u) \cos \varphi(u) + (12.31 - 4u^2).$$

Let $\cos \varphi(u) = t$, $0.87 \leqq t \leqq 1$, and write

$$F(u, t) = 8u^2t^2 - 8u(3 - u)\, t + (12.31 - 4u^2).$$

Since

$$\frac{\partial F}{\partial t} = 16u^2t - 8u(3 - u) \leqq 16u^2 - 8u(3 - u) = 24u(u - 1) \leqq 0,$$

we have

$$F(u, t) \geqq F(u, 1) = 12u^2 - 24u + 12.31 \geqq 12 - 24 + 12.31 = 0.31 > 0$$

for $0 \leqq u \leqq 1$. Thus for $|\varphi(u)| \leqq 1/2$ we have

$$Q(u) > 0.$$

From (14.9.4) we have

$$q_\tau = \int_0^1 Q(u) \sin [\alpha(u) - \tau]\, du$$

and so

$$q_\tau \leqq 0 \qquad \text{for} \qquad \tau = c + 0.08697$$

since $-0.17394 \leqq \alpha(u) - \tau = \alpha - c - 0.08697 \leqq 0$; similarly

$$q_\tau \geqq 0 \qquad \text{for} \qquad \tau = c - 0.08697$$

since $0 \leqq \alpha(u) - \tau = \alpha - c + 0.08697 \leqq 0.17394$. However, q_τ is a continuous function of τ, and it follows that there is a τ, $\tau = \tau_0$ say, $c - 0.08697 \leqq \tau_0 \leqq c + 0.08697$, such that

$$q_{\tau_0} = 0.$$

But by (14.9.5)

$$q_{\tau_0} = -a_4 \sin 3\tau_0$$

and so $\sin 3\tau_0 = 0$, $\tau_0 \equiv 0, 2\pi/3, 4\pi/3 \pmod{2\pi}$. We then have

$$a_4 = p_{\tau_0} e^{3i\tau_0} = p_{\tau_0}.$$

Since $|\varphi(u)| = |\alpha(u) - \tau_0| \leqq |\alpha(u) - c| + |c - \tau_0| \leqq 0.17394$, we have

$$a_4 = \text{Re}\,(a_4) \leqq 4$$

by (14.8.3) (with φ in place of θ).

Lemma XXXII establishes the desired local maximum property of the functions (14.1.2).

14.10. In **14.3** we proved that the maximum of Re (a_4) must occur on that portion $\mathbf{\Pi}_1$ of the boundary of V_4 which corresponds to functions Q with one double zero and no other zero on $|z| = 1$. Any such Q plainly determines a unique point of $\mathbf{\Pi}_1$. We shall now find further necessary conditions on Q in order that it should define a function

$$w = f(z) = z + a_2 z^2 + a_3 z^3 + a_4 z^4 + \cdots \tag{14.10.1}$$

with maximal Re (a_4).

If Re (a_4) is maximal, we have seen that $Q(z)$ must have the form (14.2.9). Hence we may suppose that

$$B_3 = \bar{B}_3 = A_3 = 1. \tag{14.10.2}$$

Since Q vanishes at only one point of $|z| = 1$ and is non-negative there, we have

$$Q = \sum_{\nu=-3}^{3} \frac{B_\nu}{z^\nu} = c\,\frac{(z - e^{i\theta})^2}{-ze^{i\theta}} \cdot \frac{(z - re^{i\varphi})(z - (1/r)e^{i\varphi})}{-ze^{i\varphi}} \cdot \frac{(z - \rho e^{i\psi})(z - (1/\rho)e^{i\psi})}{-ze^{i\psi}}$$

for some constant c, which must be positive. Since $B_3 = 1$, it follows that

$$-ce^{-i(\theta+\varphi+\psi)} = 1;$$

so $c = 1$, $e^{i\theta} = -e^{-i(\varphi+\psi)}$. Writing

$$p = \frac{1}{2}\left(r + \frac{1}{r}\right), \qquad q = \frac{1}{2}\left(\rho + \frac{1}{\rho}\right), \tag{14.10.3}$$

we see that $p > 1$, $q > 1$, and

$$Q = \frac{1}{z^3}\,(z + e^{-i(\varphi+\psi)})^2(z^2 - 2pze^{i\varphi} + e^{2i\varphi})(z^2 - 2qze^{i\psi} + e^{2i\psi}).$$

Comparing like powers of z, we then obtain the equations

$$\begin{aligned} B_0 &= 4\{\cos(\varphi - \psi) + 2pq - p\cos(\varphi + 2\psi) - q\cos(2\varphi + \psi)\}, \\ B_1 &= e^{2i(\varphi+\psi)} + e^{-2i\varphi} + e^{-2i\psi} + 4pqe^{-i(\varphi+\psi)} - 4pe^{i\psi} - 4qe^{i\varphi}, \\ B_2 &= 2\{e^{i(\varphi+\psi)} - pe^{-i\varphi} - qe^{-i\psi}\}. \end{aligned} \tag{14.10.4}$$

A further restriction on B_0, B_1, B_2 is provided by (14.2.9), namely that

$$B_0 = 3a_4, \qquad B_1 = 3a_3, \qquad B_2 = 2a_2, \tag{14.10.5}$$

where a_2, a_3, a_4 are the first, second, and third coefficients of a function of class $\mathfrak{S}$. We therefore impose on B_0, B_1, B_2 the further restriction that the numbers a_2, a_3, a_4 in (14.10.5) satisfy the inequality (14.3.2) with $\alpha = 3/4$. By writing

$$\mu = a_3 - \frac{11}{16} a_2^2, \tag{14.10.6}$$

this inequality becomes

$$15 |a_2|^2 + 39 |\mu|^2 + 63 \left| a_4 - \frac{11}{8} a_2 \mu - \frac{77}{192} a_2^3 \right|^2 \leqq 64. \tag{14.10.7}$$

Another restriction is provided by the inequality $|a_2| \leqq 2$.

14.11. We observe that the formulas (14.10.4) involve four real parameters p, q, φ, ψ, and it is therefore desirable to restrict the range of variation of these parameters as much as possible. We now indicate a way in which these parameters can be restricted.

The relations which we use are the following:

$$a_2 = e^{i(\varphi+\psi)} - qe^{-i\psi} - pe^{-i\varphi}, \tag{14.11.1}$$

$$3a_3 = e^{2i(\varphi+\psi)} - 4pe^{i\psi} - 4qe^{i\varphi} + e^{-2i\varphi} + e^{-2i\psi} + 4pqe^{-i(\varphi+\psi)}, \tag{14.11.2}$$

$$p \geqq 1, \qquad q \geqq 1, \tag{14.11.3}$$

$$|a_2| \leqq 2, \tag{14.11.4}$$

$$15 |a_2|^2 + 39 |\mu|^2 + 63 \left| a_4 - \frac{11}{8} a_2 \mu - \frac{77}{192} a_2^3 \right|^2 \leqq 64, \tag{14.11.5}$$

$$\mu = a_3 - \frac{11}{16} a_2^2, \tag{14.11.6}$$

$$a_4 > 0. \tag{14.11.7}$$

We have shown in preceding sections that there are other relations between these variables, but for the sake of flexibility we shall use only relations (14.11.1)–(14.11.7) at present.

Given a_2^* and a_3^*, there are p^*, q^*, φ^*, ψ^* such that (14.11.1), (14.11.2), and (14.11.3) are true. For let b_4^* be such that

$$Q^*(z) = z^3 + \frac{1}{z^3} + \frac{2a_2^*}{z^2} + 2\bar{a}_2^* z_2 + \frac{3a_3^*}{z} + 3\bar{a}_3^* z + 3b_4^*$$

is non-negative on $|z| = 1$ with at least one zero there. Then $Q^*(z)$ can be factored in the form

$$Q^*(z) = \frac{1}{z^3}[z + e^{-i(\varphi^*+\psi^*)}]^2(z - r^*e^{i\varphi^*})\left(z - \frac{1}{r^*}e^{i\varphi^*}\right)(z - \rho^*e^{i\psi^*})\left(z - \frac{1}{\rho^*}e^{i\psi^*}\right).$$

Expanding and equating coefficients of like powers of z, we obtain a_2^* and a_3^* in the forms (14.11.1) and (14.11.2) where

$$p^* = \frac{1}{2}\left(r^* + \frac{1}{r^*}\right) \geqq 1, \qquad q^* = \frac{1}{2}\left(\rho^* + \frac{1}{\rho^*}\right) \geqq 1.$$

Let a_2^*, a_3^*, and $\epsilon > 0$ be given, and let p^*, q^*, φ^*, ψ^* correspond to a_2^* and a_3^*. We show that there is a $\delta > 0$ such that for all a_2 and a_3 which satisfy $| a_2 - a_2^* | + | a_3 - a_3^* | \leqq \delta$, there are p, q, φ, ψ such that (14.11.1), (14.11.2), and (14.11.3) are true; and

$$| p - p^* | + | q - q^* | + | \varphi - \varphi^* | + | \psi - \psi^* | \leqq \epsilon.$$

Given a_2 and a_3 such that $| a_2 - a_2^* | + | a_3 - a_3^* | \leqq \delta$, let b_4 be chosen so that

$$Q(z) = z^3 + \frac{1}{z^3} + \frac{2a_2}{z^2} + 2\bar{a}_2 z^2 + \frac{3a_3}{z} + 3\bar{a}_3 z + 3b_4$$

is non-negative on $| z | = 1$ with at least one zero there. Then

$$Q^*(z) - Q(z) = 2\,\frac{a_2^* - a_2}{z^2} + 2(\bar{a}_2^* - \bar{a}_2)z^2$$
$$+ 3\,\frac{a_3^* - a_3}{z} + 3(\bar{a}_3^* - \bar{a}_3)z + 3(b_4^* - b_4).$$

At a zero of Q^* on $| z | = 1$ we have

$$0 \geqq -Q = 2\,\frac{a_2^* - a_2}{z^2} + \cdots + 3(b_4^* - b_4),$$

and so

$$3(b_4^* - b_4) - 4\,| a_2^* - a_2 | - 6\,| a_3^* - a_3 | \leqq 0$$

or

$$3(b_4^* - b_4) \leqq 6\{| a_2^* - a_2 | + | a_3^* - a_3 |\} \leqq 6\delta,$$
$$b_4^* - b_4 \leqq 2\delta.$$

At a zero of Q on $| z | = 1$,

$$0 \leqq Q^* \leqq 4\,| a_2^* - a_2 | + 6\,| a_3^* - a_3 | + 3(b_4^* - b_4);$$

so

$$b_4^* - b_4 \geqq -2\delta.$$

Thus

$$-2\delta \leqq b_4^* - b_4 \leqq 2\delta.$$

Now

$$Q(z) = \frac{1}{z^3}[z + e^{-i(\varphi+\psi)}]^2(z - re^{i\varphi})\left(z - \frac{1}{r}e^{i\varphi}\right)(z - \rho e^{i\psi})\left(z - \frac{1}{\rho}e^{i\psi}\right).$$

Given $\eta > 0$, the roots of Q can be made to lie within a distance η of those of Q^* by choosing δ small. Then by a proper lettering of the roots we have

$$|(\varphi + \psi) - (\varphi^* + \psi^*)| \leqq \eta, \quad |re^{i\varphi} - r^*e^{i\varphi^*}| \leqq \eta, \quad |\rho e^{i\psi} - \rho^* e^{i\psi^*}| \leqq \eta.$$

Since $r^* > 0$, $\rho^* > 0$, it follows that for all small η

$$|p - p^*| + |q - q^*| + |\varphi - \varphi^*| + |\psi - \psi^*| \leqq \epsilon.$$

Here φ and ψ are properly chosen modulo 2π. If $p^* = 1$, then which zero of $Q^*(z)$ on $|z| = 1$ is designated by $e^{i\varphi}$ may depend on a_2, a_3; and if $q^* = 1$, likewise.

Relations (14.11.1)–(14.11.7) implicitly contain a bound for p and q. For multiply (14.11.1) by $4e^{i(\varphi+\psi)}$ and subtract from (14.11.2). Then

$$(14.11.8) \qquad 3a_3 - 4a_2 e^{i(\varphi+\psi)} = -3e^{2i(\varphi+\psi)} + e^{-2i\varphi} + e^{-2i\psi} + 4pqe^{-i(\varphi+\psi)}.$$

Now (14.11.4) and (14.11.5) give bounds for $|a_2|$ and $|a_3|$. From (14.11.5) we have

$$|\mu| = \left|a_3 - \frac{11}{16}a_2^2\right| \leqq \left(\frac{64}{39}\right)^{1/2} < \left(\frac{64}{36}\right)^{1/2} = \frac{4}{3}$$

and so

$$|a_3| \leqq \frac{4}{3} + \frac{11}{16}|a_2|^2 \leqq \frac{4}{3} + \frac{11}{4} = \frac{49}{12} < 5.$$

Thus (14.11.8) gives

$$4\,pq \leqq 15 + 8 + 3 + 1 + 1 = 28.$$

Since $p \geqq 1$, $q \geqq 1$, we obtain

$$(14.11.9) \qquad p \leqq 7, \qquad q \leqq 7.$$

The range of variation of p, q, φ, ψ may be divided into two parts. Let γ_1 and γ_2 be given numbers which satisfy

$$(14.11.10) \qquad 1 < \gamma_1 \leqq 1.1, \qquad 1.25 \leqq \gamma_2 < \infty,$$

and let $D_1 = D_1(\gamma_1, \gamma_2)$ be the set where $\min(p, q) \leqq \gamma_1$ or $\max(p, q) \geqq \gamma_2$, $D_2 = D_2(\gamma_1, \gamma_2)$ the set where $\gamma_1 \leqq p \leqq \gamma_2$ and $\gamma_1 \leqq q \leqq \gamma_2$. The sets D_1 and D_2

intersect in the set where $p = \gamma_1$ or γ_2, $\gamma_1 \leqq q \leqq \gamma_2$, or $q = \gamma_1$ or γ_2, $\gamma_1 \leqq p \leqq \gamma_2$. We call this set the boundary of D_1. We shall show that if the maximum of a_4 in $D_1(\gamma_1, \gamma_2)$ is equal to or greater than 3.6, then this maximum must occur on the boundary of D_1. It will follow that if values of γ_1 and γ_2 are chosen, the calculation can be based on three rather than four real parameters, for it will be necessary to consider only the case in which p or q is equal to γ_1 or γ_2. By (14.11.9) we may suppose that $p \leqq 7$, $q \leqq 7$.

Suppose that the maximum of a_4 in D_1 occurs at a point $p^*, q^*, \varphi^*, \psi^*$ where $\min(p^*, q^*) < \gamma_1$ or $\max(p^*, q^*) > \gamma_2$, and that this maximum is equal to or greater than 3.6. First a_2 can be moved slightly and then a_3 can be moved locally in an arbitrary manner. Hence a_2 and μ can be varied in an arbitrary manner locally such that the point p, q, φ, ψ remains in D_1. If we write

$$\lambda = \frac{11}{8} a_2 \mu + \frac{77}{192} a_2^3, \qquad R = \left\{\frac{64 - 15|a_2|^2 - 39|\mu|^2}{63}\right\}^{1/2},$$

then inequality (14.11.5) states that a_4 lies in or on a circle of radius R with center at λ. For maximum a_4 it is clear that a_4 is a point where the circumference of this circle intersects the positive real axis and (14.11.5) is an equality. If the circle intersects the positive axis at two points, then a_4 is at the intersection lying to the right. Now

$$R \leqq \left(\frac{64}{63}\right)^{1/2} < 2;$$

so, under the assumption that $\max(a_4) > 3.6$ in D_1, $\operatorname{Re}(\lambda^*) > 0$. If $\operatorname{Im}(\lambda^*) \neq 0$, let $\mu = \mu^* e^{2i\epsilon}$, $a_2 = a_2^* e^{i\epsilon}$. Then $\lambda = \lambda^* e^{3i\epsilon}$ and $R = R^*$; so by choosing the real number ϵ properly, $|\arg(\lambda)|$ is decreased and a_4 is increased. But a_4 is maximal; so $\operatorname{Im}(\lambda^*) = 0$, $\lambda^* > 0$. Now rotate a_2 and μ slightly, keeping the absolute value of each fixed. Let

$$a_2 = a_2^* e^{i(\epsilon/3)}, \qquad \mu = \mu^* e^{i(\delta - \epsilon/3)},$$

where ϵ and δ are real numbers. Then $R = R^*$ and

$$\lambda = \frac{11}{8} a_2^* \mu^* e^{i\delta} + \frac{77}{192}(a_2^*)^3 e^{i\epsilon} = \frac{11}{8} a_2^* \mu^*(1 + i\delta) + \frac{77}{192}(a_2^*)^3(1 + i\epsilon) + O(\epsilon^2 + \delta^2)$$

$$= \lambda^* + a\delta e^{iu} + b\epsilon e^{iv} + O(\epsilon^2 + \delta^2)$$

where

$$ae^{iu} = \frac{11}{8} a_2^* \mu^* i, \qquad be^{iv} = \frac{77}{192}(a_2^*)^3 i.$$

Then

$$\lambda = \lambda^* + a\delta \cos u + b\epsilon \cos v + i\{a\delta \sin u + b\epsilon \sin v\} + O(\epsilon^2 + \delta^2).$$

The equations

$$a\delta \cos u + b\epsilon \cos v = x_1\,,$$

$$a\delta \sin u + b\epsilon \sin v = x_2$$

have determinant $ab \sin (v - u)$. If this determinant is not zero, let $x_1 > 0$ and $x_2 = 0$. Then for small x_1 we see that $\lambda = \lambda^* + x_1 + O(x_1^2)$; so a_4 can be increased. Thus the determinant is zero. If $ab = 0$, then $a_2^* = 0$ or $\mu^* = 0$. If $a_2^* = 0$, then

$$a_4 = \left(\frac{64 - 39\,|\,\mu^*\,|^2}{63}\right)^{1/2} < 2,$$

and so $a_2^* \neq 0$. If $\mu^* = 0$, then, since $\lambda^* > 0$, we have $(a_2^*)^3 > 0$, and so

$$a_4 = \frac{77}{192}\,(a_2^*)^3 + \frac{1}{(63)^{1/2}}\,\{64 - 15\,|\,a_2^*\,|^2\}^{1/2} < 3.59.$$

Since max $(a_4) > 3.6$, it follows that $\mu^* \neq 0$. Thus $\sin (v - u) = 0$. But

$$\frac{be^{i(v-u)}}{a} = \frac{77}{192}\,\frac{8}{11}\,\frac{(a_2^*)^3}{a_2^*\mu^*} = \frac{7}{24}\,\frac{(a_2^*)^2}{\mu^*}\,;$$

so

$$\frac{b}{a}\,\sin (v - u) = \operatorname{Im}\left(\frac{7}{24}\,\frac{(a_2^*)^2}{\mu^*}\right) = 0.$$

Dropping the superscript (*), we see that a_2^2/μ is real. Now

$$\lambda = \frac{11}{8}\,a_2\,\mu\left\{1 + \frac{7}{24}\,\frac{a_2^2}{\mu}\right\} > 0;$$

so $a_2\mu$ is also real, and then a_2^3 is real. If $a_2\mu \leqq 0$, we have

$$a_4 \leqq \frac{77}{192}\,a_2^3 + \frac{1}{(63)^{1/2}}\,\{64 - 15\,|\,a_2\,|^2 - 39\,|\,\mu\,|^2\}^{1/2} < 3.59.$$

Thus $a_2\mu > 0$. If $a_2^3 \leqq 0$,

$$a_4 \leqq \frac{11}{8}\,a_2\,\mu + \frac{1}{(63)^{1/2}}\,\{64 - 15\,|\,a_2\,|^2 - 39\,|\,\mu\,|^2\}^{1/2}.$$

Varying μ to obtain a maximum, we have

$$a_4 \leqq \left\{\frac{(64 - 15\,|a_2\,|^2)((121/64)\,|\,a_2\,|^2 + 39/63)}{39}\right\}^{1/2} < 1.96,$$

and it follows that $a_2^3 > 0$.

Keeping p and q fixed, we now increase both φ and ψ by $2\pi/3$ or by $4\pi/3$. Then a_2 is multiplied by $e^{4\pi i/3}$ or $e^{2\pi i/3}$ and a_3 is multiplied by $e^{2\pi i/3}$ or $e^{4\pi i/3}$ respectively. Hence $a_2\,\mu$ and a_2^3 are unchanged. Without loss of generality we may therefore suppose that $a_2 > 0$. Since $a_2\,\mu > 0$, we also have $\mu > 0$. Then

$$a_4 = \frac{11}{8} a_2 \mu + \frac{77}{192} a_2^3 + \frac{1}{(63)^{1/2}} \{64 - 15a_2^2 - 39\mu^2\}^{1/2}.$$

If $a_2 < 2$, a_2 and μ can each be varied locally in the neighborhood of the point where the maximum occurs and

$$\frac{\partial a_4}{\partial a_2} = \frac{\partial a_4}{\partial \mu} = 0.$$

Thus

$$\frac{11}{8} \mu + \frac{77}{64} a_2^2 = \frac{15a_2}{(63)^{1/2}\{64 - 15a_2^2 - 39\mu^2\}^{1/2}}, \tag{14.11.11}$$

$$\frac{11}{8} a_2 = \frac{39\mu}{(63)^{1/2}\{64 - 15a_2^2 - 39\mu^2\}^{1/2}}. \tag{14.11.12}$$

Dividing one of these equations by the other, we obtain

$$a_2^2 = \frac{104\mu^2}{40 - 91\mu}.$$

On substituting into (14.11.12) and simplifying,

$$3{,}006{,}003\,\mu^3 - 3{,}503{,}864\,\mu^2 - 4{,}175{,}808\,\mu + 2{,}001{,}920 = 0.$$

Since $\mu > 0$, $0 < a_2 \leqq 2$, we obtain

$$\mu = 0.393383, \qquad a_2 = 1.957018,$$

and

$$a_3 = \mu + \frac{11}{16} a_2^2 = 3.026452.$$

Thus if max a_4 occurs for min $(p, q) < \gamma_1$ or max $(p, q) > \gamma_2$, then either

(a) $a_2 = 1.957018$ and $a_3 = 3.026452$

or

(b) $a_2 = 2$.

In case (b) we have

$$a_4 = \frac{11}{4} \mu + \frac{77}{24} + \left\{\frac{4 - 39\mu^2}{63}\right\}^{1/2}.$$

In this case it is still possible to vary μ; so

$$\frac{\partial a_4}{\partial \mu} = \frac{11}{4} - \frac{39}{(63)^{1/2}} \frac{\mu}{(4 - 39\mu^2)^{1/2}} = 0$$

and we have

$$\mu = 0.307902.$$

Case (b) therefore gives

$$a_3 = \frac{11}{16} a_2^2 + \mu = 3.057902.$$

Finally, it will be shown that neither case (a) nor case (b) can occur in the region D_1, and we consider case (a) first. If all roots of the function Q are real, then we have one of the four following cases:

$$\text{(14.11.13)} \qquad \begin{array}{ll} \text{(i)} \quad \varphi = \psi = 0; & \text{(ii)} \quad \varphi = 0, \psi = \pi; \\ \text{(iii)} \quad \varphi = \pi, \psi = 0; & \text{(iv)} \quad \varphi = \psi = \pi. \end{array}$$

Case (i) gives $a_2 = 1 - p - q$ and this is obviously impossible since $p \geqq 1$, $q \geqq 1$. Case (ii) gives $a_2 = -1 - p + q$, $q = a_2 + p + 1$, and

$$3a_3 = 3 + 4p - 4q - 4pq = 3 + 4p - 4(a_2 + p + 1) - 4p(a_2 + p + 1),$$

that is,

$$4p^2 + 4p(a_2 + 1) + 4a_2 + 3a_3 + 1 = 0.$$

Here all terms are positive and we obtain a contradiction. Case (iii) is impossible by symmetry and case (iv) gives $a_2 = p + q + 1$, which is excluded because $p \geqq 1$, $q \geqq 1$, and $a_2 = 1.957 \cdots$ by hypothesis. Thus in case (a) there is at least one root of Q which is not real.

The roots of Q are

$$e^{i(\pi - \varphi - \psi)}, \qquad re^{i\varphi}, \qquad \frac{1}{r} e^{i\varphi}, \qquad \rho e^{i\psi}, \qquad \frac{1}{\rho} e^{i\psi},$$

the first of these roots being a double root. Here φ and ψ are real and $0 < r \leqq 1$, $0 < \rho \leqq 1$, $p = (r^2 + 1)/(2r)$, $q = (\rho^2 + 1)/(2\rho)$. There are the following cases (since the roots are symmetrical about the real axis):

$$\text{(14.11.14)} \qquad \begin{array}{llll} \text{(i)} & p = q = 1, & & \psi = \varphi + \psi - \pi, \\ \text{(ii)} & p = q = 1, & & \varphi = \varphi + \psi - \pi, \\ \text{(iii)} & p = q = 1, & & \varphi = -\psi, \\ \text{(iv)} & p = 1, & q > 1, & \psi = 0 \text{ or } \pi, \\ \text{(v)} & p > 1, & q = 1, & \varphi = 0 \text{ or } \pi, \\ \text{(vi)} & p > 1, & q > 1, & \varphi = -\psi \not\equiv 0 \pmod{\pi}. \end{array}$$

Case (i) gives $p = q = 1$, $\varphi = \pi$. Then

$$a_2 = 1 - e^{i\psi} - e^{-i\psi} = 1 - 2\cos\psi;$$

so

$$\cos \psi = \frac{1 - a_2}{2}.$$

But then

$$3a_3 = 5 - 8 \cos \psi + 4 \cos^2 \psi - 2$$

or

$$3(a_3 - 1) = -(1 - a_2)\cdot(3 + a_2).$$

This is impossible since $a_2 = 1.957 \cdots$, $a_3 = 3.026 \cdots$. Case (ii) is excluded by symmetry. In case (iii) we have

$$a_2 = 1 - e^{-i\psi} - e^{i\psi} = 1 - 2 \cos \psi;$$

so

$$\cos \psi = \frac{1 - a_2}{2},$$

$$3a_3 = 5 - 8 \cos \psi + 4 \cos^2 \psi - 2,$$

and this case is out as in (i). If $\psi = 0$ in case (iv), then

$$a_2 = e^{i\varphi} - q - e^{-i\varphi} = -q + 2i \sin \varphi;$$

so

$$a_2 + q = 2i \sin \varphi,$$

which is impossible since the left side is positive. Hence $\psi = \pi$, and

$$a_2 = q - 2 \cos \varphi, \qquad q = a_2 + 2 \cos \varphi,$$

$$3a_3 = 3 - 12 \cos^2 \varphi - 8a_2 \cos \varphi,$$

$$12 \cos^2 \varphi + 8a_2 \cos \varphi + 3a_3 - 3 = 0.$$

The roots of this equation are seen to be imaginary. Case (v) is also impossible by symmetry. Finally, in case (vi), $\varphi = -\psi \not\equiv 0 \pmod{\pi}$ since the case of all roots real has been excluded. Then $p = q$ and we have

$$a_2 = 1 - pe^{i\psi} - pe^{-i\psi} = 1 - 2p \cos \psi, \qquad 2p \cos \psi = 1 - a_2,$$

$$3a_3 + 5 - 4a_2 = 4 \cos^2 \psi + \frac{(1 - a_2)^2}{\cos^2 \psi}.$$

Then

$$6.25128 = 4 \cos^2 \psi + \frac{0.915883}{\cos^2 \psi},$$

giving

$$\cos^2 \psi = 0.16365 \quad \text{or} \quad 1.39917,$$

$$\cos \psi = \pm 0.40453,$$

$$p = \frac{1 - a_2}{2 \cos \psi} = 1.1829.$$

Thus $1.1 < p = q < 1.25$, which is contrary to hypothesis.

Now consider case (b). If all roots are real, we have one of the cases (14.11.13). In case (i), $a_2 = 2 = 1 - p - q$, which is impossible. Cases (ii), (iii), (iv) are excluded as in (a). If at least one root is not real, we obtain the six cases (14.11.14). Case (i) gives $5 < 3a_3 - 3 = -(1 - a_2)(3 + a_2)$, which is impossible. Cases (ii) and (iii) are seen to be excluded. In case (iv) we have, assuming $\psi = 0$, $2 = -q + 2i \sin \varphi$, which is impossible. If $\psi = \pi$, we obtain an equation for $\cos \varphi$ with discriminant equal to $16\{4a_2^2 - 9(a_3 - 1)\} = 16\{25 - 9a_3\} < 0$. Finally, in case (vi), $\varphi = -\psi \not\equiv 0 \pmod{\pi}$ and we obtain the equation

$$3a_3 + 5 - 4a_2 = 4 \cos^2 \psi + \frac{(1 - a_2)^2}{\cos^2 \psi};$$

so

$$6.1737 = 4 \cos^2 \psi + \frac{1}{\cos^2 \psi},$$

giving

$$\cos \psi = \pm 0.42882, \qquad p = \frac{1 - a_2}{2 \cos \psi} = 1.1660.$$

Thus $1.1 < p = q < 1.25$, contrary to assumption.

We thus see that the only possible maxima of a_4 in D_1 at which $a_4 > 3.6$ occur for $\max(p, q) = \gamma_2$ or $\min(p, q) = \gamma_1$. Therefore if, for particular γ_1 and γ_2, $1 \leqq \gamma_1 \leqq 1.1$, $1.25 \leqq \gamma_2 < \infty$, we show that the relations (14.11.1)—(14.11.7) imply $a_4 < 4$ when $\max(p, q) = \gamma_2$ or $\min(p, q) = \gamma_1$, it will follow that these same relations imply $a_4 < 4$ throughout D_1. Furthermore, by symmetry, we may assume without loss of generality either that $p \leqq q$ or that $p \geqq q$. We observe that we can add $2\pi/3$ or $4\pi/3$ to both φ and ψ without changing the maximum of a_4 and we can also replace φ and ψ by $-\varphi$ and $-\psi$. Hence, in order to exclude the region D_1 from further consideration it is sufficient to show that (14.11.1)–(14.11.7) imply $a_4 < 4$ when

$$\begin{aligned} &p = \gamma_1, \quad \gamma_1 \leqq q \leqq \gamma_2, \quad 0° \leqq \varphi \leqq 60°, \\ &p = \gamma_2, \quad \gamma_1 \leqq q \leqq \gamma_2, \quad 0° \leqq \varphi \leqq 60°. \end{aligned} \tag{14.11.15}$$

Now suppose that the parameters lie in the part of D_2 for which

$$\gamma_1 \leqq p \leqq q \leqq \gamma_2, \qquad 0^\circ \leqq \varphi \leqq 60^\circ. \tag{14.11.16}$$

14.12. The final step in the calculation is to integrate the system of differential equations described in Chapters IX and XI when the starting values satisfy (14.11.16). Writing $z_1 = re^{i\varphi}$, $z_2 = \rho e^{i\psi}$, and separating the equations (9.2.17) into their real and imaginary parts, we obtain the four equations

$$\begin{aligned} u\frac{dr}{du} &= r\frac{1 - r^2}{1 - 2r\cos(\alpha + \varphi) + r^2}, \\ u\frac{d\rho}{du} &= \rho\frac{1 - \rho^2}{1 - 2\rho\cos(\alpha + \psi) + \rho^2}, \\ u\frac{d\varphi}{du} &= 2r\frac{\sin(\alpha + \varphi)}{1 - 2r\cos(\alpha + \varphi) + r^2}, \\ u\frac{d\psi}{du} &= 2\rho\frac{\sin(\alpha + \psi)}{1 - 2\rho\cos(\alpha + \psi) + \rho^2}. \end{aligned} \tag{14.12.1}$$

Here

$$e^{-i\alpha}, \qquad re^{i\varphi}, \qquad \frac{1}{r}e^{i\varphi}, \qquad \rho e^{i\psi}, \qquad \frac{1}{\rho}e^{i\psi}$$

denote the roots of the function

$$Q(z, u) = \sum_{\nu=-3}^{3} \frac{B_\nu(u)}{z^\nu},$$

where at $u = 1$ the quantities $B_\nu(u)$ have the values (14.10.5). Since $B_3(u) = B_{-3}(u) = u^3$, we see that

$$-z_1 z_2 e^{-i\alpha} > 0,$$

and so

$$\alpha = \varphi + \psi - \pi. \tag{14.12.2}$$

Writing, as in (9.2.8),

$$C_k(u) = B_k(u)u^{-k}, \tag{14.12.3}$$

we have by (9.2.9),

$$u\frac{dC_k}{du} = \sum_{\nu=1}^{3-k} (k - \nu)C_{k+\nu}u^\nu e^{i\nu\alpha}, \qquad k = 0, 1, 2. \tag{14.12.4}$$

Write

$$p = \frac{1}{2}\left(r + \frac{1}{r}\right), \qquad q = \frac{1}{2}\left(\rho + \frac{1}{\rho}\right). \tag{14.12.5}$$

Introducing these quantities into (14.12.1) and using (14.12.2), we obtain

$$(14.12.1)' \qquad \begin{aligned} u\frac{dp}{du} &= -\frac{p^2-1}{p-\cos(\alpha+\varphi)}, & u\frac{dq}{du} &= -\frac{q^2-1}{q-\cos(\alpha+\psi)}, \\ u\frac{d\varphi}{du} &= \frac{\sin(\alpha+\varphi)}{p-\cos(\alpha+\varphi)}, & u\frac{d\psi}{du} &= \frac{\sin(\alpha+\psi)}{q-\cos(\alpha+\psi)}. \end{aligned}$$

Thus, in particular,

$$(14.12.6) \qquad u\frac{d\alpha}{du} = \frac{\sin(\alpha+\varphi)}{p-\cos(\alpha+\varphi)} + \frac{\sin(\alpha+\psi)}{q-\cos(\alpha+\psi)}.$$

To distinguish the starting values of these equations from the variable quantities we attach the subscript 1. By (14.11.16) we have

$$(14.12.7) \qquad \gamma_1 \leqq p_1 \leqq q_1 \leqq \gamma_2, \qquad 0° \leqq \varphi_1 \leqq 60°.$$

Since

$$-(p+1) \leqq u\frac{dp}{du} \leqq -(p-1), \qquad \frac{p+1}{p_1+1} \leqq \frac{1}{u} \leqq \frac{p-1}{p_1-1},$$

and similarly for q by (14.12.1)$'$, we obtain from (14.12.6)

$$(14.12.8) \qquad u\left|\frac{d\alpha}{du}\right| \leqq \frac{1}{p-1} + \frac{1}{q-1} \leqq u\left\{\frac{1}{p_1-1} + \frac{1}{q_1-1}\right\} \leqq \frac{2u}{\gamma_1-1}.$$

Hence

$$(14.12.9) \qquad \left|\frac{d\alpha}{du}\right| \leqq \frac{2}{\gamma_1-1}.$$

The equations (14.12.1)$'$ constitute a closed system and can be integrated to obtain $\alpha(u)$ as a function of u. Then using Löwner's formula

$$(14.12.10) \qquad \begin{aligned} a_4 = {} & 2\int_0^1 u^2 e^{3i\alpha(u)}\,du - 12\int_0^1 ue^{2i\alpha(u)}\,du\cdot\int_0^1 e^{i\alpha(u)}\,du \\ & + 4\int_{v=0}^1\int_{u=0}^v ve^{i[\alpha(u)+2\alpha(v)]}\,du\,dv + 8\left(\int_0^1 e^{i\alpha(u)}\,du\right)^3, \end{aligned}$$

we obtain a value for a_4 corresponding to a boundary point of $\mathbf{\Pi}_1$. However, it may be simpler to integrate the equations (14.12.4) by using the relations

$$(14.12.11) \qquad \begin{aligned} Q(e^{-i\alpha}, u) &= \sum_{\nu=-3}^{3} C_\nu(u)u^\nu e^{i\nu\alpha} = 0, \\ -\left[z\frac{\partial}{\partial z}Q(z,u)\right]_{z=e^{-i\alpha}} &= \sum_{\nu=-3}^{3} \nu C_\nu u^\nu e^{i\nu\alpha} = 0 \end{aligned}$$

which express the fact that $Q(z, u)$ has a double zero at the point $z = e^{-i\alpha}$. By (9.2.13) we have

$$C_1(0) = A_1, \qquad C_2(0) = A_2, \qquad C_3(0) = 1,$$

and hence a_4 can be calculated from (14.2.5) with

$$\begin{aligned} &B_0 = C_0(1), \qquad B_1 = C_1(1), \qquad B_2 = C_2(1), \qquad B_3 = 1, \\ &A_1 = C_1(0), \qquad A_2 = C_2(0), \qquad A_3 = 1. \end{aligned} \tag{14.12.12}$$

We begin at $u = 1$ with values corresponding to the range (14.12.7) and whenever $\alpha(u)$ deviates from a constant c by an amount not exceeding 0.08697, we apply Lemma XXXII. Since values of $\alpha(u)$ will be obtained at a discrete set of points, we use (14.12.9) as a means of estimating intermediate values of $\alpha(u)$.

We remark that the values of a_2 and a_3 given by (14.2.5) should also be calculated since each point (a_2, a_3, a_4) obtained in this way is a point on $\mathbf{\Pi}_1$ no matter how we choose the starting values (so long as $p_1 > 1$, $q_1 > 1$).

CHAPTER XV

THE REGION OF VALUES OF THE DERIVATIVE OF A SCHLICHT FUNCTION

BY

ARTHUR GRAD

15.1. Let z_0 be a fixed point in $|z| < 1$. As $f(z)$ varies over the class $\mathfrak{S}$, the set of values $f'(z_0)$ covers a region in the complex plane which has been called the region of variability of the derivative of a schlicht function (with respect to the point z_0). Our object is to obtain this region, which we shall denote by $\boldsymbol{R}(z_0)$.[1] First, however, we deduce some geometrical properties of $\boldsymbol{R}(z_0)$ which can be obtained without exact knowledge of the region.

The region $\boldsymbol{R}(z_0)$ is essentially the solution to a more general problem. Let

$$f(w) = w + a_2 w^2 + \cdots$$

be regular and schlicht in a simply-connected domain D containing the origin. Let $w = w(z)$ be the function belonging to class $\mathfrak{S}$ which maps the unit circle on D. If we write $f(w) = f[w(z)] = F(z)$, then $F(z)$ also belongs to class $\mathfrak{S}$. If $w_0 = w(z_0)$, then

$$F'(z_0) = f'(w_0) \cdot w'(z_0).$$

Since $w'(z_0)$ depends only on D and z_0, not on f, it follows that the region of variability with respect to the point w_0 of a function regular and schlicht in D is simply a euclidean magnification and rotation of $\boldsymbol{R}(z_0)$, the magnification factor being $z'(w_0)$.

The point z_0 can be chosen to be real and positive without any loss of generality, because if $f(z)$ belongs to class $\mathfrak{S}$, then so does $F(z) = e^{-i\theta} f(e^{i\theta} z)$, and $F'(z) = f'(e^{i\theta} z)$. We shall therefore take $z_0 = r$, $0 \leqq r < 1$.

Since the class $\mathfrak{S}$ is compact, $\boldsymbol{R}(r)$ is closed. It is also bounded, the precise bounds being given by the distortion and rotation theorems, (1.1.1) and (1.1.1)′.

The region $\boldsymbol{R}(r)$ has two symmetries. It is symmetric with respect to the real axis since if $f(z)$ belongs to class $\mathfrak{S}$, then so does $\bar{f}(\bar{z})$. The other symmetry is less obvious. Let $f(z)$ belong to class $\mathfrak{S}$, and let

$$z = \frac{w + r}{1 + rw}.$$

Then $f(z) = \varphi(w)$ is regular and schlicht in $|w| < 1$, and

$$\varphi'(w) = \frac{1 - r^2}{(1 + rw)^2} f'(z),$$

[1] In a paper not yet published, H. Grunsky investigated the region by integrating Löwner's differential equation. He has been kind enough to inform the writer of the general nature of his results, which include some of those stated here but do not characterize the boundary functions.

$$\varphi'(0) = (1 - r^2)f'(r).$$

Now let

$$F(w) = \frac{1}{\varphi'(0)} [\varphi(w) - \varphi(0)].$$

The function $F(w)$ belongs to class $\mathfrak{S}$, and

$$F'(w) = \frac{\varphi'(w)}{\varphi'(0)} = \frac{f'(z)}{(1 + rw)^2 f'(r)}.$$

Taking $w = -r$, we obtain

$$F'(-r) \cdot f'(r) = \frac{1}{(1 - r^2)^2}.$$

Thus $\boldsymbol{R}(r)$ is symmetric by inversion in the circle about the origin of radius $1/(1 - r^2)$. In the logarithmic plane, $\boldsymbol{R}(r)$ is symmetric with respect to the line

$$\text{Re}\,[\log f'(r)] = \log \frac{1}{1 - r^2}.$$

Clearly $\boldsymbol{R}(r)$ varies continuously with r. For if we let $0 < r < \rho_0 < 1$, then given $\epsilon > 0$, there exists a δ, depending only on ρ_0 and ϵ, such that $| f'(z_1) - f'(z_2) | < \epsilon$ for all points z_1 and z_2 satisfying $| z_1 | \leqq \rho_0$, $| z_2 | \leqq \rho_0$, $| z_1 - z_2 | < \delta$, and for all functions $f(z)$ belonging to class $\mathfrak{S}$. In particular, $| f'(r + \Delta r) - f'(r) | < \epsilon$ for $-\delta < \Delta r < \delta$; so there is a point of $\boldsymbol{R}(r + \Delta r)$ arbitrarily near any point of $\boldsymbol{R}(r)$ for Δr sufficiently small, and each point of $\boldsymbol{R}(r + \Delta r)$ lies near some point of $\boldsymbol{R}(r)$.

The region $\boldsymbol{R}(r)$ expands with increasing r; that is, $\boldsymbol{R}(r_1) \subset \boldsymbol{R}(r_2)$ for $r_1 < r_2$. To show this, write $r_1 = \rho r_2$, $0 < \rho < 1$, let $f(z)$ belong to class $\mathfrak{S}$, and let $g(z) = f(\rho z)/\rho$. Then $g(z)$ also belongs to class $\mathfrak{S}$, and $g'(r_2) = f'(\rho r_2) = f'(r_1)$. Thus any point $f'(r_1)$ of $\boldsymbol{R}(r_1)$ is a point $g'(r_2)$ of $\boldsymbol{R}(r_2)$.

We remark that if $f'(r)$ is an interior point of $\boldsymbol{R}(r)$, then $f'(r)$ belongs to $\boldsymbol{R}(r - \Delta r)$ provided that Δr is sufficiently small. Hence there is a function $g(z)$ of class $\mathfrak{S}$ such that $g'(r - \Delta r) = f'(r)$, that is, $g'(\rho r) = f'(r)$ where $\rho = (r - \Delta r)/r$. Writing

$$h(z) = \frac{1}{\rho}\, g(\rho z),$$

we have $h'(r) = f'(r)$ where $h(z)$ is bounded in $| z | < 1$. Conversely, if a function $f(z)$ belonging to the point $f'(r)$ of $\boldsymbol{R}(r)$ is bounded or satisfies the weaker condition that the map of $| z | < 1$ by $w = f(z)$ has an exterior point w_0, then $f'(r)$ is an interior point of $\boldsymbol{R}(r)$. This follows from an argument similar to the one given in the proof of Lemma I. If $f(z)$ belongs to the point $f'(r)$ of $\boldsymbol{R}(r)$, the function $f(\rho z)/\rho$ belongs to the point $f'(\rho r)$. If $0 \leqq \rho < 1$, the point $f'(\rho r)$ is an interior point of $\boldsymbol{R}(r)$ since $f(\rho z)/\rho$ is bounded, and it follows that $\boldsymbol{R}(r)$ is a closed domain. However, when considered as a schlicht domain in the plane of the values

$f'(r)$, $\boldsymbol{R}(r)$ is not simply-connected for $r > 1/2^{1/2}$. This is owing to the fact that $\arg f'(r)$, $\arg f'(0) = 0$, can exceed π in magnitude when $r > 1/2^{1/2}$. For this reason it is better to consider the set of values $\log f'(r)$ where $\log f'(r)$ is the value at $z = r$ of that branch of $\log f'(z)$ which vanishes at $z = 0$. Since for any function $f(z)$ of class $\mathfrak{S}$ the derivative $f'(z)$ does not vanish for $|z| < 1$, we see that any branch of $\log f'(z)$ is single-valued in $|z| < 1$. Let $\boldsymbol{L}(r)$ be the domain of variability of the branch of $\log f'(z)$ at $z = r$ which vanishes at $z = 0$.

Finally, we note that the curve $f'(re^{i\theta})$, where $f(z)$ belongs to class $\mathfrak{S}$ and θ is real, lies in $\boldsymbol{R}(r)$. Hence at a point where the boundary of $\boldsymbol{R}(r)$ has a tangent, $f''(r)$ defines a vector normal to the boundary.

For convenience we refer to functions $f(z)$ belonging to class $\mathfrak{S}$, for which $\log f'(r)$ lies on the boundary of $\boldsymbol{L}(r)$, as boundary functions. Here and henceforth, without explicit statement to the contrary, we interpret $\log f'(z)$ as that branch defined in $|z| < 1$ which vanishes at $z = 0$. By the principal value of the logarithm of a complex number we shall mean that value whose imaginary part exceeds $-\pi$ and is not greater than π. The value of $\log f'(z)$ will in general not be principal. By the principal value of the argument of a complex number, we mean the value which exceeds $-\pi$ and is not greater than π.

15.2. The region $\boldsymbol{L}(r)$ will be found in terms of three real parameters φ, ψ, and ρ, $0 < \rho \leqq 1$, which are connected by the relation

$$-e^{2i(\varphi+\psi)}(1 - r^2)^2 = (e^{i\psi} - r)^2(r - \rho e^{i\varphi})(r - \frac{1}{\rho} e^{i\varphi}). \tag{15.2.1}$$

Let

$$\sigma = \frac{r - \rho e^{i\varphi}}{r - (1/\rho)e^{i\varphi}}.$$

It will be shown from (15.2.1) that if $0 < \rho < 1$, then σ cannot be positive; we therefore take $\sigma^{1/2}$ to be the branch of the square root which has positive imaginary part $(0 < \rho < 1)$. We summarize the results to be proved.

For $0 < r < 1/2^{1/2}$, the region $\boldsymbol{L}(r)$ is bounded by a closed curve whose equation is

$$\begin{aligned}\log f'(r) = 2i \log \frac{\rho - \sigma^{1/2}}{\rho + \sigma^{1/2}} + 2 \log \frac{1 - re^{-i\psi}}{1 - r^2}\\ + e^{-i(\varphi+\psi)} \left\{ \log \left[\frac{1 - \rho}{1 + \rho} \cdot \frac{\rho - |\sigma|}{\rho + |\sigma|} \right] + 2 \arg \frac{1 - \sigma^{1/2}}{1 + \sigma^{1/2}} \right\} - \pi.\end{aligned} \tag{15.2.2}$$

Here each logarithm on the right side and the argument have principal values, and the logarithm on the left is as defined above. This is actually the equation of a curve since there is but one real independent variable; indeed we shall show that (when $0 < r < 1/2^{1/2}$) for each ψ there are unique ρ and $e^{i\varphi}$, $0 < \rho < 1$, which satisfy (15.2.1). Any boundary function $w = f(z)$ maps $|z| < 1$ onto the exterior of a single analytic slit in the w-plane extending to $w = \infty$. For $r = 1/2^{1/2}$, the corresponding boundary curve is given by (15.2.2), and, in this case, we have $0 < \rho < 1$ except at $\psi = \pm\pi/4$ where $\rho = 1$.

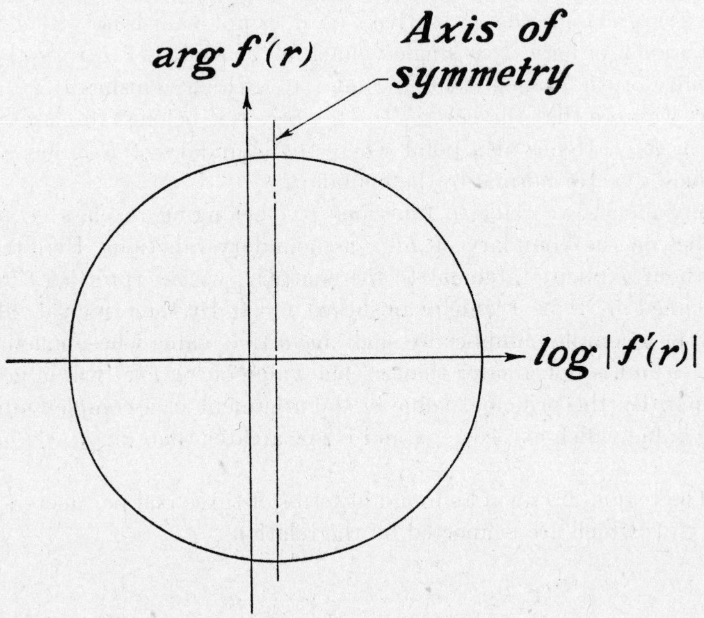

Region $L(r)$ $(r = \frac{1}{2})$

FIG. 16

If $1/2^{1/2} < r < 1$, the region $L(r)$ is bounded by two arcs alternating with two straight-line segments. The arcs are given by (15.2.2), one for ψ in the interval $-\pi/4 + \tau < \psi < \pi/4 - \tau$ and the other for ψ in the interval $\pi/4 + \tau < \psi < 2\pi - (\pi/4 + \tau)$. Here $\tau = \tau(r)$ is given by

$$\tau = \text{arc cos}\left[\frac{1}{2^{1/2}r}\right], \qquad 0 < \tau < \pi/4.$$

In each of these intervals ρ and $e^{i\varphi}$ are unique functions of ψ defined by (15.2.1) and $0 < \rho < 1$. For ψ in either of these intervals, the corresponding boundary function $w = f(z)$ maps $|z| < 1$ onto the exterior of a single analytic slit in the w-plane extending to $w = \infty$. The straight-line segments on the boundary of $L(r)$ are given by

$$C_0(r) \leqq \text{Re}\,[\log f'(r)] \leqq C_1(r), \qquad \text{Im}\,[\log f'(r)] = \pm\left(\log\frac{r^2}{1 - r^2} + \pi\right)$$

where

$$C_0(r) = -\log (r^2 + (2r^2 - 1)^{1/2}) + 2\tau,$$
$$C_1(r) = -\log (r^2 - (2r^2 - 1)^{1/2}) - 2\tau.$$

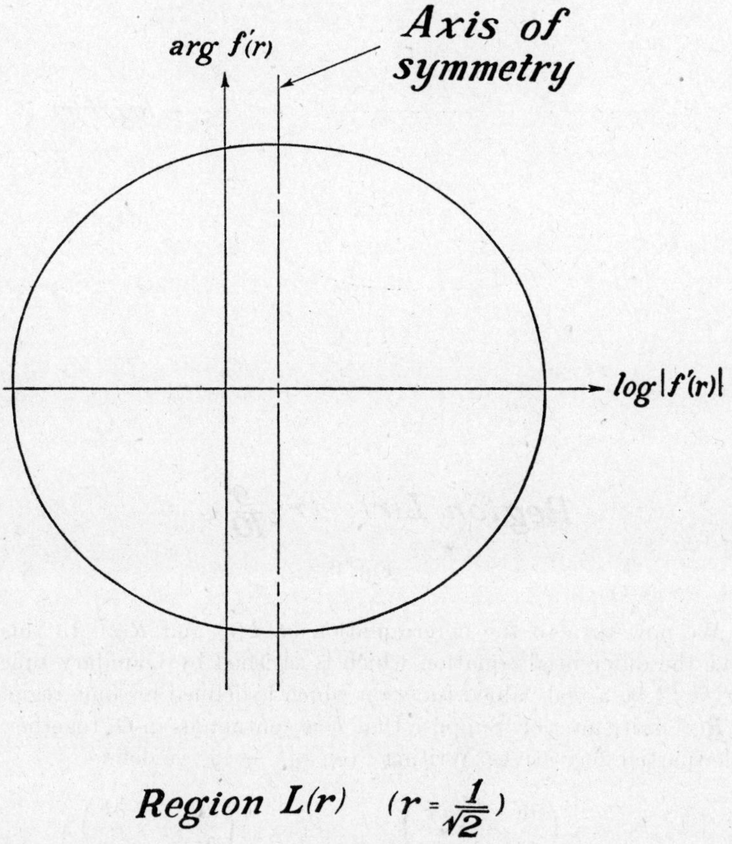

FIG. 17

The functions $w = f(z)$ corresponding to points on these line segments map $|z| < 1$ onto the w-plane minus a straight-line segment and two arcs that meet this straight-line segment. At the end points of these straight-line segments of the boundary of $L(r)$, one of the two arcs in the w-plane degenerates to a point.

The boundary curves of the region $L(r)$ corresponding to $r = 1/2$, $r = 1/2^{1/2}$, and $r = 9/10$ are shown in Figures 16, 17, and 18. We remark that for $1/2^{1/2} < r < 1$, there is a continuum of boundary functions for which $|\arg f'(r)|$ is maximal, a result which complements Golusin's result (1.1.1)′.

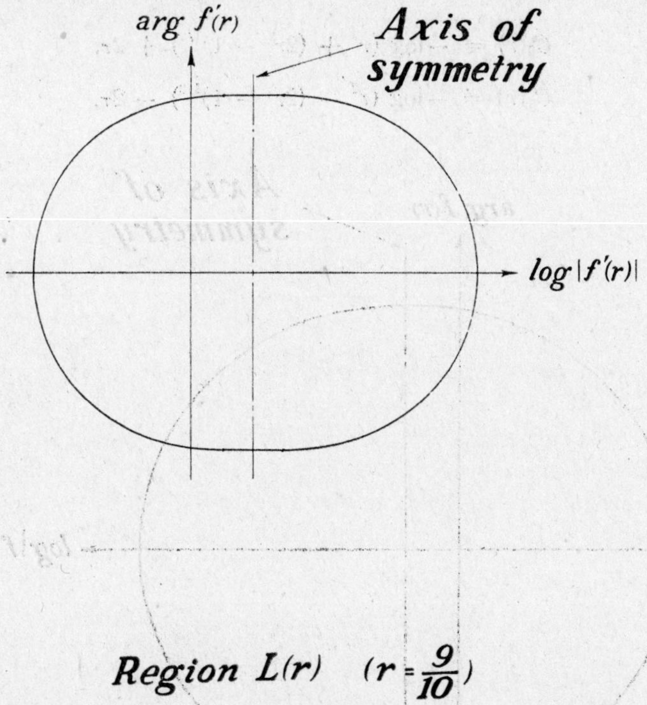

FIG. 18

15.3. We now turn to the determination of $L(r)$ and $R(r)$. In this section we derive the differential equation which is satisfied by boundary functions.

Let $F(f', \bar{f}')$ be a real-valued function which is defined in some region O containing $R(r)$ in its interior. Suppose that F is continuous in O, together with its first order partial derivatives. Writing $f'(r) = x + iy$, we define

$$F_1 = \frac{1}{2}\left(\frac{\partial F}{\partial x} - i\,\frac{\partial F}{\partial y}\right), \qquad \bar{F}_1 = \frac{1}{2}\left(\frac{\partial F}{\partial x} + i\,\frac{\partial F}{\partial y}\right),$$

and we suppose that $F_1 \neq 0$ in O.

Given any f belonging to class $\mathfrak{S}$, there exists for all sufficiently small complex numbers ϵ a neighboring function $f^*(z) = f_\epsilon(z)$ defined by (2.2.22), namely

$$f_\epsilon(z) = f(z) + \frac{\epsilon}{2\pi i}\int_\alpha^\beta \frac{p(u)}{2u^2}\left\{\left(u\,\frac{f'(u)}{f(u)}\right)^2 \frac{2f(z)^2}{f(u) - f(z)} - zf'(z)\,\frac{u+z}{u-z} + f(z)\right\} du$$
$$+ \frac{\bar{\epsilon}}{2\pi i}\int_\alpha^\beta \frac{\overline{p(u)}}{2\bar{u}^2}\left\{zf'(z)\,\frac{1+\bar{u}z}{1-\bar{u}z} - f(z)\right\} d\bar{u} + o(\epsilon).$$

Differentiating this equation with respect to z, we obtain

$$f'_\epsilon(z) = f'(z) + \frac{\epsilon}{2\pi i}\int_\alpha^\beta \frac{p(u)}{2u^2}\left\{\left(u\frac{f'(u)}{f(u)}\right)^2 \frac{2f(u) - f(z)}{[f(u) - f(z)]^2} 2f(z)f'(z)\right.$$

$$\left. + \frac{z - 2u}{(u - z)^2} 2zf'(z) - \frac{u + z}{u - z} zf''(z)\right\} du \tag{15.3.1}$$

$$+ \frac{\bar{\epsilon}}{2\pi i}\int_\alpha^\beta \overline{\frac{p(u)}{2\bar{u}^2}}\left\{\frac{\bar{u}(2 - \bar{u}z)}{(1 - \bar{u}z)^2} 2zf'(z) + \frac{1 + \bar{u}z}{1 - \bar{u}z} zf''(z)\right\} d\bar{u} + o(\epsilon).$$

Now let $f(z)$, belonging to class $\mathfrak{S}$, maximize F in $\boldsymbol{R}(r)$. Since the gradient of F is non-vanishing in $\boldsymbol{O}$, the maximum of F is attained at a boundary point. Writing

$$\Delta F = F(f'_\epsilon, \bar{f}'_\epsilon) - F(f', \bar{f}')$$

and

$$\Delta f' = f'_\epsilon - f', \qquad f' = f'(r), \qquad f'_\epsilon = f'_\epsilon(r),$$

we have

$$\Delta F = 2 \text{ Re } (F_1 \Delta f') + o(\epsilon).$$

Substituting the value of $\Delta f'$ from (15.3.1) and replacing the second term by its conjugate, we obtain

$$\Delta F = 2 \text{ Re}\left\{\frac{\epsilon}{2\pi i}\int_\alpha^\beta \frac{p(u)}{2u^2}\left[Q\left(u\frac{f'(u)}{f(u)}\right)^2 \frac{2f(u) - f(r)}{[f(u) - f(r)]^2} - M\frac{u + r}{u - r}\right.\right.$$

$$\left.\left. - \bar{M}\frac{1 + ur}{1 - ur} - N\frac{2u - r}{(u - r)^2} - \bar{N}\frac{u(2 - ur)}{(1 - ur)^2}\right] du\right\} + o(\epsilon)$$

where $Q = 2F_1 f(r) f'(r)$, $M = F_1 r f''(r)$, and $N = 2F_1 r f'(r)$. Since F attains its maximum for $f' = f'(r)$, we have for all sufficiently small ϵ,

$$\Delta F \leqq 0,$$

and it follows that f satisfies the differential equation

$$Q\left(\frac{zf'(z)}{f(z)}\right)^2 \frac{2f(z) - f(r)}{[f(z) - f(r)]^2}$$

$$= M\frac{z + r}{z - r} + \bar{M}\frac{1 + zr}{1 - zr} + N\frac{2z - r}{(z - r)^2} + \bar{N}\frac{z(2 - zr)}{(1 - zr)^2}. \tag{15.3.2}$$

The right side of (15.3.2) is real on $|z| = 1$. We shall show that it is non-negative there. The function defined by (2.3.6)′, namely

$$f_\epsilon(z) = f(z) + \epsilon\left\{f(z) - z\frac{1 + e^{-i\theta} z}{1 - e^{-i\theta} z} f'(z)\right\} + o(\epsilon), \qquad \epsilon > 0,$$

belongs to class $\mathfrak{S}$. Differentiating, we have

$$f'_\epsilon(z) = f'(z) - \epsilon \left\{ 2zf'(z) \frac{e^{-i\theta}(2 - e^{-i\theta} z)}{(1 - e^{-i\theta} z)^2} + zf''(z) \frac{1 + e^{-i\theta} z}{1 - e^{-i\theta} z} \right\} + o(\epsilon).$$

Applying this variation to the boundary function f, we obtain

$$\Delta F = -2\mathrm{Re} \left\{ \epsilon \left[N \frac{e^{-i\theta}(2 - e^{-i\theta} r)}{(1 - e^{-i\theta} r)^2} + M \frac{1 + e^{-i\theta} r}{1 - e^{-i\theta} r} \right] \right\} + o(\epsilon) \leqq 0.$$

This inequality implies that the expression

$$G(z) = M \frac{z + r}{z - r} + \bar{M} \frac{1 + zr}{1 - zr} + N \frac{2z - r}{(z - r)^2} + \bar{N} \frac{z(2 - zr)}{(1 - zr)^2}$$

is non-negative on $|z| = 1$.

Placing the four terms in $G(z)$ over a common denominator, we obtain a rational fraction in z with numerator of fourth degree. The degree of the numerator is precisely 4 since the coefficient of z^4 in the numerator is $r^2(M - \bar{M}) - r\bar{N}$ and this is equal to the conjugate of the constant term in the numerator. If the constant term were zero, then $G(z)$ would be equal to zero at $z = 0$, but the left side of (15.3.2) does not vanish at $z = 0$. Hence the right side of (15.3.2) has precisely four zeros, and since it is real on $|z| = 1$, the zeros not on $|z| = 1$ must come in pairs at inverse points with respect to $|z| = 1$. But the left side has at most one zero inside $|z| < 1$, which is attained when $f(z) = (1/2)f(r)$. If $G(z)$ has no zero on $|z| = 1$, it has two zeros in $|z| < 1$ and the differential equation (15.3.2) shows that this is impossible since $f'(z)$ does not vanish in $|z| < 1$. Thus $G(z)$ has at least one zero on $|z| = 1$, and this zero must be of second or higher order. On writing $w = f(z)$, it follows that (15.3.2) can be written in the form

$$A \left(\frac{z}{w} \frac{dw}{dz} \right)^2 \frac{2w - \alpha}{(w - \alpha)^2} = e^{-i(\varphi+\psi)} \frac{(z - e^{i\psi})^2(z - \rho e^{i\varphi})(z - (1/\rho)\cdot e^{i\varphi})}{(z - r)^2(z - 1/r)^2}.$$

Here A, φ, ψ, ρ are constants, φ, ψ, ρ are real, $0 < \rho \leqq 1$, and $\alpha = w(r)$.

Recalling that $w = z + O(z^2)$ and letting z approach zero, we have

$$A = -\alpha e^{i(\varphi+\psi)}.$$

Letting z approach r and noting that $(w - \alpha)/(z - r)$ approaches $w'(r)$, we obtain

$$A = \frac{\alpha e^{-i(\varphi+\psi)}}{(1 - r^2)^2} (r - e^{i\psi})^2(r - \rho e^{i\varphi}) \left(r - \frac{1}{\rho} e^{i\varphi} \right).$$

It follows that

$$-\alpha e^{i(\varphi+\psi)} \left(\frac{z}{w} \frac{dw}{dz} \right)^2 \frac{2w - \alpha}{(w - \alpha)^2} \tag{15.3.3}$$
$$= e^{-i(\varphi+\psi)} \frac{(z - e^{i\psi})^2(z - \rho e^{i\varphi})(z - (1/\rho)e^{i\varphi})}{(z - r)^2(z - 1/r)^2}$$

where the right side is non-negative on $|z| = 1$ and

$$-e^{2i(\varphi+\psi)}(1 - r^2)^2 = (r - e^{i\psi})^2(r - \rho e^{i\varphi})(r - \frac{1}{\rho} e^{i\varphi}). \tag{15.3.4}$$

The proof that the map of the unit circle by a boundary function has no exterior points parallels the argument given in Lemma I and so will be omitted. By the method of Lemma VIII, it can be shown that to each point of the boundary of $\boldsymbol{R}(r)$ there is at least one function of class $\mathfrak{S}$ which satisfies a differential equation of the form (15.3.3) with the right side non-negative on $|z| = 1$.

We remark for future reference that if ψ and r are given, $0 < r < 1$, then there is no more than one set of values $e^{i\varphi}$ and ρ which satisfy (15.3.4) where $0 < \rho \leqq 1$. To show this, let r be fixed and let

$$g(\psi) = \frac{(1 - r^2)^2}{(1 - re^{-i\psi})^2}, \qquad p = \frac{1}{2}\left(\rho + \frac{1}{\rho}\right).$$

Then equation (15.3.4) states that

$$r^2 e^{-2i\varphi} - 2pre^{-i\varphi} + 1 + g(\psi) = 0. \tag{15.3.5}$$

If $g(\psi) = g_1 + ig_2$, then multiplying (15.3.5) by $e^{i\varphi}$, we have

$$2pr = r^2 e^{-i\varphi} + (1 + g_1 + ig_2)e^{i\varphi}.$$

Taking real and imaginary parts, we obtain

$$2pr = (r^2 + 1 + g_1) \cos \varphi - g_2 \sin \varphi, \tag{15.3.6}$$

$$0 = (1 + g_1 - r^2) \sin \varphi + g_2 \cos \varphi. \tag{15.3.7}$$

The coefficients $(1 + g_1 - r^2)$ and g_2 in the second equation do not both vanish, for if they did, it would imply that $g(\psi) = r^2 - 1$ and, therefore, by the definition of $g(\psi)$, that $re^{-i\psi} = 1 \pm i(1 - r^2)^{1/2}$. This is impossible since $0 < r < 1$. Thus (15.3.7) has precisely two roots modulo 2π, and if φ_1 is one of them, then $\varphi_1 + \pi$ is the other. Substituting φ_1 and $\varphi_1 + \pi$ into (15.3.6), we see that p is positive in one case and negative in the other. Thus, with fixed r and ψ, $0 < r < 1$, there is no more than one set of solutions p, $e^{i\varphi}$ of (15.3.5) with $p \geqq 1$. Indeed, there is no more than one such set with $p > 0$.

We now determine the solutions of (15.3.5) in case $p = 1$, r fixed. In this case $(re^{-i\varphi} - 1)^2 = -g(\psi)$; so

$$re^{-i\varphi} = 1 + \delta i \frac{1 - r^2}{1 - re^{-i\psi}}, \qquad \delta = \pm 1. \tag{15.3.8}$$

As ψ varies through real values, the right side of (15.3.8) moves over a circle of radius r with center at $1 + \delta i$. Thus if $0 < r < 1/2^{1/2}$, there is no solution of (15.3.8). If $r = 1/2^{1/2}$, then ($\psi = \varphi = \pi/4$) and ($\psi = \varphi = -\pi/4$) satisfy (15.3.8) but there are no other solutions modulo 2π. If $1/2^{1/2} < r < 1$, then the four points ($\psi = -\pi/4 - \tau, \varphi = -\pi/4 + \tau$), ($\psi = -\pi/4 + \tau, \varphi = -\pi/4 - \tau$), ($\psi =$

$\pi/4 - \tau, \varphi = \pi/4 + \tau$), and ($\psi = \pi/4 + \tau, \varphi = \pi/4 - \tau$) satisfy (15.3.8) but there are no other solutions modulo 2π. Here and henceforth

$$\tau = \arccos\left(\frac{1}{2^{1/2}r}\right), \qquad 0 < \tau < \pi/4. \tag{15.3.9}$$

The two solutions for which $\varphi + \psi = \pi/2$ satisfy (15.3.8) with $\delta = -1$, and the two solutions for which $\varphi + \psi = -\pi/2$ satisfy (15.3.8) with $\delta = +1$.

Next, we determine the solutions of (15.3.5) when $\psi = 0$ or π. If $\psi = 0$, then

$$r^2e^{-2i\varphi} - 2pre^{-i\varphi} + 2 + 2r + r^2 = 0.$$

By taking imaginary parts it is clear, since $p \geqq 1$, that $\sin\varphi = 0$, and then it follows that $e^{-i\varphi} = 1$, $p = (1 + r + r^2)/r > 1$. If $\psi = \pi$, then $e^{-i\varphi} = 1$, $p = (1 - r + r^2)/r > 1$.

Returning to equation (15.3.5), we note that this equation implies an upper bound for p which depends only on r. Write

$$f(\rho, \varphi) = x + iy = r^2e^{-2i\varphi} - 2pre^{-i\varphi} + 1.$$

Then

$$\frac{\partial f}{\partial\varphi} = 2ire^{-i\varphi}(p - re^{-i\varphi}), \qquad \frac{\partial f}{\partial p} = -2re^{-i\varphi},$$

and so

$$\operatorname{Im}\left(\frac{\partial f/\partial\varphi}{\partial f/\partial p}\right) = -p + r\cos\varphi \neq 0.$$

It follows that the Jacobian of the pairs of real variables (p, φ) and (x, y) is not zero for $0 < r < 1$, $p > r$, and, in particular, for $0 < r < 1$, $p \geqq 1$.

With r fixed, $1/2^{1/2} < r < 1$, let ψ increase or decrease from the initial value $\psi = 0$. Then for $-\pi/4 + \tau \leqq \psi \leqq \pi/4 - \tau$, there is precisely one set of values of p and $e^{i\varphi}$ which satisfy (15.3.5) with $p \geqq 1$, and $p > 1$ except at the end points of this interval, where it is unity. For p is bounded above and the non-vanishing of the Jacobian shows that p varies continuously so long as $p \geqq 1$. We have shown that $p \neq 1$ for $-\pi/4 + \tau < \psi < \pi/4 - \tau$. Consider the end point $\psi = \pi/4 - \tau$, where τ is given by (15.3.9). Then $p = 1$, $\varphi = \pi/4 + \tau$, $0 < \tau < \pi/4$. We show now that if ψ is increased slightly, then the value of p is decreased. We note from (15.3.5) that

$$\frac{\partial f}{\partial\varphi}\,d\varphi + \frac{\partial f}{\partial p}\,dp + g'(\psi)\,d\psi = 0. \tag{15.3.10}$$

Since

$$g'(\psi) = -\frac{2ire^{-i\psi}}{1 - re^{-i\psi}}\,g(\psi),$$

we have from (15.3.5), with $p = 1$,

$$g'(\psi) = \frac{2ire^{-i\psi}}{1 - re^{-i\psi}}(1 - re^{-i\varphi})^2.$$

Substituting in (15.3.10), we obtain

$$(15.3.11) \qquad d\varphi + \frac{i}{1 - re^{-i\varphi}}\,dp + e^{i(\varphi-\psi)}\frac{1 - re^{-i\varphi}}{1 - re^{-i\psi}}\,d\psi = 0.$$

Taking imaginary parts, we have

$$(15.3.12) \qquad \frac{r \sin \psi}{1 - r^2}\,dp + \frac{\sin 2\tau + r(\sin \psi - \sin \varphi)}{|\,e^{i\psi} - r\,|^2}\,d\psi = 0.$$

Since

$$\sin 2\tau + r(\sin \psi - \sin \varphi) = 2(\sin \tau)(\cos \tau - r \cos \pi/4) > 0$$

and since $d\psi > 0$, it follows that $dp < 0$. Thus if ψ is increased slightly from $\pi/4 - \tau$, then p is decreased from 1. On the other hand, there is a solution with p slightly less than 1 and this solution is the unique positive solution. Since equation (15.3.5) implies an upper bound for p which depends only on r, the above arguments show that if there is a solution of (15.3.5) with $p \geqq 1$ in the open interval $\pi/4 - \tau < \psi < \pi/4 + \tau$, then there is a least value of ψ for which there is a solution with $p \geqq 1$. We have shown that there is no solution in this open interval with $p = 1$. Thus, if there is a solution with $p \geqq 1$, we must have $p > 1$. The non-vanishing of the Jacobian of the pairs of real variables (p, φ) and (x, y) shows that there would then be a solution in the open interval corresponding to a smaller value of ψ with $p \geqq 1$. This contradiction shows that there is no solution of (15.3.5) with $p \geqq 1$ in the interval $\pi/4 - \tau < \psi < \pi/4 + \tau$. Likewise, there is no solution of (15.3.5) with $p \geqq 1$ in the interval $-\pi/4 - \tau < \psi < -\pi/4 + \tau$.

With r fixed, $1/2^{1/2} < r < 1$, let ψ increase or decrease from the initial value $\psi = \pi$. Then for $\pi/4 + \tau \leqq \psi \leqq 2\pi - (\pi/4 + \tau)$ there is precisely one set of values p and $e^{i\varphi}$ which satisfies (15.3.5) with $p \geqq 1$.

To sum up, if r is fixed, $1/2^{1/2} < r < 1$, then for each ψ in $-\pi/4 + \tau \leqq \psi \leqq \pi/4 - \tau$ or in $\pi/4 + \tau \leqq \psi \leqq 2\pi - (\pi/4 + \tau)$ there is a unique set of values ρ and $e^{i\varphi}$ satisfying (15.3.4) with $0 < \rho \leqq 1$. At interior points of these intervals $0 < \rho < 1$. Outside these two intervals (modulo 2π), there is no solution with $0 < \rho \leqq 1$.

If $r = 1/2^{1/2}$, then for each ψ there is a unique set of values ρ, $e^{i\varphi}$ satisfying (15.3.4) with $0 < \rho \leqq 1$. In this case $0 < \rho < 1$ except for $\psi = \pm\pi/4$ where $\rho = 1$.

If $0 < r < 1/2^{1/2}$, then for each ψ there is a unique solution of (15.3.4) with $0 < \rho \leqq 1$ and in this case $0 < \rho < 1$.

For future reference we remark that $\rho e^{i\varphi} \neq r$, for if $\rho e^{i\varphi} = r$, then $g(\psi) = 0$ by (15.3.4) and this is impossible. We also remark that if $0 < r < 1$ and $0 < \rho < 1$, then the quantity $(r - \rho e^{i\varphi})/(r - e^{i\varphi}/\rho)$ never lies on the positive real axis. For if $(r - \rho e^{i\varphi})/(r - e^{i\varphi}/\rho)$ is real, then clearly $\sin \varphi = 0$ and $(r - \rho e^{i\varphi})/(r - e^{i\varphi}/\rho)$ has the same sign as $r^2 - 2pr \cos \varphi + 1$. The equation (15.3.5) shows that under these circumstances the function $g(\psi)$ is real and has the sign opposite to $r^2 - 2pr \cos \varphi + 1$. But if $g(\psi)$ is real, then by its definition it is positive and $e^{i\psi} = \pm 1$. From (15.3.5) it follows that $r^2 - 2pr \cos \varphi + 1 < 0$ and $\cos \varphi = +1$.

15.4. By making the substitution $w = \alpha v$, the differential equation (15.3.3) becomes

$$-e^{i(\varphi+\psi)} \left(\frac{z}{v}\frac{dv}{dz}\right)^2 \frac{2v - 1}{(v - 1)^2} = \Psi(z) \tag{15.4.1}$$

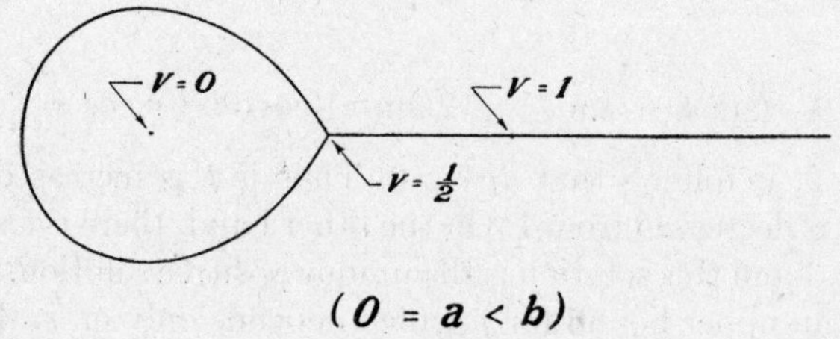

FIG. 19a

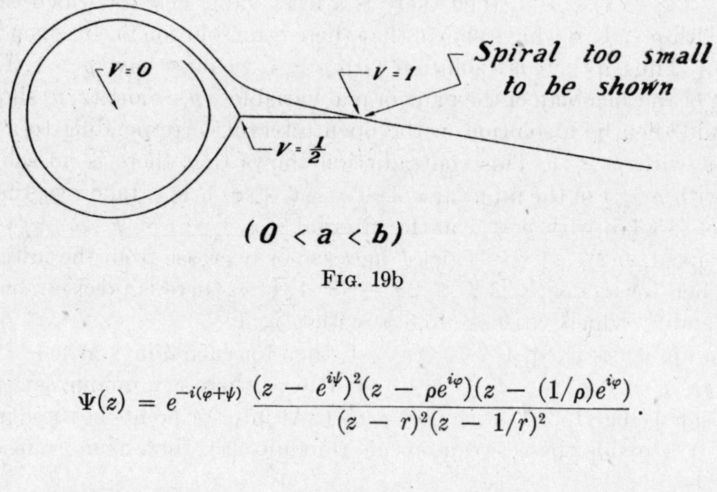

FIG. 19b

where

$$\Psi(z) = e^{-i(\varphi+\psi)} \frac{(z - e^{i\psi})^2(z - \rho e^{i\varphi})(z - (1/\rho)e^{i\varphi})}{(z - r)^2(z - 1/r)^2}. \tag{15.4.2}$$

Writing

$$\Phi(v) = A \frac{2v - 1}{v^2(v - 1)^2}, \qquad A = -e^{i(\varphi+\psi)},$$

let

$$\zeta = \int (\Phi(v)\, dv)^{1/2}, \qquad A^{1/2} = a + ib, \tag{15.4.3}$$

where

$$a = -\sin \frac{\varphi + \psi}{2}, \qquad b = \cos \frac{\varphi + \psi}{2}.$$

Let $\mathbf{\Gamma}_v$ be the set of all loci Re (ζ) = constant which have one or both end points at $v = 1/2$ or at $v = \infty$. This set is described in Lemma XX and the various configurations are shown in Figures 19a–19e. By Lemma XXII we see that $\mathbf{\Gamma}_v = \mathbf{\Gamma}_v(A)$ depends continuously on A unless A is real, and the manner in which the configurations change is illustrated in the sequence of diagrams (Figures 19a–19e) beginning with the case $0 = a < b$ and ending with the case $0 < a = b$. The other possible configurations are obtained by reflection on the lines Re $(v) = 1/2$, Im $(v) = 0$.

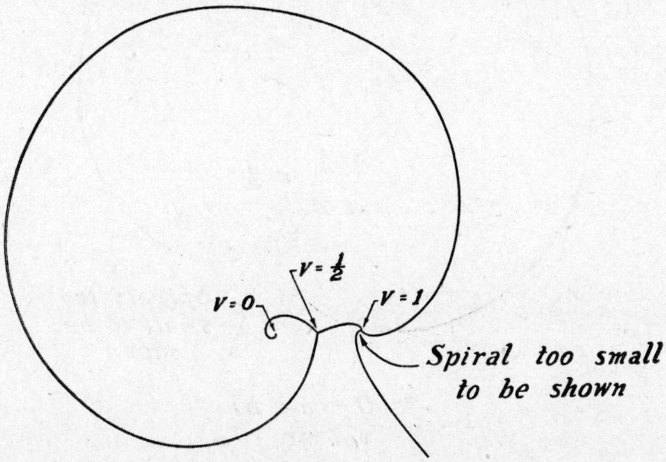

$(0 < a < b)$

FIG. 19c

In order to calculate these loci we make the substitution $u^2 = 2v - 1$, and we obtain

$$\zeta = \pm(a + ib)\left(\log\frac{u - 1}{u + 1} + i\log\frac{u + i}{u - i}\right).$$

Writing

$$\log\frac{u - 1}{u + 1} = \lambda + i\mu, \qquad \log\frac{u + i}{u - i} = \sigma + i\tau,$$

we have

$$\pm\text{Re}\,(\zeta) = a(\lambda - \tau) - b(\mu + \sigma).$$

Let μ_0, σ_0, λ_0, τ_0 correspond to a point on $\mathbf{\Gamma}_v$. Then on this arc of $\mathbf{\Gamma}_v$ we have

$$\frac{(\mu - \mu_0) + (\sigma - \sigma_0)}{(\lambda - \lambda_0) - (\tau - \tau_0)} = \frac{a}{b}. \tag{15.4.4}$$

Choosing principal values for the logarithms, we may take $\mu_0 = \sigma_0 = \lambda_0 = \tau_0 = 0$

at $u = \infty$ and $\lambda_0 = \sigma_0 = 0$, $\mu_0 = \pm\pi$, $\tau_0 = \pm\pi$ at $u = 0$. Formula (15.4.4) was used in computing $\mathbf{\Gamma}_v$ for several different values of a, b.

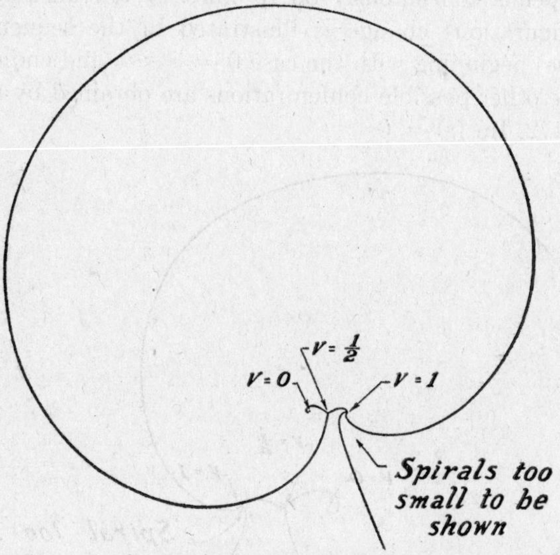

FIG. 19d

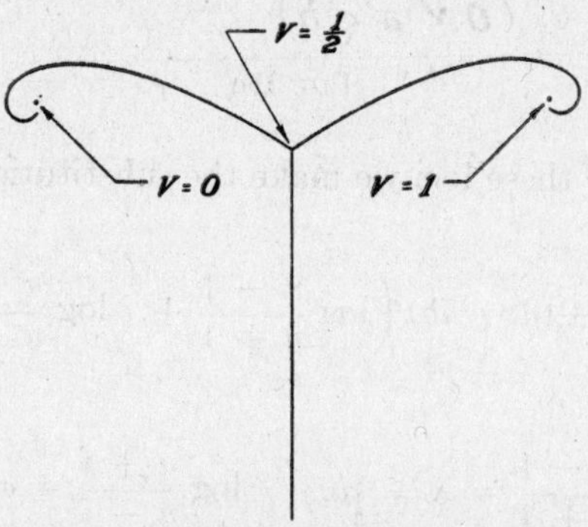

FIG. 19e

15.5. We now turn to the actual determination of $L(r)$ and we suppose first that $0 < \rho < 1$. Let $w = f(z)$ be a function of class $\mathfrak{S}$ which satisfies the differential equation (15.3.3) where (15.3.4) is true with $0 < \rho < 1$.

As in **15.4,** we write

(15.5.1) $$w = \alpha v, \qquad u = (2v - 1)^{1/2}.$$

Also let

(15.5.2) $$\zeta = \frac{z - \rho e^{i\varphi}}{z - (1/\rho)e^{i\varphi}}, \qquad s = \zeta^{1/2},$$

where ζ is not to be confused with the function ζ of (15.4.3). Under the transformation (15.5.2) the unit circle $|z| < 1$ is mapped onto the circle $|\zeta| < \rho$, and $|\zeta| < \rho$, cut along the negative real axis from $\zeta = 0$ to $\zeta = -\rho$, is mapped onto the semi-circle $|s| < \rho^{1/2}$, Re $(s) > 0$. The mappings (15.5.2), $w = f(z)$, and (15.5.1) define a function $u(s)$ which is regular and schlicht in $|s| < \rho^{1/2}$, Re $(s) \geqq 0$, and maps $s = 0$ into $u = 0$. The relation $u(s) = -u(-s)$, which is true on the portion of the imaginary axis in $|s| < \rho^{1/2}$, defines a continuation of $u(s)$ throughout the circle $|s| < \rho^{1/2}$. The function $u(s)$ is therefore regular, schlicht, and odd in $|s| < \rho^{1/2}$. We remark that the function $-u(s)$ is obtained from a composition of the same mappings by using a different branch of the square root in (15.5.1). Conversely, either of the odd functions $u(s)$ or $-u(s)$ together with the mappings (15.5.1) and (15.5.2) defines the given function $w = f(z)$ of class $\mathfrak{S}$. Hence, since $s = \rho$ maps into $u = \pm i$, we consider the function $u(s)$ for which $s = \rho$ maps into $u = i$. Let

$$s_1 = \sigma^{1/2} = \left(\frac{r - \rho e^{i\varphi}}{r - (1/\rho)e^{i\varphi}}\right)^{1/2}.$$

According to the remark near the end of **15.3,** the function $(r - \rho e^{i\varphi})/(r - e^{i\varphi}/\rho)$ is never positive for $0 < \rho < 1$, and so we suppose that

$$\text{Im}\,(s_1) > 0.$$

We have the following correspondence of points

(15.5.3) $$\begin{bmatrix} s = 0 \\ u = 0 \\ z = \rho e^{i\varphi} \\ w = \alpha/2 \end{bmatrix} \quad \begin{bmatrix} s = \rho \\ u = i \\ z = 0 \\ w = 0 \end{bmatrix} \quad \begin{bmatrix} s = s_1 \\ u = \pm 1 \\ z = r \\ w = \alpha \end{bmatrix}.$$

Using the substitutions (15.5.1) and (15.5.2), we transform the differential equation (15.3.3) into an equation involving the variables u and s. Since

$$\frac{dw}{dz} = \frac{dw}{du}\frac{du}{ds}\frac{ds}{dz},$$

we see that

$$\frac{dw}{dz} = \frac{\alpha\rho e^{-i\varphi}}{2(\rho^2 - 1)}\,\frac{u}{s}\,(1 - s^2)^2\,\frac{du}{ds},$$

and we obtain

(15.5.4) $$e^{2i(\varphi+\psi)}\frac{4u^4}{(u^4-1)^2}\left(\frac{du}{ds}\right)^2 = K^2\frac{s^4\{((e^{i\varphi}/\rho)-e^{i\psi})s^2-(e^{i\varphi}\rho-e^{i\psi})\}^2}{(s^2-1)^2(s^2-\rho^2)^2(s^2-\sigma)^2(s^2-\sigma_1)^2}$$

where

(15.5.5) $$K = \frac{ie^{i\varphi}(1-\rho^2)^2}{\rho(e^{i\varphi}/\rho - r)(e^{i\varphi}/\rho - 1/r)},$$

and

$$\sigma_1 = \frac{1/r - \rho e^{i\varphi}}{1/r - (1/\rho)e^{i\varphi}} = \frac{\rho^2}{\bar\sigma}.$$

Taking square roots of both sides of (15.5.4), we have

(15.5.5)′ $$\delta_1 e^{i(\varphi+\psi)}\frac{2u^2}{u^4-1}\frac{du}{ds} = K\frac{s^2\{((e^{i\varphi}/\rho)-e^{i\psi})s^2-(e^{i\varphi}\rho-e^{i\psi})\}}{(s^2-1)(s^2-\rho^2)(s^2-\sigma)(s^2-\sigma_1)}$$

where $\delta_1 = \pm 1$. The right side is a function of s^2; so it can be written in the form

$$\frac{A}{s^2-1}+\frac{B}{s^2-\rho^2}+\frac{C}{s^2-\sigma}+\frac{D}{s^2-\sigma_1}.$$

Comparing residues and simplifying, we find that

$$A = i, \qquad B = -i\rho e^{i(\varphi+\psi)},$$

(15.5.6) $$C = -i\frac{(e^{i\psi}-r)(r-\rho e^{i\varphi})}{1-r^2}, \qquad D = -i\frac{(1-re^{i\psi})r(1/r-\rho e^{i\varphi})}{1-r^2}.$$

Using relation (15.3.4), namely,

(15.5.7) $$-e^{2i(\varphi+\psi)}(1-r^2)^2 = (r-e^{i\psi})^2\sigma(r-\frac{1}{\rho}e^{i\varphi})^2,$$

we have

(15.5.8) $$ie^{i(\varphi+\psi)}\delta_2(1-r^2) = (r-e^{i\psi})(r-\frac{1}{\rho}e^{i\varphi})s_1,$$

where $\delta_2 = \pm 1$. From (15.5.8) we find that

(15.5.9) $$C = -\delta_2 e^{i(\varphi+\psi)}s_1, \qquad \bar D = \frac{-\delta_2\rho}{s_1}.$$

Integrating (15.5.5)′, we obtain

(15.5.10) $$\begin{aligned}\delta_1 e^{i(\varphi+\psi)}\left\{\log\frac{u-1}{u+1}+i\log\frac{u+i}{u-i}\right\} &= i\log\frac{1-s}{1+s} \\ &- ie^{i(\varphi+\psi)}\log\frac{s-\rho}{s+\rho}+\delta_2\left\{e^{i(\varphi+\psi)}\log\frac{s+s_1}{s-s_1}+\log\frac{\rho+\bar s_1 s}{\rho-\bar s_1 s}\right\} \\ &+ \text{constant.}\end{aligned}$$

Since we have chosen the function $u(s)$ such that $s = \rho$ maps into $u = i$, it follows that $\delta_1 = 1$. Then it is clear that $s = s_1$ maps into $u = -\delta_2$. This shows that the $\pm$ sign of (15.5.3) is equal to $-\delta_2$. Equation (15.5.10) can be written

$$e^{i(\varphi+\psi)} \log \left[\frac{u-1}{u+1}\left(\frac{s-s_1}{s+s_1}\right)^{\delta_2}\right] + ie^{i(\varphi+\psi)} \log \left[\frac{u+i}{u-i}\frac{s-\rho}{s+\rho}\right] \tag{15.5.11}$$

$$= i \log \frac{1-s}{1+s} + \delta_2 \log \frac{\rho + \bar{s}_1 s}{\rho - \bar{s}_1 s} + \text{constant}.$$

Here each logarithm is regular and single-valued in $|s| < \rho^{1/2}$. Choosing the principal values at the point $s = 0$, $u = 0$, we see that each logarithm is zero there; so the constant term is zero.

We are going to construct a family of functions $f(z, x)$, $0 \leqq x \leqq 1$, which belong to class $\mathbb{S}$, which satisfy a family of corresponding differential equations (15.3.3), and which lead continuously from the given boundary function $w = f(z)$ at $x = 1$ to a Koebe function at $x = 0$ of the form

$$f(z, 0) = \frac{z}{(1 \pm z)^2}. \tag{15.5.12}$$

This construction will remove some ambiguities that remain in the equation (15.5.11). We shall suppose that the given boundary function $w = f(z)$ satisfies a differential equation of the form (15.3.3) where $0 < \rho < 1$.

If $1 > r > 1/2^{1/2}$, then, changing ψ by some multiple of 2π, we know from relation (15.3.4) that either $|\psi| < \pi/4 - \tau$ or $|\psi - \pi| < 3\pi/4 - \tau$, where τ is defined by (15.3.9), $0 < \tau < \pi/4$. If $r = 1/2^{1/2}$, then (after reduction modulo 2π) either $|\psi| < \pi/4$ or $|\psi - \pi| < 3\pi/4$. If $0 < r < 1/2^{1/2}$, then ψ can have any real value. Thus for given r, $0 < r < 1$, let $\psi(x)$ be defined by

$$\psi(x) = \begin{cases} x\psi, & |\psi| \leqq \pi/4, \\ \pi + x(\psi - \pi), & |\psi - \pi| < 3\pi/4. \end{cases} \tag{15.5.13}$$

Then by the remarks of **15.3**, there are unique real functions $\varphi(x)$ and $\rho(x)$, $0 < \rho(x) < 1$, which satisfy (15.3.4) for $0 \leqq x \leqq 1$, and they are continuous functions of x.

Let

$$\Psi(z, x) = e^{-i[\varphi(x)+\psi(x)]} \frac{(z - e^{i\psi(x)})^2(z - \rho(x)e^{i\varphi(x)})(z - (1/\rho(x))e^{i\varphi(x)})}{(z-r)^2(z-1/r)^2}. \tag{15.5.14}$$

For fixed x, $0 \leqq x \leqq 1$, let the unit circumference $|z| = 1$ be divided into two disjoint arcs L and U such that one of the common end points is at $\exp(i\psi(x))$ and the other is at a point which we call $\exp(i\beta(x))$. We suppose that a point moving in a counterclockwise direction on $|z| = 1$ passes from L to U at the point $\exp(i\psi(x))$. Let

$$\int_{e^{i\psi(x)}}^{z'} (\Psi(z, x))^{1/2} \frac{dz}{z} = \int_{e^{i\psi(x)}}^{z''} (\Psi(z, x))^{1/2} \frac{dz}{z} \tag{15.5.15}$$

where z' is a point of U, z'' a point of L, and $\Psi^{1/2}$ is positive on L, negative on U. Let $\exp(i\beta(x))$ be such that as z' describes U in the counterclockwise sense, the point z'' describes L in the clockwise sense, each point beginning at $\exp(i\psi(x))$ and ending at $\exp(i\beta(x))$. The method of Chapter VIII shows that there is a function

$$f(z, x) = z + b_2(x)z^2 + b_3(x)z^3 + \cdots$$

of class $\mathfrak{S}$ which realizes the identification defined by (15.5.15), that is,

$$f(z', x) = f(z'', x).$$

Moreover, $w = f(z, x)$ satisfies a differential equation of the form

$$A(x)\,\frac{2w - \chi(x)}{(w - \alpha(x))^2}\left(\frac{z}{w}\frac{dw}{dz}\right)^2 = e^{-i(\varphi(x)+\psi(x))}\,\frac{(z - e^{i\psi(x)})^2(z - \rho(x)e^{i\varphi(x)})(z - (1/\rho(x))e^{i\varphi(x)})}{(z - r)^2(z - 1/r)^2}$$

where

$$\alpha(x) = f(r, x), \qquad \chi(x) = 2f(\rho(x)e^{i\varphi(x)}, x),$$

and $A(x)$ depends on x.

Since $w = z + \cdots$, it follows that

$$A(x)\chi(x) = -\alpha(x)^2 e^{i(\varphi(x)+\psi(x))}.$$

Letting z approach r, we see that w approaches $\alpha(x)$ and $(w - \alpha(x))/(z - r)$ approaches dw/dz. It follows that

$$A(x)\,\frac{2\alpha(x) - \chi(x)}{\alpha(x)^2}\,r^2 = e^{-i(\varphi(x)+\psi(x))}\,\frac{(r - e^{i\psi(x)})^2(r - \rho(x)e^{i\varphi(x)})(r - (1/\rho(x))e^{i\varphi(x)})}{(r - 1/r)^2}.$$

But $\varphi(x)$ and $\rho(x)$ were chosen such that (15.3.4) is satisfied for each x; so these relations imply that $\chi(x) = \alpha(x)$. Thus for $0 \leqq x \leqq 1$, the function $w = f(z, x)$ satisfies the differential equation

$$-\alpha(x)e^{i(\varphi(x)+\psi(x))}\,\frac{2w - \alpha(x)}{(w - \alpha(x))^2}\left(\frac{z}{w}\frac{dw}{dz}\right)^2 = e^{-i(\varphi(x)+\psi(x))}\,\frac{(z - e^{i\psi(x)})^2(z - \rho(x)e^{i\varphi(x)})(z - (1/\rho(x))e^{i\varphi(x)})}{(z - r)^2(z - 1/r)^2}$$

where

$$-e^{2i(\varphi(x)+\psi(x))}(1 - r^2)^2 = (r - e^{i\psi(x)})^2(r - \rho(x)e^{i\varphi(x)})\left(r - \frac{1}{\rho(x)}\,e^{i\varphi(x)}\right).$$

Let

$$s_1(x) = \left(\frac{r - \rho(x)e^{i\varphi(x)}}{r - (1/\rho(x))e^{i\varphi(x)}}\right)^{1/2}, \qquad \operatorname{Im}(s_1(x)) > 0.$$

We make the substitutions (15.5.1) and (15.5.2) and obtain a function $u(s, x)$ which is regular, schlicht, and odd in $|s| < (\rho(x))^{1/2}$. Moreover, from (15.5.3),

$$u(0, x) = 0, \qquad u(\rho(x), x) = i, \qquad u(s_1(x), x) = -\delta_2 .$$

The function $u(s, x)$ satisfies

$$(15.5.16) \qquad \begin{aligned} &e^{i(\varphi(x)+\psi(x))} \log \left[\frac{u-1}{u+1}\left(\frac{s-s_1(x)}{s+s_1(x)}\right)^{\delta_2}\right] \\ &\quad + ie^{i(\varphi(x)+\psi(x))} \log \left[\frac{u+i}{u-i}\,\frac{s-\rho(x)}{s+\rho(x)}\right] \\ &= i \log \frac{1-s}{1+s} + \delta_2 \log \frac{\rho(x) + \bar{s}_1(x)s}{\rho(x) - \bar{s}_1(x)s} . \end{aligned}$$

It is clear that $\alpha(x)$, $\varphi(x)$, $\psi(x)$, and $\rho(x)$ are continuous functions of x for $0 \leqq x \leqq 1$. Since, by **15.3**,

$$(15.5.17) \qquad \frac{r - \rho(x)e^{i\varphi(x)}}{r - (1/\rho(x))e^{i\varphi(x)}}$$

is not positive and $\operatorname{Im}(s_1(x)) > 0$, it follows that $s_1(x)$ is also a continuous function of x. Since $u(s_1(x), x) = -\delta_2$ and $\delta_2 = \pm 1$, it follows that δ_2 is independent of x. At $x = 0$ we have $\exp(i\psi(0)) = \pm 1$. Then from (15.3.4) we see that $\exp(i^{\varphi(0)}) = 1$ and $0 < \rho(0) < 1$. The identification defined by (15.5.15) is such that $z'' = \bar{z}' = 1/z'$. Then $f(z, 0) = f(1/z, 0)$ for z on U, and hence by continuation everywhere. It follows that $f(z, 0)$ is a Koebe function of the form

$$(15.5.18) \qquad f(z) = \frac{z}{(1 \pm z)^2} .$$

The plus sign corresponds to $\exp(i\psi(0)) = 1$, the minus sign to $\exp(i\psi(0)) = -1$. If $\exp(i\psi(0)) = 1$, then the mappings (15.5.1) and (15.5.2) show that $|s| < (\rho(0))^{1/2}$ is mapped by $u(s, 0)$ onto the u-plane cut along two segments of the real axis extending to infinity, one along the negative real axis and the other along the positive real axis. Since $u(s, 0)$ is an odd function which maps $s = \rho(0)$ into $u = i$, it follows that

$$(15.5.19) \qquad u(s, 0) = i(\rho(0) - 1)\,\frac{s}{s^2 - \rho(0)} .$$

Since the expression (15.5.17) cannot be positive and since at $x = 0$ it is real, it follows that $s_1(0) = i\,|s_1(0)|$. Substituting into (15.5.19), we see that $u(s_1(0), 0) < 0$. But $u(s_1(x), x) = -\delta_2$, and this shows that $\delta_2 = 1$. In a similar fashion, if $\exp(i\psi(0)) = -1$, then

$$f(z, 0) = \frac{z}{(1-z)^2}, \qquad u(s, 0) = i(\rho(0) + 1)\,\frac{s}{s^2 + \rho(0)} .$$

Since $|s_1(0)| < \rho^{1/2}$, $s_1(0) = i\,|s_1(0)|$, we see that $u(s_1(0), 0) < 0$ and it again follows that $\delta_2 = 1$. Since δ_2 is independent of x, it follows that $\delta_2 = 1$ for all x. Since the given boundary function $w = f(z)$ may be any function with $0 < \rho < 1$, we conclude that $\delta_1 = 1$, $\delta_2 = 1$ in all cases where $0 < \rho < 1$. From equation (15.5.11) we have

$$\log\left[\frac{u-1}{u+1}\frac{s-s_1}{s+s_1}\right] + i\log\left[\frac{u+i}{u-i}\frac{s-\rho}{s+\rho}\right] = e^{-i(\varphi+\psi)}\left\{i\log\frac{1-s}{1+s} + \log\frac{\rho+\bar{s}_1 s}{\rho-\bar{s}_1 s}\right\}, \tag{15.5.20}$$

where each logarithm has its principal value at $s = 0$, $u = 0$.

We recall that

$$\frac{dw}{dz} = \frac{dw}{du}\frac{du}{ds}\frac{ds}{dz} = \frac{\alpha}{2}\frac{\rho(1-s^2)^2}{(\rho^2-1)e^{i\varphi}}\cdot\frac{u}{s}\cdot\frac{du}{ds} \tag{15.5.21}$$

where $\alpha = f(r)$, $\alpha/2 = f(\rho e^{i\varphi})$. If the point s beginning at $s = 0$ approaches ρ along a radius, then u, z, w approach i, 0, 0 respectively and dw/dz approaches 1. Hence from (15.5.20) and (15.5.21) we obtain

$$\begin{aligned}\log f(r) &= \log\frac{2\rho}{1-\rho^2} + i\varphi + i\log\frac{\rho-s_1}{\rho+s_1} \\ &\quad + e^{-i(\varphi+\psi)}\left\{\log\frac{1-\rho}{1+\rho} + i\log\frac{1-\bar{s}_1}{1+\bar{s}_1}\right\} \\ &\quad + (2n-1/2)\pi + 2m\pi i,\end{aligned} \tag{15.5.22}$$

where, here and subsequently, m and n are positive, negative, or zero integers, not necessarily the same each place where they occur. If s begins at the origin and approaches $s = s_1$ along a radius, then u, z, w approach -1, r, α respectively and dw/dz approaches $f'(r)$. Thus (15.5.20) and (15.5.21) show that

$$\begin{aligned}\log f'(r) &= \log f(r) + \log\frac{2\rho}{1-\rho^2} - i\varphi + \log\frac{(1-s_1^2)^2}{4s_1^2} \\ &\quad - i\log\frac{s_1+\rho}{s_1-\rho} + e^{-i(\varphi+\psi)}\left\{i\log\frac{1+s_1}{1-s_1} + \log\frac{\rho-|s_1|^2}{\rho+|s_1|^2}\right\} \\ &\quad + \left(2n+\frac{1}{2}\right)\pi + (2m+1)\pi i.\end{aligned} \tag{15.5.23}$$

From (15.5.22) and (15.5.23) we obtain

$$\begin{aligned}\log f'(r) &= 2i\log\frac{\rho-s_1}{\rho+s_1} + 2\log\left[\frac{1-s_1^2}{s_1}\cdot\frac{\rho}{1-\rho^2}\right] \\ &\quad + e^{-i(\varphi+\psi)}\left\{\log\left[\frac{1-\rho}{1+\rho}\cdot\frac{\rho-|s_1|^2}{\rho+|s_1|^2}\right] + i\log\frac{1+s_1}{1-s_1} + i\log\frac{1-\bar{s}_1}{1+\bar{s}_1}\right\} \\ &\quad + (2n+1)\pi + (2m+1)\pi i\end{aligned} \tag{15.5.24}$$

where each logarithm has its principal value. If $f(z, x)$, $0 \leqq x \leqq 1$, is the family of functions of class $\mathfrak{S}$ discussed above, then formulas (15.5.22) and (15.5.24) are valid where ρ, φ, ψ, and s_1 are continuous functions of x and each logarithm has its principal value. In formula (15.5.22) we take φ to lie between $-\pi$ and π. Since σ is not positive, it is clear that $e^{i\varphi} \neq -1$ for $0 < \rho < 1$. This determination of $\varphi(x)$ is also a continuous function of x. Since we have $\operatorname{Im}(s_1) > 0$, the functions $(\rho - s_1)/(\rho + s_1)$ and $(1 - s_1^2)/s_1$ have negative imaginary parts while the functions $(1 + s_1)/(1 - s_1)$ and $(1 - \bar{s}_1)/(1 + \bar{s}_1)$ have positive imaginary parts. We also note that $0 < | s_1 |^2 < \rho < 1$. It is then clear that the principal part of each logarithm is a continuous function of x. Letting x approach zero, we know that $e^{i\varphi}$ approaches 1 and $e^{i\psi}$ approaches δ where $\delta = \pm 1$. Then

$$f(z, 0) = \frac{z}{(1 + \delta z)^2}, \qquad \alpha = \frac{r}{(1 + \delta r)^2} \qquad \frac{\alpha}{2} = \frac{\rho}{(1 + \delta\rho)^2}, \qquad s_1^2 = \frac{r - \rho}{r - 1/\rho}.$$

We now let r tend to zero, keeping the same function $f(z, 0)$. Then $f'(r, 0)$ clearly approaches 1 while

$$\rho = r/2 + o(r), \qquad s_1 = ir/2 + o(s).$$

Substituting into (15.5.24) with $x = 0$, we see, since n and m are independent of r, that $n = -1$ and $m = 0$. Likewise the n and m of (15.5.22) are each zero.

Since

$$s_1 = \sigma^{1/2} = \left(\frac{r - \rho e^{i\varphi}}{r - (1/\rho)e^{i\varphi}}\right)^{1/2}, \qquad \operatorname{Im}(\sigma^{1/2}) > 0,$$

it follows that

$$\text{(15.5.25)} \qquad \begin{aligned} \log f'(r) = 2i \log \frac{\rho - \sigma^{1/2}}{\rho + \sigma^{1/2}} + 2 \log \left[\frac{1 - \sigma}{\sigma^{1/2}} \cdot \frac{\rho}{1 - \rho^2}\right] \\ + e^{-i(\varphi+\psi)} \left\{\log \left[\frac{1 - \rho}{1 + \rho} \frac{\rho - | \sigma |}{\rho + | \sigma |}\right] + 2 \arg \frac{1 - \sigma^{1/2}}{1 + \sigma^{1/2}}\right\} - \pi + \pi i \end{aligned}$$

and

$$\text{(15.5.26)} \qquad \begin{aligned} \log f(r) = \log \frac{2\rho e^{i\varphi}}{1 - \rho^2} + i \log \frac{\rho - \sigma^{1/2}}{\rho + \sigma^{1/2}} \\ + e^{-i(\varphi+\psi)} \left\{\log \frac{1 - \rho}{1 + \rho} + i \log \frac{1 - \bar{\sigma}^{1/2}}{1 + \bar{\sigma}^{1/2}}\right\} - \frac{\pi}{2} \end{aligned}$$

for all functions of class $\mathfrak{S}$ which satisfy (15.3.3) with $0 < \rho < 1$, $0 < r < 1$. Here the argument and each logarithm have their principal values.

In order to simplify one of the terms in (15.5.25), we note that

$$\frac{1 - \sigma}{\sigma^{1/2}} \frac{\rho}{1 - \rho^2} = \frac{1 - \dfrac{r - \rho e^{i\varphi}}{r - (1/\rho)e^{i\varphi}}}{\sigma^{1/2}} \frac{\rho}{1 - \rho^2}$$

$$= -\frac{e^{i\varphi}}{(r - (1/\rho)e^{i\varphi})\sigma^{1/2}}.$$

On using relation (15.3.4) or (15.5.7), this can be written

$$\pm i \frac{1 - re^{-i\psi}}{1 - r^2},$$

and since $\operatorname{Im}(\sigma^{1/2}) > 0$, it follows that the sign is minus. Thus (15.5.25) can be written

$$\log f'(r) = 2i \log \frac{\rho - \sigma^{1/2}}{\rho + \sigma^{1/2}} + 2 \log \frac{1 - re^{-i\psi}}{1 - r^2}$$

(15.5.27)

$$+ e^{-i(\varphi+\psi)} \left\{ \log \left[\frac{1 - \rho}{1 + \rho} \frac{\rho - |\sigma|}{\rho + |\sigma|} \right] + 2 \arg \frac{1 - \sigma^{1/2}}{1 + \sigma^{1/2}} \right\} - \pi,$$

where again each logarithm and the argument have the principal value.

For fixed r, equation (15.5.27) is actually the parametric equation of a curve or curves since it involves only one independent real variable. For we have shown in **15.3** that for each r and ψ there is at most one pair of numbers ρ and $e^{i\varphi}$ for which (15.3.4) is satisfied. Thus (15.5.27) together with

$$-e^{2i(\varphi+\psi)} = \frac{(r - e^{i\psi})^2}{(1 - r^2)^2} (r - \rho e^{i\varphi}) \left(r - \frac{1}{\rho} e^{i\varphi} \right) \qquad (15.5.28)$$

defines the part of the boundary of $L(r)$, r fixed, $0 < r < 1$, for which $0 < \rho < 1$.

We now wish to find the limit of $\log f'(r)$ for fixed r, $1/2^{1/2} \leqq r < 1$, as ψ approaches a boundary point of one of its intervals. Thus let ψ approach $-\pi/4 \pm \tau$. Then φ approaches $-\pi/4 \mp \tau$ and $\varphi + \psi$ approaches $-\pi/2$. Since ρ tends toward 1, we write $\rho = 1 - t$, $t > 0$. Then

$$p = \frac{1}{2}\left(\rho + \frac{1}{\rho}\right) = 1 + O(t^2);$$

so by (15.3.12) and (15.3.11),

$$\psi = -\frac{\pi}{4} \pm \tau + O(t^2), \qquad \varphi = -\frac{\pi}{4} \mp \tau + O(t^2).$$

Then

$$\sigma = \frac{r - \rho e^{i\varphi}}{r - (1/\rho)e^{i\varphi}} = 1 - \frac{2t}{1 - re^{-i\varphi}} + O(t^2).$$

This shows that $\operatorname{Im}(\sigma) < 0$, and so, since $\operatorname{Im}(\sigma^{1/2}) > 0$, it follows that

$$\sigma^{1/2} = -1 + \frac{t}{1 - re^{-i\varphi}} + O(t^2), \qquad |\sigma| = 1 - 2t \frac{1 - r\cos\varphi}{|e^{i\varphi} - r|^2} + O(t^2).$$

Substituting into (15.5.25), we have, since $e^{-i(\varphi+\psi)} = i + O(t^2)$,

$$\log f'(r) = i \log \frac{1 - r^2}{r^2} + 2i \log \frac{e^{i\varphi} - r}{|e^{i\varphi} - r|}$$

$$- 2 \log \left[\frac{1 - re^{-i\varphi}}{-1} \right] + 2i \arg (1 - re^{-i\varphi}) - \pi + \pi i,$$

or

$$\log f'(r) = i \log \frac{1 - r^2}{r^2}$$

(15.5.29)

$$- 2 \log | e^{i\varphi} - r | + 2i \log \frac{e^{i\varphi} - r}{| e^{i\varphi} - r |} - \pi - \pi i.$$

Likewise

(15.5.30) $$\log f(r) = \log \frac{1}{2(1 - re^{i\varphi})} + i \log \frac{e^{i\varphi} - r}{r} + i\varphi - \frac{\pi}{2}.$$

If ψ approaches $\pi/4 \pm \tau$, then φ approaches $\pi/4 \mp \tau$ and $\varphi + \psi$ approaches $\pi/2$. In this case, if $\rho = 1 - t$, we have

$$\sigma = 1 - \frac{2t}{1 - re^{-i\varphi}} + O(t^2),$$

$$\sigma^{1/2} = 1 - \frac{t}{1 - re^{-i\varphi}} + O(t^2), \qquad | \sigma | = 1 - 2t \frac{1 - r \cos \varphi}{| e^{i\varphi} - r |^2} + O(t^2).$$

Hence

$$\log f'(r) = i \log \frac{r^2}{1 - r^2} - 2i \log \frac{e^{i\varphi} - r}{| e^{i\varphi} - r |}$$

$$- 2 \log (1 - re^{-i\varphi}) + 2i \arg (1 - re^{-i\varphi}) - \pi + \pi i,$$

or

$$\log f'(r) = i \log \frac{r^2}{1 - r^2}$$

(15.5.31)

$$- 2i \log \frac{e^{i\varphi} - r}{| e^{i\varphi} - r |} - 2 \log | e^{i\varphi} - r | - \pi + \pi i.$$

Likewise,

(15.5.32) $$\log f(r) = \log \frac{1}{2(1 - re^{i\varphi})} + i \log \frac{r}{e^{i\varphi} - r} + i\varphi - \frac{\pi}{2}.$$

15.6. In the preceding section we considered the case $0 < \rho < 1$, and in that case it followed that $f(\rho e^{i\varphi}) = \alpha/2$. In the present section we discuss the case $\rho = 1$, and here we do not necessarily have $f(\rho e^{i\varphi}) = \alpha/2$.

From equation (15.3.3) we obtain, in the case $\rho = 1$,

$$-\alpha e^{i(\varphi+\psi)} \frac{2w - \alpha}{(w - \alpha)^2} \left(\frac{z}{w} \frac{dw}{dz}\right)^2 = e^{-i(\varphi+\psi)} \frac{(z - e^{i\psi})^2(z - e^{i\varphi})^2}{(z - r)^2(z - 1/r)^2}.$$

Setting $w = \alpha v$ gives the equation

(15.6.1) $$-e^{i(\varphi+\psi)} \frac{2v - 1}{(v - 1)^2} \left(\frac{z}{v} \frac{dv}{dz}\right)^2 = e^{-i(\varphi+\psi)} \frac{(z - e^{i\psi})^2(z - e^{i\varphi})^2}{(z - r)^2(z - 1/r)^2}.$$

The function $v(z)$ is regular and schlicht in $|z| < 1$ and $v(0) = 0$, but it does not in general belong to class $\mathcal{S}$. Since we suppose that $\rho = 1$, equation (15.3.4) has a solution only if $r \geqq 1/2^{1/2}$; so we shall consider the case $r > 1/2^{1/2}$. Then the case $r = 1/2^{1/2}$ will be obtained as a limit by letting r approach $1/2^{1/2}$. Also, we know from (15.3.4) that $e^{i(\varphi+\psi)} = \pm i$, and we shall consider first the case $e^{i(\varphi+\psi)} = -i$. Then there are two solutions ($\varphi = -\pi/4 - \tau, \psi = -\pi/4 + \tau$) and ($\varphi = -\pi/4 + \tau, \psi = -\pi/4 - \tau$), modulo 2π, and since the roles of φ and ψ are interchangeable in (15.6.1), let $\varphi = -\pi/4 - \tau, \psi = -\pi/4 + \tau$. Then we have

$$i\,\frac{2v - 1}{(v - 1)^2}\left(\frac{z}{v}\frac{dv}{dz}\right)^2 = i\,\frac{(z - e^{i\psi})^2(z - e^{i\varphi})^2}{(z - r)^2(z - 1/r)^2};$$

so

$$(15.6.2)\quad (1 + i)\int \frac{(2v - 1)^{1/2}}{v(v - 1)}\,dv = \pm(1 + i)\int \frac{(z - e^{i\psi})(z - e^{i\varphi})}{(z - r)(z - 1/r)}\frac{dz}{z}.$$

This is the case $0 < a = b$ discussed in **15.4**. The unit circle $|z| < 1$ is mapped by $v(z)$ into the v-plane cut by the straight-line segment Re $(v) = 1/2$, Im $(v) \leqq 0$ and, in general, by parts of the other two arcs of $\mathbf{\Gamma}_v$ that emerge from $v = 1/2$. The points $e^{i\varphi}$ and $e^{i\psi}$ map into the tips of these arcs, and we denote the arcs by L_φ and L_ψ respectively. One of the arcs, however, may degenerate to a point. As the point z describes the unit circumference $|z| = 1$ in a counterclockwise direction, the point v describes the boundary of its domain with area to the left. Thus there is a point $e^{i\lambda}$, $-\pi/4 - \tau \leqq \lambda \leqq -\pi/4 + \tau$, of $|z| = 1$ which maps into $v = 1/2$.

By an arc (α, β) of $|z| = 1$ we mean the open arc consisting of points $e^{i\theta}$ where $\alpha < \theta < \beta$. Using the metric

$$(15.6.3)\qquad ds = \left|\frac{(z - e^{i\psi})(z - e^{i\varphi})}{(z - r)(z - 1/r)}\frac{dz}{z}\right|,$$

we define the distance of two points on $|z| = 1$ in this metric as the minimum of the integral of ds over the two arcs of $|z| = 1$ which join these two points. Then it is clear that the distance between two points on the upper half of $|z| = 1$ is greater than the distance between their conjugate points. Thus, if $e^{i\beta}$ is the point on $|z| = 1$ which is mapped into infinity by $v(z)$ or $w(z)$, then $-\pi/4 + \tau < \beta < 2\pi - (\pi/4 + \tau)$, modulo 2π.

We now show that λ can have any value in its interval. Thus let $\varphi \leqq \lambda \leqq \psi$ where $\varphi = -\pi/4 - \tau, \psi = -\pi/4 + \tau$. Let the arc $I_1' = (\lambda, \psi)$ be identified with an arc $I_1'' = (\psi, \psi_1)$ where ψ_1 is chosen such that I_1' and I_1'' have the same length in the metric (15.6.3). Likewise let the arc $I_2' = (\varphi, \lambda)$ be identified with the arc $I_2'' = (\varphi_1, \varphi)$ where φ_1 is such that I_2' and I_2'' have the same length in the metric. Since I_1' plus I_2' has a total length in the metric (15.6.3) less than that of the arc $(\pi/4 - \tau, \pi/4 + \tau)$, it follows that I_1'' and I_2'' are disjoint, and so there is an arc (ψ_1, φ_1) complementary to I_1', I_1'', I_2', I_2'', and their end points. Let I_3' and I_3'' be two open disjoint arcs whose sum, together with the common end point, is the arc (ψ_1, φ_1), and let each have metrical length half that of

(ψ_1, φ_1). Let I_3' and I_3'' be identified by using the metric (15.6.3). Then $|z| = 1$ is the sum of the open arcs $I_1', I_1'', I_2', I_2'', I_3', I_3''$, and their end points, and the arcs are identified according to the metric (15.6.3). If z' describes I_ν' clockwise, then the identified point z'' describes I_ν'' counterclockwise, $\nu = 1, 2, 3$. The method of Chapter VIII shows that there is a function

$$f(z) = z + b_2 z^2 + \cdots$$

of class $\mathfrak{S}$ which realizes the given identification.

The function $w = f(z)$ satisfies a differential equation of the form

$$-A \frac{2w - \chi}{(w - \alpha)^2} \left(\frac{z}{w} \frac{dw}{dz}\right)^2 = i \frac{(z - e^{i\psi})^2 (z - e^{i\varphi})^2}{(z - r)^2 (z - 1/r)^2}$$

where $\alpha = f(r)$, and χ and A are certain constants. As z approaches zero, it follows that

$$A\chi = -i\alpha^2,$$

and as z approaches r, it follows that

$$-A(2\alpha - \chi)(r^2 - 1)^2 = i\alpha^2 (r - e^{i\psi})^2 (r - e^{i\varphi})^2.$$

Since $\varphi = -\pi/4 - \tau$, $\psi = -\pi/4 + \tau$, we have from these relations

$$\alpha = \chi, \qquad A = -i\alpha.$$

Thus $w = f(z)$ satisfies the differential equation

$$i\alpha \frac{2w - \alpha}{(w - \alpha)^2} \left(\frac{z}{w} \frac{dw}{dz}\right)^2 = i \frac{(z - e^{i\psi})^2 (z - e^{i\varphi})^2}{(z - r)^2 (z - 1/r)^2}.$$

This shows that λ can take all values in the range $\varphi \leqq \lambda \leqq \psi$. In particular, we obtain a family of functions $f(z, \lambda) = z + \cdots$ of class $\mathfrak{S}$, and $\log f'(r, \lambda)$ is a continuous function of λ in the interval $-\pi/4 - \tau \leqq \lambda \leqq -\pi/4 + \tau$. If $-\pi/4 - \tau < \lambda < -\pi/4 + \tau$, then it is clear that L_φ and L_ψ are analytic arcs, each containing more than one point. By tracing out $|z| = 1$ with internal area to the left and using Lemma XX, it follows that L_φ is a subarc of the arc of $\mathbf{\Gamma}_v$ issuing from $v = 1/2$ and tending to $v = 0$ while L_ψ is a subarc of the arc issuing from $v = 1/2$ and tending to $v = 1$. If $\lambda = \varphi = -\pi/4 - \tau$, then L_φ degenerates to a point; so the tree in the v-plane which is the map of $|z| = 1$ is a Jordan arc having a critical point at $v = 1/2$. If $\lambda = \psi = -\pi/4 + \tau$, then L_ψ is a single point and the tree in the v-plane is a Jordan arc with a finite critical point.

Returning now to (15.6.2), we see that

$$(1 + i)\left\{\log \frac{(2v - 1)^{1/2} - 1}{(2v - 1)^{1/2} + 1} + i \log \frac{(2v - 1)^{1/2} + i}{(2v - 1)^{1/2} - i}\right\} = \pm(1 + i)$$

$$\cdot \left\{-i \log z - \frac{(r - e^{i\psi})(r - e^{i\varphi})}{1 - r^2} \log (z - r)\right.$$

$$\left. + \frac{(1 - re^{i\varphi})(1 - re^{i\psi})}{1 - r^2} \log \left(z - \frac{1}{r}\right)\right\} + C.$$

According to (15.3.4),

$$\frac{(r - e^{i\varphi})(r - e^{i\psi})}{1 - r^2} = \pm 1, \tag{15.6.4}$$

and the sign is minus for the present choice of φ and ψ. Taking conjugates, we have

$$\frac{(1 - re^{i\varphi})(1 - re^{i\psi})}{1 - r^2} = i. \tag{15.6.5}$$

The function $v(z)$ maps $|z| < 1$ onto a portion of the v-plane complementary to the line segment $\operatorname{Re}(v) = 1/2$, $\operatorname{Im}(v) \leqq 0$, and we shall suppose that the v-plane is cut by this line segment. Then the function $(2v - 1)^{1/2}$ is single-valued in the cut v-plane. Let it have the value 1 at $v = 1$. Then $(2v - 1)^{1/2} = i$ at $v = 0$. The points $z = 0$, $z = r$ map respectively into $v = 0$, $v = 1$; so, using (15.6.4) and (15.6.5), we obtain

$$\log \frac{(2v - 1)^{1/2} - 1}{(2v - 1)^{1/2} + 1} \cdot \frac{1}{z - r} + i \log \frac{(2v - 1)^{1/2} + i}{(2v - 1)^{1/2} - i}\, z = i \log (1 - zr) + C_1.$$

Since

$$v = \frac{1}{\alpha}(z + \cdots)$$

near $z = 0$, we see that the last equation may be written in the form

$$\log \frac{ir}{z - r}\,\frac{(2v - 1)^{1/2} - 1}{(2v - 1)^{1/2} + 1} + i \log \frac{i + (2v - 1)^{1/2}}{i - (2v - 1)^{1/2}} \cdot \frac{z}{2\alpha} = i \log (1 - rz),$$

where each logarithm is regular and single-valued in $|z| < 1$ and vanishes at $z = 0$.

The function $(2v - 1)^{1/2}$ is equal to zero when $z = e^{i\lambda}$ where $-\pi/4 - \tau \leqq \lambda \leqq -\pi/4 + \tau$. Thus setting $v = 1/2$, $z = e^{i\lambda}$, we have, since $\alpha = f(r)$,

$$\log f(r) = -i \log \frac{r}{e^{i\lambda} - r} + \log \frac{e^{i\lambda}}{2(1 - re^{i\lambda})} + \left(2n - \frac{1}{2}\right)\pi + 2m\pi i, \tag{15.6.6}$$

where m and n are positive or negative integers or zero (not necessarily the same everywhere they occur).

Now let z approach r. Then v approaches 1 and we obtain

$$\begin{aligned} \log f'(r) = {} & (1 + i) \log f(r) - (1 + i) \log (r/2) \\ & + i \log (1 - r^2) + (2n - 1/2)\pi + (2m - 1/2)\pi i. \end{aligned} \tag{15.6.7}$$

Combining (15.6.6) and (15.6.7), we have

$$\begin{aligned} \log f'(r) = {} & -2 \log |e^{i\lambda} - r| - 2 \arg (e^{i\lambda} - r) \\ & + i \log \frac{1 - r^2}{r^2} + (2n - 1)\pi + (2m - 1)\pi i. \end{aligned} \tag{15.6.8}$$

If $\lambda = -\pi/4 - \tau$, then we obtain a function $w = f(z)$ which maps $|z| < 1$ onto the complement of a single Jordan arc in the w-plane with a finite critical point. This is the limit of the case discussed in **15.4** in which ψ approaches $-\pi/4 + \tau$ and φ approaches $-\pi/4 - \tau$. Thus comparing (15.5.29) and (15.6.8) with $\varphi = +\lambda = -\pi/4 - \tau$, we see that $n = 0, m = 0$. These two equations also agree, as they must, at $\varphi = \lambda = -\pi/4 + \tau$. Thus the equation

$$\log f'(r) = i \log \frac{1 - r^2}{r^2} - 2 \log | e^{i\lambda} - r | \qquad (15.6.9)$$
$$- 2 \arg (e^{i\lambda} - r) - \pi - \pi i,$$

where $\arg (e^{i\lambda} - r)$ has its principal value and each logarithm is real, holds for all r and λ satisfying $1/2^{1/2} \leqq r < 1, -\pi/4 - \tau \leqq \lambda \leqq -\pi/4 + \tau$. Comparing (15.6.6) and (15.5.30) with $\lambda = \varphi = -\pi/4 \pm \tau$, we see that in (15.6.6) we have $n = m = 0$. Thus

$$\log f(r) = i \log \frac{e^{i\lambda} - r}{r} + \log \frac{e^{i\lambda}}{2(1 - re^{i\lambda})} - \pi/2 \qquad (15.6.10)$$

for $-\pi/4 - \tau \leqq \lambda \leqq -\pi/4 + \tau$.

We have considered the case of (15.6.1) in which $e^{i(\varphi+\psi)} = -i$. Now we suppose that $e^{i(\varphi+\psi)} = i$ and that $\varphi = \pi/4 + \tau, \psi = \pi/4 - \tau$. Equation (15.6.1) remains valid, but in place of (15.6.2) we have

$$(1 - i) \int \frac{(2v - 1)^{1/2}}{v(v - 1)}\, dv = \pm(1 - i) \int \frac{(z - e^{i\psi})(z - e^{i\varphi})}{(z - r)(z - 1/r)} \frac{dz}{z}. \qquad (15.6.11)$$

This is the case $0 < a = -b$ mentioned in **15.4.** The unit circle $|z| < 1$ is mapped by $v(z)$ onto the v-plane cut by the straight-line segment $\operatorname{Re}(v) = 1/2$, $\operatorname{Im}(v) \geqq 0$ and, in general, by parts of the other two arcs of $\mathbf{\Gamma}_v$ that emerge from $v = 1/2$. The points $z = e^{i\varphi}$ and $z = e^{i\psi}$ map into the tips of these slits. The $\mathbf{\Gamma}_v$-structure in this case is the conjugate of the $\mathbf{\Gamma}_v$-structure in the case previously discussed. From (15.3.3) we have

$$\pm \int i\alpha^{1/2} e^{i(\varphi+\psi)/2} \frac{(2w - \alpha)^{1/2}}{w - \alpha} \frac{dw}{w} = e^{-i(\varphi+\psi)/2} \int \frac{(z - e^{i\psi})(z - e^{i\varphi})}{(z - r)(z - 1/r)} \frac{dz}{z},$$

and this shows that the identification of points of $|z| = 1$ in this case is conjugate to the identification obtained in the case previously discussed. Hence if $f(z)$ is the function corresponding to $e^{i(\varphi+\psi)} = i$ and $f_1(z)$ is the function corresponding to $e^{i(\varphi+\psi)} = -i$, then

$$f(z) = \overline{f_1(\bar{z})}.$$

Then $f'(r) = \overline{f'(r)}$; so from (15.6.9) and (15.6.10),

$$\log f'(r) = i \log \frac{r^2}{1 - r^2} -2 \log | e^{i\lambda} - r | \qquad (15.6.12)$$
$$+ 2 \arg (e^{i\lambda} - r) - \pi + \pi i;$$

$$\text{(15.6.13)} \qquad \log f(r) = -i \log \frac{e^{i\lambda} - r}{r} + \log \frac{e^{i\lambda}}{2(1 - re^{i\lambda})} - \pi/2$$

where arg $(e^{i\lambda} - r)$ has its principal value; each logarithm is real; and

$$1/2^{1/2} \leqq r \leqq 1, \qquad \pi/4 - \tau \leqq \lambda \leqq \pi/4 + \tau.$$

These equations agree with (15.5.31) and (15.5.32) in the cases $\lambda = \varphi = \pi/4 - \tau$ and $\lambda = \varphi = \pi/4 + \tau$.

Equations (15.6.9) and (15.6.12) show that if r is fixed and λ varies over the appropriate interval, then $\log f'(r)$ moves over a straight-line segment. Let

$$\text{(15.6.14)} \qquad \begin{aligned} C_0(r) &= -2 \log \frac{1 + (2r^2 - 1)^{1/2}}{2^{1/2}} + 2\tau, \\ C_1(r) &= -2 \log \frac{1 - (2r^2 - 1)^{1/2}}{2^{1/2}} - 2\tau. \end{aligned}$$

If r is fixed and λ moves over the interval $-\pi/4 - \tau \leqq \lambda \leqq -\pi/4 + \tau$, then $\log f'(r)$ moves over the line segment

$$C_0 \leqq \operatorname{Re} (\log f'(r)) \leqq C_1, \qquad \operatorname{Im} (\log f'(r)) = \log \frac{1 - r^2}{r^2} - \pi.$$

If r is fixed and λ moves over the interval $+ \pi/4 - \tau \leqq \lambda \leqq \pi/4 + \tau$, then $\log f'(r)$ moves over the straight-line segment

$$C_0 \leqq \operatorname{Re} (\log f'(r)) \leqq C_1, \qquad \operatorname{Im} (\log f'(r)) = -\log \frac{1 - r^2}{r^2} + \pi.$$

15.7. If $1/2^{1/2} < r < 1$, we have seen that the region $L(r)$ is bounded by two arcs alternating with two straight-line segments. We now determine the internal angles at the four points where the straight-line segments meet the arcs. By symmetry, these four angles are equal. It can be verified from (15.5.25) that $L(r)$ has definite interior angles at the ends of the straight-line segments.

Let

$$f(z) = z + a_2 z^2 + \cdots$$

belong to class $\mathfrak{S}$, and let

$$\varphi(z) = f\left(\frac{z + \eta}{1 + \bar{\eta} z}\right)$$

where η is a small complex number. Then

$$\varphi'(z) = f'\left(\frac{z + \eta}{1 + \bar{\eta} z}\right) \frac{1 - |\eta|^2}{(1 + \bar{\eta} z)^2}, \qquad \varphi'(0) = (1 - |\eta|^2) f'(\eta).$$

The function

$$F(z) = \frac{1}{\varphi'(0)} [\varphi(z) - \varphi(0)]$$

clearly belongs to class $\mathfrak{S}$. Differentiating, we obtain

$$F'(r) = \frac{f'\left(\dfrac{r + \eta}{1 + \bar{\eta} r}\right)}{(1 + \bar{\eta} r)^2 f'(\eta)}$$

$$= f'(r) + [f''(r) - 2a_2 f'(r)]\eta - [r^2 f''(r) + 2rf'(r)]\bar{\eta} + o(\eta).$$

Now let $f(z)$ extremalize

$$\operatorname{Im}[e^{-i\tau} f'(r)],$$

where τ is some fixed real number. Then

$$\operatorname{Im}\{e^{-i\tau} F'(r)\} = \operatorname{Im}\{e^{-i\tau} f'(r) + e^{-i\tau}[f''(r) - 2a_2 f'(r)]\eta + e^{i\tau} r[r\overline{f''(r)} + 2\overline{f'(r)}]\eta + o(\eta)\},$$

and it readily follows that

(15.7.1) $$e^{-i\tau}[f''(r) - 2a_2 f'(r)] + e^{i\tau} r[r\overline{f''(r)} + 2\overline{f'(r)}] = 0.$$

Let $f(z)$ be the boundary function belonging to one of the points where the straight-line segment meets the arc. If $f''(r) \neq 0$, the curve $f'(re^{i\theta})$, θ real, lies in $\boldsymbol{R}(r)$ and has a tangent at the point where the straight line meets the arc. Hence the interior angle at the point is not less than π. If $f''(r) = 0$, then by (15.7.1)

(15.7.2) $$a_2 e^{-i\tau} f'(r) = re^{i\tau}\overline{f'(r)}.$$

Since $f'(r) \neq 0$, we see that $a_2 \neq 0$. If the internal angle were less than π, there would be infinitely many values of τ satisfying (15.7.2) for the same function $f(z)$ (and so for the same values a_2 and $f'(r)$). This is impossible.

Hence the internal angle at the point is not less than π. But the internal angle cannot exceed π since arg $f'(r)$ is maximal at the point; therefore the internal angle is precisely equal to π.

APPENDIX

BOUNDARY OF V_3

TABLE I (a_2 REAL)

Part I (The portion Π_1 corresponding to unforked functions)

ρ	a_2	Re (a_3)	Im (a_3)	ρ	a_2	Re (a_3)	Im (a_3)
	$\varphi = 15°$				$\varphi = 60°$		
0.1	1.9994	2.9989	0.0278	0.1	1.9970	2.9938	0.0627
0.2	1.9984	2.9974	0.0454	0.2	1.9867	2.9718	0.1312
0.3	1.9975	2.9966	0.0544	0.3	1.9687	2.9340	0.1979
0.4	1.9971	2.9966	0.0565	0.4	1.9449	2.8870	0.2560
0.5	1.9971	2.9972	0.0532	0.5	1.9186	2.8395	0.3014
0.6	1.9974	2.9980	0.0460	0.6	1.8926	2.7982	0.3327
0.7	1.9979	2.9988	0.0363	0.7	1.8690	2.7658	0.3508
0.8	1.9984	2.9994	0.0258	0.8	1.8490	2.7424	0.3584
0.9	1.9987	2.9997	0.0163	0.9	1.8338	2.7274	0.3592
1.0	1.9989	2.9998	0.0111	1.0	1.8264	2.7208	0.3580
	$\varphi = 30°$				$\varphi = 67.5°$		
0.1	1.9980	2.9962	0.0514	0.1	1.9978	2.9954	0.0536
0.2	1.9938	2.9894	0.0891	0.2	1.9892	2.9760	0.1179
0.3	1.9895	2.9837	0.1132	0.3	1.9717	2.9354	0.1878
0.4	1.9862	2.9811	0.1250	0.4	1.9450	2.8745	0.2560
0.5	1.9841	2.9812	0.1266	0.5	1.9118	2.8017	0.3161
0.6	1.9832	2.9833	0.1205	0.6	1.8757	2.7283	0.3644
0.7	1.9831	2.9860	0.1090	0.7	1.8407	2.6628	0.3999
0.8	1.9833	2.9886	0.0948	0.8	1.8096	2.6101	0.4234
0.9	1.9834	2.9905	0.0810	0.9	1.7856	2.5728	0.4371
1.0	1.9834	2.9912	0.0732	1.0	1.7739	2.5556	0.4425
	$\varphi = 45°$				$\varphi = 75°$		
0.1	1.9968	2.9936	0.0654	0.1	1.9988	2.9975	0.0393
0.2	1.9882	2.9776	0.1235	0.2	1.9936	2.9852	0.0905
0.3	1.9768	2.9585	0.1693	0.3	1.9812	2.9539	0.1522
0.4	1.9651	2.9422	0.2010	0.4	1.9586	2.8954	0.2204
0.5	1.9546	2.9311	0.2189	0.5	1.9252	2.8089	0.2884
0.6	1.9460	2.9255	0.2248	0.6	1.8839	2.7045	0.3503
0.7	1.9393	2.9239	0.2212	0.7	1.8397	2.5972	0.4024
0.8	1.9342	2.9245	0.2115	0.8	1.7980	2.5009	0.4429
0.9	1.9304	2.9257	0.1998	0.9	1.7645	2.4268	0.4713
1.0	1.9284	2.9264	0.1925	1.0	1.7478	2.3909	0.4842

ρ	a_2	Re (a_3)	Im (a_3)	ρ	a_2	Re (a_3)	Im (a_3)
	$\varphi = 1.353000$ rad.				$\varphi = 1.492000$ rad.		
0.1	1.9992	2.9981	0.0334	0.1	1.9999	2.9997	0.0127
0.2	1.9953	2.9888	0.0780	0.2	1.9993	2.9983	0.0304
0.3	1.9854	2.9636	0.1336	0.3	1.9976	2.9936	0.0543
0.4	1.9663	2.9122	0.1977	0.4	1.9935	2.9816	0.0854
0.5	1.9359	2.8291	0.2648	0.5	1.9846	2.9539	0.1246
0.6	1.8957	2.7203	0.3285	0.6	1.9672	2.8974	0.1707
0.7	1.8504	2.6012	0.3840	0.7	1.9387	2.8027	0.2189
0.8	1.8062	2.4891	0.4287	0.8	1.9017	2.6791	0.2617
0.9	1.7700	2.4000	0.4607	0.9	1.8654	2.5576	0.2928
1.0	1.7518	2.3559	0.4755	1.0	1.8452	2.4894	0.3064
	$\varphi = 1.400000$ rad.				$\varphi = 1.518000$ rad.		
0.1	1.9995	2.9988	0.0268	0.1	1.9999	2.9999	0.0085
0.2	1.9969	2.9926	0.0632	0.2	1.9997	2.9992	0.0205
0.3	1.9901	2.9747	0.1101	0.3	1.9989	2.9971	0.0367
0.4	1.9757	2.9348	0.1666	0.4	1.9969	2.9913	0.0583
0.5	1.9507	2.8628	0.2294	0.5	1.9925	2.9773	0.0862
0.6	1.9143	2.7575	0.2922	0.6	1.9828	2.9452	0.1211
0.7	1.8702	2.6313	0.3492	0.7	1.9641	2.8805	0.1610
0.8	1.8250	2.5048	0.3962	0.8	1.9351	2.7788	0.1993
0.9	1.7867	2.3998	0.4304	0.9	1.9028	2.6650	0.2275
1.0	1.7672	2.3465	0.4461	1.0	1.8835	2.5963	0.2394
	$\varphi = 1.431000$ rad.				$\varphi = 1.546000$ rad.		
0.1	1.9996	2.9992	0.0221	0.1	2.0000	3.0000	0.0040
0.2	1.9978	2.9948	0.0526	0.2	1.9999	2.9998	0.0097
0.3	1.9929	2.9818	0.0926	0.3	1.9997	2.9993	0.0174
0.4	1.9822	2.9512	0.1422	0.4	1.9993	2.9980	0.0277
0.5	1.9619	2.8913	0.1996	0.5	1.9983	2.9947	0.0414
0.6	1.9301	2.7951	0.2596	0.6	1.9957	2.9861	0.0595
0.7	1.8887	2.6704	0.3158	0.7	1.9895	2.9640	0.0831
0.8	1.8443	2.5379	0.3629	0.8	1.9756	2.9127	0.1107
0.9	1.8054	2.4238	0.3972	0.9	1.9533	2.8294	0.1341
1.0	1.7853	2.3645	0.4129	1.0	1.9368	2.7673	0.1437
	$\varphi = 1.466000$ rad.				$\varphi = 1.560000$ rad.		
0.1	1.9998	2.9995	0.0168	0.1	2.0000	3.0000	0.0017
0.2	1.9987	2.9970	0.0401	0.2	2.0000	3.0000	0.0042
0.3	1.9958	2.9889	0.0711	0.3	2.0000	2.9999	0.0076
0.4	1.9890	2.9695	0.1109	0.4	1.9999	2.9996	0.0121
0.5	1.9752	2.9272	0.1592	0.5	1.9997	2.9990	0.0181
0.6	1.9508	2.8501	0.2130	0.6	1.9992	2.9973	0.0262
0.7	1.9153	2.7368	0.2660	0.7	1.9978	2.9924	0.0374
0.8	1.8738	2.6046	0.3114	0.8	1.9937	2.9770	0.0528
0.9	1.8356	2.4835	0.3445	0.9	1.9824	2.9337	0.0703
1.0	1.8151	2.4183	0.3593	1.0	1.9698	2.8847	0.0786

Part II (The portion Π_2 corresponding to forked functions)

$\varphi = 0.523599$ rad.

μ	a_2	Re (a_3)	Im (a_3)
0.0000000	1.9812	2.9936	0.0000
0.0931004	1.9834	2.9912	0.0732

$\varphi = 0.785398$ rad.

μ	a_2	Re (a_3)	Im (a_3)
0.0000000	1.9043	2.9334	0.0000
0.1500000	1.9102	2.9312	0.1010
0.3034926	1.9284	2.9264	0.1925

$\varphi = 0.916000$ rad.

μ	a_2	Re (a_3)	Im (a_3)
0.0000000	1.8220	2.8424	0.0000
0.1600000	1.8291	2.8413	0.1087
0.3200000	1.8499	2.8405	0.2037
0.4706602	1.8819	2.8460	0.2703

$\varphi = 0.982000$ rad.

μ	a_2	Re (a_3)	Im (a_3)
0.0000000	1.7640	2.7692	0.0000
0.2000000	1.7753	2.7681	0.1398
0.4000000	1.8088	2.7710	0.2531
0.5724916	1.8546	2.7888	0.3136

$\varphi = 1.047198$ rad.

μ	a_2	Re (a_3)	Im (a_3)
0.0000000	1.6931	2.6736	0.0000
0.1700000	1.7017	2.6726	0.1282
0.3400000	1.7269	2.6729	0.2391
0.5100000	1.7683	2.6842	0.3185
0.6848541	1.8264	2.7208	0.3580

$\varphi = 1.090000$ rad.

μ	a_2	Re (a_3)	Im (a_3)
0.0000000	1.6383	2.5963	0.0000
0.1500000	1.6451	2.5951	0.1205
0.3000000	1.6655	2.5937	0.2285
0.4500000	1.6989	2.5985	0.3131
0.6000000	1.7447	2.6186	0.3672
0.7650357	1.8081	2.6704	0.3871

$\varphi = 1.132000$ rad.

μ	a_2	Re (a_3)	Im (a_3)
0.0000000	1.5771	2.5081	0.0000
0.1600000	1.5852	2.5060	0.1374
0.3200000	1.6092	2.5028	0.2586
0.4800000	1.6485	2.5077	0.3504
0.6400000	1.7020	2.5328	0.4046
0.8486667	1.7909	2.6171	0.4147

$\varphi = 1.174000$ rad.

μ	a_2	Re (a_3)	Im (a_3)
0.0000000	1.5078	2.4067	0.0000
0.1900000	1.5197	2.4022	0.1753
0.3800000	1.5549	2.3959	0.3218
0.5700000	1.6119	2.4068	0.4186
0.7500000	1.6840	2.4540	0.4560
0.9371865	1.7753	2.5611	0.4402

$\varphi = 1.208000$ rad.

μ	a_2	Re (a_3)	Im (a_3)
0.0000000	1.4451	2.3144	0.0000
0.1900000	1.4575	2.3080	0.1887
0.3800000	1.4942	2.2969	0.3461
0.5700000	1.5534	2.3022	0.4501
0.7600000	1.6327	2.3496	0.4921
1.0124021	1.7644	2.5151	0.4580

$\varphi = 1.242000$ rad.

μ	a_2	Re (a_3)	Im (a_3)
0.0000000	1.3758	2.2129	0.0000
0.1500000	1.3840	2.2070	0.1636
0.3000000	1.4082	2.1932	0.3094
0.4500000	1.4476	2.1818	0.4235
0.6000000	1.5010	2.1867	0.4982
0.7500000	1.5670	2.2217	0.5314
1.0907697	1.7558	2.4698	0.4722

$\varphi = 1.275000$ rad.

μ	a_2	Re (a_3)	Im (a_3)
0.0000000	1.3017	2.1054	0.0000
0.1200000	1.3072	2.1001	0.1439
0.2400000	1.3236	2.0861	0.2773
0.3600000	1.3505	2.0688	0.3913
0.4800000	1.3874	2.0557	0.4797
0.6000000	1.4333	2.0554	0.5398
0.7200000	1.4875	2.0756	0.5713
1.1698113	1.7501	2.4284	0.4813

$\varphi = 1.308997$ rad.

μ	a_2	Re (a_3)	Im (a_3)
0.0000000	1.2173	1.9857	0.0000
0.1800000	1.2305	1.9695	0.2350
0.3600000	1.2694	1.9329	0.4304
0.5400000	1.3317	1.9043	0.5611
0.7200000	1.4143	1.9163	0.6211
0.9000000	1.5139	1.9941	0.6183
1.0800000	1.6273	2.1531	0.5669
1.1700000	1.6884	2.2655	0.5274
1.2542651	1.7478	2.3909	0.4842

$\varphi = 1.331000$ rad.

μ	a_2	Re (a_3)	Im (a_3)
0.0000000	1.1579	1.9035	0.0000
0.1500000	1.1675	1.8892	0.2123
0.3000000	1.1961	1.8535	0.3979
0.4500000	1.2422	1.8139	0.5377
0.6000000	1.3041	1.7923	0.6242
0.7500000	1.3795	1.8082	0.6593
1.3105349	1.7488	2.3710	0.4818

μ	a_2	Re (a_3)	Im (a_3)
$\varphi = 1.353000$ rad.			
0.0000000	1.0943	1.8179	0.0000
0.2000000	1.1124	1.7872	0.3003
0.4000000	1.1651	1.7202	0.5323
0.6000000	1.2480	1.6724	0.6626
0.8000000	1.3555	1.6962	0.6961
1.0000000	1.4824	1.8228	0.6568
1.1000000	1.5516	1.9288	0.6180
1.2000000	1.6240	2.0640	0.5702
1.3680436	1.7518	2.3559	0.4755
$\varphi = 1.377000$ rad.			
0.0000000	1.0196	1.7211	0.0000
0.1500000	1.0306	1.6981	0.2516
0.3000000	1.0628	1.6399	0.4662
0.4500000	1.1145	1.5734	0.6202
0.6000000	1.1831	1.5285	0.7079
0.7500000	1.2658	1.5293	0.7366
0.9000000	1.3600	1.5903	0.7192
1.4321799	1.7581	2.3465	0.4634
$\varphi = 1.400000$ rad.			
0.0000000	0.9423	1.6255	0.0000
0.1900000	0.9613	1.5789	0.3453
0.3700000	1.0124	1.4798	0.5939
0.5600000	1.0962	1.3869	0.7387
0.7500000	1.2044	1.3676	0.7758
0.9300000	1.3240	1.4465	0.7415
1.0200000	1.3887	1.5245	0.7077
1.1200000	1.4637	1.6412	0.6615
1.2200000	1.5416	1.7888	0.6087
1.3100000	1.6137	1.9474	0.5576
1.4949916	1.7672	2.3465	0.4461
$\varphi = 1.422000$ rad.			
0.0000000	0.8625	1.5321	0.0000
0.1200000	0.8708	1.5068	0.2494
0.2400000	0.8952	1.4396	0.4681
0.3600000	0.9346	1.3519	0.6355
0.4800000	0.9870	1.2687	0.7449
0.6000000	1.0506	1.2112	0.8009
0.7200000	1.1235	1.1930	0.8137
0.8400000	1.2039	1.2213	0.7945
0.9600000	1.2905	1.2983	0.7534
1.0800000	1.3821	1.4236	0.6979
1.2000000	1.4778	1.5953	0.6337
1.5562835	1.7793	2.3568	0.4238
$\varphi = 1.444000$ rad.			
0.0000000	0.7759	1.4378	0.0000

μ	a_2	Re (a_3)	Im (a_3)
$\varphi = 1.444000$ rad.			
0.1300000	0.7867	1.3983	0.3035
0.2600000	0.8183	1.2970	0.5560
0.3900000	0.8684	1.1736	0.7296
0.5200000	0.9340	1.0685	0.8230
0.6500000	1.0122	1.0091	0.8516
0.7800000	1.1002	1.0088	0.8347
0.9100000	1.1959	1.0708	0.7889
1.0400000	1.2975	1.1934	0.7262
1.1700000	1.4039	1.3731	0.6548
1.3000000	1.5139	1.6059	0.5795
1.6187370	1.7951	2.3797	0.3951
$\varphi = 1.466000$ rad.			
0.0000000	0.6815	1.3438	0.0000
0.1000000	0.6888	1.3107	0.2726
0.1900000	0.7075	1.2325	0.4889
0.2800000	0.7368	1.1270	0.6598
0.3700000	0.7755	1.0154	0.7796
0.4600000	0.8222	0.9157	0.8515
0.5600000	0.8821	0.8336	0.8853
0.6500000	0.9418	0.7927	0.8848
0.7500000	1.0134	0.7868	0.8603
0.8400000	1.0817	0.8169	0.8242
0.9400000	1.1611	0.8878	0.7746
1.0400000	1.2434	0.9956	0.7194
1.1200000	1.3111	1.1067	0.6733
1.1800000	1.3627	1.2036	0.6383
1.2400000	1.4149	1.3117	0.6033
1.3100000	1.4767	1.4514	0.5627
1.3700000	1.5302	1.5823	0.5284
1.4400000	1.5931	1.7476	0.4890
1.5000000	1.6476	1.8997	0.4559
1.5450000	1.6886	2.0199	0.4315
1.5900000	1.7299	2.1453	0.4074
1.6350000	1.7714	2.2758	0.3837
1.6823271	1.8151	2.4183	0.3593
$\varphi = 1.483000$ rad.			
0.0000000	0.6022	1.2724	0.0000
0.1400000	0.6183	1.1879	0.4233
0.2800000	0.6641	0.9899	0.7273
0.4200000	0.7342	0.7843	0.8792
0.5600000	0.8224	0.6443	0.9145
0.7000000	0.9234	0.5954	0.8812
0.8400000	1.0336	0.6367	0.8143
0.9800000	1.1502	0.7587	0.7340
1.1200000	1.2716	0.9515	0.6506
1.2600000	1.3965	1.2069	0.5691
1.4000000	1.5240	1.5188	0.4916
1.7322273	1.8339	2.4613	0.3261

μ	a_2	Re (a_3)	Im (a_3)
	$\varphi = 1.500000$ rad.		
0.0000000	0.5162	1.2034	0.0000
0.1400000	0.5349	1.0846	0.4905
0.2800000	0.5873	0.8226	0.8070
0.4200000	0.6655	0.5753	0.9295
0.5600000	0.7616	0.4250	0.9274
0.7000000	0.8698	0.3850	0.8659
0.8400000	0.9859	0.4428	0.7822
0.9800000	1.1076	0.5826	0.6943
1.1200000	1.2332	0.7918	0.6095
1.2600000	1.3616	1.0614	0.5304
1.4000000	1.4921	1.3852	0.4576
1.7827784	1.8560	2.5180	0.2874
	$\varphi = 1.518000$ rad.		
0.0000000	0.4160	1.1347	0.0000
0.0800000	0.4237	1.0694	0.3645
0.1600000	0.4457	0.9012	0.6569
0.2400000	0.4803	0.6902	0.8452
0.3200000	0.5249	0.4896	0.9381
0.4000000	0.5771	0.3287	0.9626
0.4800000	0.6352	0.2167	0.9447
0.5600000	0.6976	0.1524	0.9036
0.6400000	0.7633	0.1306	0.8515
0.7200000	0.8316	0.1454	0.7952
0.7800000	0.8840	0.1774	0.7528
0.8400000	0.9374	0.2254	0.7112
0.9100000	1.0006	0.2994	0.6644
0.9700000	1.0555	0.3770	0.6262
1.0400000	1.1201	0.4826	0.5838
1.1000000	1.1760	0.5853	0.5493
1.1600000	1.2323	0.6984	0.5167
1.2200000	1.2890	0.8215	0.4856
1.2800000	1.3459	0.9542	0.4561
1.4700000	1.5277	1.4336	0.3718
1.6500000	1.7016	1.9663	0.3026
1.8369980	1.8835	2.5963	0.2394
	$\varphi = 1.531000$ rad.		
0.0000000	0.3361	1.0894	0.0000
0.1000000	0.3507	0.9363	0.5404
0.2000000	0.3911	0.6048	0.8657
0.3000000	0.4505	0.2898	0.9728
0.4000000	0.5225	0.0735	0.9563
0.5000000	0.6025	−0.0424	0.8892
0.6000000	0.6877	−0.0777	0.8077
0.7000000	0.7765	−0.0510	0.7272
0.8000000	0.8677	0.0241	0.6528

μ	a_2	Re (a_3)	Im (a_3)
	$\varphi = 1.531000$ rad.		
0.9000000	0.9607	0.1387	0.5860
1.0000000	1.0550	0.2867	0.5264
1.1000000	1.1502	0.4644	0.4732
1.2000000	1.2462	0.6690	0.4256
1.3000000	1.3428	0.8988	0.3828
1.4000000	1.4398	1.1524	0.3441
1.8765924	1.9065	2.6670	0.1989
	$\varphi = 1.546000$ rad.		
0.0000000	0.2329	1.0439	0.0000
0.0400000	0.2363	0.9882	0.3321
0.0800000	0.2463	0.8390	0.6112
0.1200000	0.2620	0.6383	0.8093
0.1600000	0.2826	0.4276	0.9266
0.2000000	0.3070	0.2341	0.9798
0.2400000	0.3344	0.0703	0.9889
0.2800000	0.3642	−0.0612	0.9706
0.3200000	0.3958	−0.1626	0.9369
0.3600000	0.4288	−0.2380	0.8954
0.4000000	0.4629	−0.2915	0.8509
0.4400000	0.4978	−0.3266	0.8062
0.5100000	0.5607	−0.3528	0.7314
0.5800000	0.6250	−0.3440	0.6637
0.6400000	0.6811	−0.3147	0.6118
0.7000000	0.7377	−0.2689	0.5652
0.7700000	0.8045	−0.1978	0.5168
0.8300000	0.8621	−0.1234	0.4798
0.8900000	0.9200	−0.0381	0.4463
0.9600000	0.9879	0.0744	0.4112
1.1600000	1.1832	0.4647	0.3290
1.3500000	1.3699	0.9219	0.2686
1.5400000	1.5575	1.4579	0.2197
1.7300000	1.7456	2.0700	0.1790
1.9227228	1.9368	2.7673	0.1437
	$\varphi = 1.553000$ rad.		
0.0000000	0.1790	1.0263	0.0000
0.3000000	0.3493	−0.3596	0.8699
0.4000000	0.4382	−0.4811	0.7319
0.5000000	0.5311	−0.4977	0.6172
0.6000000	0.6261	−0.4514	0.5268
0.7000000	0.7225	−0.3622	0.4554
0.8000000	0.8198	−0.2396	0.3979
0.9000000	0.9176	−0.0888	0.3508
1.0000000	1.0159	0.0871	0.3115
1.9444109	1.9526	2.8226	0.1135

μ	a_2	Re (a_3)	Im (a_3)
	φ = 1.560000 rad.		
0.0000000	0.1194	1.0119	0.0000
0.0200000	0.1210	0.9577	0.3256
0.0400000	0.1259	0.8116	0.6018
0.0600000	0.1336	0.6121	0.8014
0.0800000	0.1437	0.3983	0.9234
0.1000000	0.1557	0.1970	0.9826
0.1200000	0.1693	0.0208	0.9976
0.1400000	0.1840	−0.1268	0.9846
0.1600000	0.1996	−0.2476	0.9554
0.1800000	0.2160	−0.3451	0.9175
0.2000000	0.2329	−0.4231	0.8761
0.2200000	0.2503	−0.4851	0.8338
0.2700000	0.2952	−0.5886	0.7340
0.3300000	0.3509	−0.6482	0.6328
0.3900000	0.4079	−0.6651	0.5514
0.4500000	0.4656	−0.6545	0.4861
0.5200000	0.5335	−0.6180	0.4250
0.5900000	0.6020	−0.5618	0.3761
0.7900000	0.7990	−0.3198	0.2785
0.9800000	0.9872	0.0011	0.2190
1.1800000	1.1860	0.4241	0.1748
1.3800000	1.3852	0.9307	0.1420
1.5700000	1.5745	1.4878	0.1173
1.7700000	1.7740	2.1534	0.0961
1.9661996	1.9698	2.8847	0.0786
	φ = 1.564000 rad.		
0.0000000	0.0814	1.0056	0.0000
0.1650000	0.1840	−0.5755	0.7916
0.2200000	0.2346	−0.7051	0.6482
0.2750000	0.2868	−0.7577	0.5408
0.3300000	0.3399	−0.7708	0.4608
0.3850000	0.3935	−0.7607	0.3997
0.4400000	0.4475	−0.7347	0.3520
0.4950000	0.5017	−0.6968	0.3137
0.5500000	0.5560	−0.6492	0.2823
0.6050000	0.6105	−0.5929	0.2562
1.9786950	1.9804	2.9240	0.0553
	φ = 1.567000 rad.		
0.0000000	0.0499	1.0021	0.0000
0.1290000	0.1383	−0.7208	0.6721
0.1720000	0.1791	−0.8130	0.5340

μ	a_2	Re (a_3)	Im (a_3)
	φ = 1.567000 rad.		
0.2150000	0.2207	−0.8494	0.4389
0.2580000	0.2628	−0.8592	0.3710
0.3010000	0.3051	−0.8538	0.3205
0.3440000	0.3476	−0.8383	0.2816
0.3870000	0.3902	−0.8154	0.2508
0.4300000	0.4329	−0.7864	0.2258
0.4730000	0.4756	−0.7522	0.2051
1.9880879	1.9887	2.9558	0.0351
	φ = 1.570000 rad.		
0.0000000	0.0130	1.0001	0.0000
0.0020000	0.0131	0.9536	0.3015
0.0040000	0.0136	0.8261	0.5637
0.0060000	0.0143	0.6470	0.7626
0.0080000	0.0152	0.4482	0.8940
0.0100000	0.0164	0.2537	0.9673
0.0120000	0.0177	0.0769	0.9970
0.0140000	0.0191	−0.0769	0.9970
0.0160000	0.0206	−0.2075	0.9781
0.0180000	0.0222	−0.3169	0.9483
0.0200000	0.0238	−0.4082	0.9126
0.0220000	0.0255	−0.4843	0.8745
0.0260000	0.0290	−0.6013	0.7984
0.0300000	0.0327	−0.6845	0.7280
0.0340000	0.0364	−0.7451	0.6655
0.0380000	0.0401	−0.7901	0.6109
0.0490000	0.0507	−0.8667	0.4942
0.0600000	0.0614	−0.9071	0.4126
0.0800000	0.0810	−0.9423	0.3155
0.1000000	0.1008	−0.9568	0.2547
0.1500000	0.1506	−0.9625	0.1712
0.2000000	0.2004	−0.9515	0.1287
0.3000000	0.3003	−0.9061	0.0857
0.4000000	0.4002	−0.8378	0.0641
0.6000000	0.6001	−0.6389	0.0422
0.8000000	0.8001	−0.3593	0.0311
1.0000000	1.0001	0.0005	0.0243
1.2000000	1.2001	0.4404	0.0197
1.4000000	1.4001	0.9603	0.0163
1.6000000	1.6001	1.5603	0.0136
1.8000000	1.8000	2.2402	0.0115
1.9974989	1.9975	2.9902	0.0098

TABLE II (a_3 REAL)

Part I (The portion $\mathbf{II}_1$ corresponding to unforked functions)

ρ	Re (a_2)	Im (a_2)	a_3	ρ	Re (a_2)	Im (a_2)	a_3
	$\varphi = 15°$				$\varphi = 60°$		
0.1	1.9994	0.0093	2.9991	0.1	1.9969	0.0209	2.9945
0.2	1.9983	0.0151	2.9977	0.2	1.9862	0.0438	2.9747
0.3	1.9975	0.0181	2.9970	0.3	1.9675	0.0663	2.9407
0.4	1.9970	0.0188	2.9971	0.4	1.9430	0.0860	2.8983
0.5	1.9970	0.0177	2.9976	0.5	1.9159	0.1014	2.8555
0.6	1.9974	0.0153	2.9984	0.6	1.8893	0.1119	2.8179
0.7	1.9979	0.0121	2.9990	0.7	1.8653	0.1178	2.7879
0.8	1.9984	0.0086	2.9995	0.8	1.8451	0.1201	2.7658
0.9	1.9987	0.0054	2.9998	0.9	1.8299	0.1200	2.7509
1.0	1.9989	0.0037	2.9999	1.0	1.8225	0.1194	2.7443
	$\varphi = 30°$				$\varphi = 67.5°$		
0.1	1.9979	0.0171	2.9967	0.1	1.9978	0.0179	2.9958
0.2	1.9936	0.0297	2.9907	0.2	1.9888	0.0394	2.9783
0.3	1.9891	0.0377	2.9859	0.3	1.9707	0.0630	2.9414
0.4	1.9857	0.0416	2.9837	0.4	1.9431	0.0864	2.8859
0.5	1.9837	0.0421	2.9839	0.5	1.9087	0.1074	2.8195
0.6	1.9828	0.0400	2.9857	0.6	1.8716	0.1245	2.7525
0.7	1.9827	0.0362	2.9880	0.7	1.8356	0.1371	2.6926
0.8	1.9830	0.0314	2.9901	0.8	1.8038	0.1454	2.6442
0.9	1.9833	0.0269	2.9916	0.9	1.7793	0.1501	2.6096
1.0	1.9832	0.0243	2.9921	1.0	1.7674	0.1519	2.5936
	$\varphi = 45°$				$\varphi = 75°$		
0.1	1.9967	0.0218	2.9943	0.1	1.9988	0.0131	2.9977
0.2	1.9878	0.0412	2.9802	0.2	1.9934	0.0302	2.9865
0.3	1.9760	0.0565	2.9634	0.3	1.9805	0.0510	2.9578
0.4	1.9640	0.0670	2.9490	0.4	1.9572	0.0744	2.9037
0.5	1.9532	0.0728	2.9393	0.5	1.9227	0.0985	2.8236
0.6	1.9446	0.0746	2.9341	0.6	1.8800	0.1213	2.7271
0.7	1.9379	0.0732	2.9323	0.7	1.8343	0.1412	2.6282
0.8	1.9329	0.0698	2.9322	0.8	1.7911	0.1574	2.5398
0.9	1.9292	0.0658	2.9325	0.9	1.7563	0.1690	2.4722
1.0	1.9273	0.0633	2.9327	1.0	1.7391	0.1743	2.4395

ρ	Re (a_2)	Im (a_2)	a_3
	$\varphi = 1.353000$ rad.		
0.1	1.9991	0.0111	2.9983
0.2	1.9951	0.0260	2.9898
0.3	1.9849	0.0447	2.9666
0.4	1.9652	0.0666	2.9189
0.5	1.9338	0.0903	2.8415
0.6	1.8922	0.1138	2.7401
0.7	1.8454	0.1355	2.6294
0.8	1.7997	0.1538	2.5257
0.9	1.7620	0.1676	2.4438
1.0	1.7432	0.1741	2.4034
	$\varphi = 1.400000$ rad.		
0.1	1.9994	0.0089	2.9989
0.2	1.9968	0.0211	2.9932
0.3	1.9897	0.0368	2.9767
0.4	1.9750	0.0560	2.9395
0.5	1.9491	0.0780	2.8720
0.6	1.9116	0.1010	2.7729
0.7	1.8661	0.1233	2.6544
0.8	1.8194	0.1430	2.5360
0.9	1.7797	0.1583	2.4381
1.0	1.7594	0.1658	2.3885
	$\varphi = 1.431000$ rad.		
0.1	1.9996	0.0074	2.9993
0.2	1.9978	0.0176	2.9953
0.3	1.9927	0.0309	2.9833
0.4	1.9816	0.0477	2.9547
0.5	1.9608	0.0676	2.8982
0.6	1.9280	0.0893	2.8072
0.7	1.8855	0.1111	2.6890
0.8	1.8396	0.1308	2.5637
0.9	1.7995	0.1465	2.4561
1.0	1.7786	0.1541	2.4002
	$\varphi = 1.466000$ rad.		
0.1	1.9998	0.0056	2.9996
0.2	1.9987	0.0134	2.9972
0.3	1.9956	0.0237	2.9898
0.4	1.9887	0.0371	2.9716
0.5	1.9745	0.0537	2.9315
0.6	1.9494	0.0727	2.8581
0.7	1.9130	0.0927	2.7497
0.8	1.8705	0.1114	2.6231
0.9	1.8312	0.1264	2.5073
1.0	1.8102	0.1337	2.4449

ρ	Re (a_2)	Im (a_2)	a_3
	$\varphi = 1.492000$ rad.		
0.1	1.9999	0.0042	2.9998
0.2	1.9993	0.0101	2.9984
0.3	1.9975	0.0181	2.9941
0.4	1.9933	0.0285	2.9828
0.5	1.9841	0.0418	2.9565
0.6	1.9663	0.0579	2.9025
0.7	1.9372	0.0755	2.8112
0.8	1.8995	0.0926	2.6919
0.9	1.8624	0.1063	2.5743
1.0	1.8417	0.1129	2.5082
	$\varphi = 1.518000$ rad.		
0.1	1.9999	0.0028	2.9999
0.2	1.9997	0.0068	2.9993
0.3	1.9988	0.0123	2.9973
0.4	1.9968	0.0194	2.9919
0.5	1.9923	0.0288	2.9786
0.6	1.9824	0.0407	2.9476
0.7	1.9633	0.0548	2.8850
0.8	1.9339	0.0692	2.7860
0.9	1.9011	0.0810	2.6747
1.0	1.8815	0.0865	2.6073
	$\varphi = 1.546000$ rad.		
0.1	2.0000	0.0013	3.0000
0.2	1.9999	0.0032	2.9998
0.3	1.9997	0.0058	2.9994
0.4	1.9993	0.0092	2.9982
0.5	1.9982	0.0138	2.9949
0.6	1.9956	0.0199	2.9867
0.7	1.9893	0.0279	2.9652
0.8	1.9752	0.0375	2.9148
0.9	1.9527	0.0462	2.8326
1.0	1.9361	0.0502	2.7710
	$\varphi = 1.560000$ rad.		
0.1	2.0000	0.0006	3.0000
0.2	2.0000	0.0014	3.0000
0.3	2.0000	0.0025	2.9999
0.4	1.9999	0.0040	2.9996
0.5	1.9997	0.0060	2.9990
0.6	1.9991	0.0087	2.9974
0.7	1.9978	0.0125	2.9926
0.8	1.9936	0.0177	2.9775
0.9	1.9823	0.0237	2.9345
1.0	1.9696	0.0268	2.8858

Part II (The portion $\mathbf{II}_2$ corresponding to forked functions)

μ	Re (a_2)	Im (a_2)	a_3
$\varphi = 0.523599$ rad.			
0.0000000	1.9812	0.0000	2.9936
0.0931004	1.9832	0.0243	2.9921
$\varphi = 0.785398$ rad.			
0.0000000	1.9043	0.0000	2.9334
0.1500000	1.9100	0.0329	2.9329
0.3034926	1.9273	0.0633	2.9327
$\varphi = 0.916000$ rad.			
0.0000000	1.8220	0.0000	2.8424
0.1600000	1.8287	0.0350	2.8434
0.3200000	1.8488	0.0662	2.8478
0.4706602	1.8797	0.0891	2.8588
$\varphi = 0.982000$ rad.			
0.0000000	1.7640	0.0000	2.7692
0.2000000	1.7747	0.0448	2.7717
0.4000000	1.8069	0.0824	2.7826
0.5724916	1.8516	0.1038	2.8064
$\varphi = 1.047198$ rad.			
0.0000000	1.6931	0.0000	2.6736
0.1700000	1.7012	0.0408	2.6756
0.3400000	1.7252	0.0770	2.6836
0.5100000	1.7652	0.1044	2.7030
0.6848541	1.8225	0.1194	2.7443
$\varphi = 1.090000$ rad.			
0.0000000	1.6383	0.0000	2.5963
0.1500000	1.6447	0.0382	2.5979
0.3000000	1.6639	0.0731	2.6037
0.4500000	1.6959	0.1018	2.6173
0.6000000	1.7404	0.1214	2.6442
0.7650357	1.8034	0.1300	2.6983
$\varphi = 1.132000$ rad.			
0.0000000	1.5771	0.0000	2.5081
0.1600000	1.5846	0.0434	2.5097
0.3200000	1.6071	0.0828	2.5162
0.4800000	1.6445	0.1143	2.5321
0.6400000	1.6966	0.1347	2.5649
0.8486667	1.7854	0.1406	2.6497

μ	Re (a_2)	Im (a_2)	a_3
$\varphi = 1.174000$ rad.			
0.0000000	1.5078	0.0000	2.4067
0.1900000	1.5187	0.0554	2.4086
0.3800000	1.5515	0.1037	2.4175
0.5700000	1.6059	0.1386	2.4429
0.7500000	1.6769	0.1545	2.4960
0.9371865	1.7689	0.1509	2.5987
$\varphi = 1.208000$ rad.			
0.0000000	1.4451	0.0000	2.3144
0.1900000	1.4563	0.0594	2.3157
0.3800000	1.4900	0.1116	2.3228
0.5700000	1.5462	0.1497	2.3458
0.7600000	1.6240	0.1682	2.4005
1.0124021	1.7573	0.1587	2.5565
$\varphi = 1.242000$ rad.			
0.0000000	1.3758	0.0000	2.2129
0.1500000	1.3830	0.0512	2.2131
0.3000000	1.4047	0.0986	2.2149
0.4500000	1.4409	0.1386	2.2226
0.6000000	1.4916	0.1678	2.2427
0.7500000	1.5562	0.1835	2.2844
1.0907697	1.7479	0.1656	2.5146
$\varphi = 1.275000$ rad.			
0.0000000	1.3017	0.0000	2.1054
0.1200000	1.3064	0.0447	2.1050
0.2400000	1.3207	0.0874	2.1045
0.3600000	1.3446	0.1260	2.1054
0.4800000	1.3783	0.1587	2.1110
0.6000000	1.4215	0.1835	2.1251
0.7200000	1.4741	0.1992	2.1528
1.1698113	1.7417	0.1709	2.4757
$\varphi = 1.308997$ rad.			
0.0000000	1.2173	0.0000	1.9857
0.1800000	1.2284	0.0730	1.9835
0.3600000	1.2618	0.1388	1.9802
0.5400000	1.3180	0.1901	1.9853
0.7200000	1.3970	0.2207	2.0144
0.9000000	1.4968	0.2267	2.0877
1.0800000	1.6139	0.2089	2.2265
1.1700000	1.6774	0.1927	2.3261
1.2542651	1.7391	0.1743	2.4395

$\varphi = 1.331000$ rad.

μ	Re (a_2)	Im (a_2)	a_3
0.0000000	1.1579	0.0000	1.9035
0.1500000	1.1657	0.0653	1.9011
0.3000000	1.1894	0.1262	1.8957
0.4500000	1.2294	0.1784	1.8919
0.6000000	1.2858	0.2175	1.8979
0.7500000	1.3585	0.2399	1.9246
1.3105349	1.7400	0.1750	2.4195

$\varphi = 1.353000$ rad.

μ	Re (a_2)	Im (a_2)	a_3
0.0000000	1.0943	0.0000	1.8179
0.2000000	1.1085	0.0925	1.8123
0.4000000	1.1520	0.1742	1.8006
0.6000000	1.2258	0.2340	1.7989
0.8000000	1.3299	0.2623	1.8335
1.0000000	1.4603	0.2551	1.9375
1.1000000	1.5330	0.2396	2.0254
1.2000000	1.6093	0.2182	2.1413
1.3680436	1.7432	0.1741	2.4034

$\varphi = 1.377000$ rad.

μ	Re (a_2)	Im (a_2)	a_3
0.0000000	1.0196	0.0000	1.7211
0.1500000	1.0278	0.0757	1.7166
0.3000000	1.0527	0.1467	1.7049
0.4500000	1.0949	0.2080	1.6912
0.6000000	1.1554	0.2546	1.6845
0.7500000	1.2340	0.2817	1.6975
0.9000000	1.3295	0.2867	1.7454
1.4321799	1.7497	0.1711	2.3918

$\varphi = 1.400000$ rad.

μ	Re (a_2)	Im (a_2)	a_3
0.0000000	0.9423	0.0000	1.6255
0.1900000	0.9557	0.1033	1.6162
0.3700000	0.9940	0.1920	1.5945
0.5600000	1.0635	0.2656	1.5714
0.7500000	1.1645	0.3073	1.5723
0.9300000	1.2870	0.3107	1.6255
1.0200000	1.3560	0.2994	1.6808
1.1200000	1.4369	0.2787	1.7695
1.2200000	1.5209	0.2517	1.8896
1.3100000	1.5981	0.2243	2.0257
1.4949916	1.7594	0.1658	2.3885

$\varphi = 1.422000$ rad.

μ	Re (a_2)	Im (a_2)	a_3
0.0000000	0.8625	0.0000	1.5321
0.1200000	0.8678	0.0713	1.5273
0.2400000	0.8842	0.1401	1.5138
0.3600000	0.9121	0.2037	1.4938

$\varphi = 1.422000$ rad.

μ	Re (a_2)	Im (a_2)	a_3
0.4800000	0.9525	0.2589	1.4712
0.6000000	1.0061	0.3026	1.4520
0.7200000	1.0736	0.3313	1.4441
0.8400000	1.1542	0.3424	1.4570
0.9600000	1.2462	0.3354	1.5011
1.0800000	1.3464	0.3123	1.5855
1.2000000	1.4515	0.2777	1.7166
1.5562835	1.7723	0.1581	2.3946

$\varphi = 1.444000$ rad.

μ	Re (a_2)	Im (a_2)	a_3
0.0000000	0.7759	0.0000	1.4378
0.1300000	0.7822	0.0839	1.4309
0.2600000	0.8016	0.1646	1.4112
0.3900000	0.8350	0.2384	1.3819
0.5200000	0.8842	0.3011	1.3487
0.6500000	0.9507	0.3475	1.3204
0.7800000	1.0351	0.3727	1.3093
0.9100000	1.1361	0.3733	1.3300
1.0400000	1.2494	0.3502	1.3970
1.1700000	1.3693	0.3098	1.5213
1.3000000	1.4913	0.2608	1.7073
1.6187370	1.7890	0.1475	2.4123

$\varphi = 1.466000$ rad.

μ	Re (a_2)	Im (a_2)	a_3
0.0000000	0.6815	0.0000	1.3438
0.1000000	0.6852	0.0705	1.3387
0.1900000	0.6949	0.1328	1.3259
0.2800000	0.7111	0.1929	1.3060
0.3700000	0.7343	0.2494	1.2801
0.4600000	0.7652	0.3008	1.2504
0.5600000	0.8098	0.3498	1.2160
0.6500000	0.8599	0.3841	1.1879
0.7500000	0.9274	0.4086	1.1659
0.8400000	0.9984	0.4162	1.1604
0.9400000	1.0872	0.4076	1.1782
1.0400000	1.1831	0.3827	1.2283
1.1200000	1.2624	0.3538	1.2954
1.1800000	1.3224	0.3289	1.3624
1.2400000	1.3822	0.3026	1.4438
1.3100000	1.4515	0.2715	1.5567
1.3700000	1.5103	0.2455	1.6682
1.4400000	1.5783	0.2167	1.8147
1.5000000	1.6362	0.1936	1.9536
1.5450000	1.6793	0.1773	2.0655
1.5900000	1.7223	0.1621	2.1836
1.6350000	1.7652	0.1478	2.3079
1.6823271	1.8102	0.1337	2.4449

μ	Re (a_2)	Im a_2	a_3
	$\varphi = 1.483000$ rad.		
0.0000000	0.6022	0.0000	1.2724
0.1400000	0.6092	0.1053	1.2611
0.2800000	0.6311	0.2069	1.2283
0.4200000	0.6700	0.3002	1.1782
0.5600000	0.7300	0.3787	1.1187
0.7000000	0.8155	0.4332	1.0635
0.8400000	0.9290	0.4529	1.0337
0.9800000	1.0663	0.4314	1.0556
1.1200000	1.2149	0.3756	1.1527
1.2600000	1.3628	0.3052	1.3344
1.4000000	1.5054	0.2376	1.5963
1.7322273	1.8299	0.1207	2.4828
	$\varphi = 1.500000$ rad.		
0.0000000	0.5162	0.0000	1.2034
0.1400000	0.5228	0.1127	1.1904
0.2800000	0.5436	0.2221	1.1524
0.4200000	0.5814	0.3239	1.0931
0.5600000	0.6410	0.4114	1.0202
0.7000000	0.7293	0.4739	0.9476
0.8400000	0.8517	0.4966	0.8989
0.9800000	1.0039	0.4681	0.9063
1.1200000	1.1675	0.3973	0.9992
1.2600000	1.3253	0.3127	1.1865
1.4000000	1.4732	0.2370	1.4588
1.7827784	1.8530	0.1054	2.5343
	$\varphi = 1.518000$ rad.		
0.0000000	0.4160	0.0000	1.1347
0.0800000	0.4180	0.0693	1.1298
0.1600000	0.4238	0.1381	1.1152
0.2400000	0.4339	0.2059	1.0912
0.3200000	0.4488	0.2721	1.0582
0.4000000	0.4694	0.3357	1.0171
0.4800000	0.4968	0.3958	0.9692
0.5600000	0.5327	0.4504	0.9164
0.6400000	0.5792	0.4972	0.8614
0.7200000	0.6387	0.5325	0.8084
0.7800000	0.6931	0.5487	0.7734
0.8400000	0.7563	0.5537	0.7461
0.9100000	0.8404	0.5431	0.7288
0.9700000	0.9188	0.5193	0.7309
1.0400000	1.0134	0.4771	0.7574
1.1000000	1.0935	0.4328	0.8027
1.1600000	1.1704	0.3859	0.8687
1.2200000	1.2433	0.3400	0.9543
1.2800000	1.3126	0.2976	1.0576

μ	Re (a_2)	Im (a_2)	a_3
	$\varphi = 1.518000$ rad.		
1.4700000	1.5155	0.1933	1.4811
1.6500000	1.6967	0.1298	1.9895
1.8369980	1.8815	0.0865	2.6073
	$\varphi = 1.531000$ rad.		
0.0000000	0.3361	0.0000	1.0894
0.1000000	0.3387	0.0907	1.0810
0.2000000	0.3468	0.1808	1.0561
0.3000000	0.3612	0.2693	1.0151
0.4000000	0.3833	0.3550	0.9592
0.5000000	0.4157	0.4361	0.8902
0.6000000	0.4624	0.5090	0.8115
0.7000000	0.5295	0.5680	0.7290
0.8000000	0.6248	0.6022	0.6533
0.9000000	0.7535	0.5960	0.6022
1.0000000	0.9070	0.5388	0.5994
1.1000000	1.0606	0.4451	0.6630
1.2000000	1.1965	0.3483	0.7929
1.3000000	1.3156	0.2685	0.9769
1.4000000	1.4247	0.2081	1.2027
1.8765924	1.9051	0.0709	2.6744
	$\varphi = 1.546000$ rad.		
0.0000000	0.2329	0.0000	1.0439
0.0400000	0.2332	0.0381	1.0425
0.0800000	0.2342	0.0763	1.0381
0.1200000	0.2358	0.1143	1.0307
0.1600000	0.2380	0.1523	1.0205
0.2000000	0.2410	0.1902	1.0074
0.2400000	0.2447	0.2279	0.9914
0.2800000	0.2493	0.2655	0.9725
0.3200000	0.2548	0.3029	0.9509
0.3600000	0.2614	0.3399	0.9265
0.4000000	0.2691	0.3766	0.8995
0.4400000	0.2782	0.4129	0.8698
0.5100000	0.2981	0.4748	0.8121
0.5800000	0.3247	0.5341	0.7476
0.6400000	0.3547	0.5814	0.6880
0.7000000	0.3940	0.6237	0.6259
0.7700000	0.4560	0.6627	0.5533
0.8300000	0.5282	0.6813	0.4954
0.8900000	0.6223	0.6776	0.4480
0.9600000	0.7581	0.6333	0.4179
1.1600000	1.1275	0.3587	0.5693
1.3500000	1.3562	0.1935	0.9602
1.5400000	1.5532	0.1164	1.4743
1.7300000	1.7440	0.0753	2.0778
1.9227228	1.9361	0.0502	2.7710

$\varphi = 1.553000$ rad.

μ	Re (a_2)	Im (a_2)	a_3
0.0000000	0.1790	0.0000	1.0263
0.3000000	0.1942	0.2904	0.9413
0.4000000	0.2080	0.3857	0.8758
0.5000000	0.2291	0.4791	0.7929
0.6000000	0.2617	0.5688	0.6938
0.7000000	0.3139	0.6508	0.5818
0.8000000	0.4034	0.7137	0.4645
0.9000000	0.5636	0.7241	0.3619
1.0000000	0.8094	0.6140	0.3234
1.9444109	1.9522	0.0392	2.8249

$\varphi = 1.560000$ rad.

μ	Re (a_2)	Im (a_2)	a_3
0.0000000	0.1194	0.0000	1.0119
0.0200000	0.1194	0.0197	1.0115
0.0400000	0.1195	0.0395	1.0103
0.0600000	0.1198	0.0592	1.0084
0.0800000	0.1201	0.0790	1.0057
0.1000000	0.1205	0.0987	1.0021
0.1200000	0.1209	0.1184	0.9978
0.1400000	0.1215	0.1382	0.9928
0.1600000	0.1222	0.1579	0.9869
0.1800000	0.1229	0.1776	0.9803
0.2000000	0.1238	0.1973	0.9729
0.2200000	0.1248	0.2170	0.9647
0.2700000	0.1277	0.2661	0.9408
0.3300000	0.1323	0.3250	0.9058
0.3900000	0.1384	0.3837	0.8640
0.4500000	0.1462	0.4420	0.8153
0.5200000	0.1583	0.5095	0.7501
0.5900000	0.1750	0.5760	0.6761
0.7900000	0.2801	0.7483	0.4241
0.9800000	0.6998	0.6964	0.2190
1.1800000	1.1634	0.2304	0.4587
1.3800000	1.3812	0.1047	0.9415
1.5700000	1.5733	0.0619	1.4925
1.7700000	1.7736	0.0396	2.1555
1.9661996	1.9696	0.0268	2.8858

$\varphi = 1.564000$ rad.

μ	Re (a_2)	Im (a_2)	a_3
0.0000000	0.0814	0.0000	1.0056
0.1650000	0.0835	0.1640	0.9787
0.2200000	0.0852	0.2186	0.9578
0.2750000	0.0875	0.2731	0.9309
0.3300000	0.0905	0.3276	0.8981
0.3850000	0.0943	0.3821	0.8593
0.4400000	0.0991	0.4364	0.8147
0.4950000	0.1053	0.4905	0.7642
0.5500000	0.1133	0.5443	0.7079
0.6050000	0.1236	0.5978	0.6459
1.9786950	1.9803	0.0187	2.9246

$\varphi = 1.567000$ rad.

μ	Re (a_2)	Im (a_2)	a_3
0.0000000	0.0499	0.0000	1.0021
0.1290000	0.0507	0.1287	0.9856
0.1720000	0.0513	0.1716	0.9727
0.2150000	0.0521	0.2145	0.9561
0.2580000	0.0532	0.2573	0.9359
0.3010000	0.0545	0.3002	0.9120
0.3440000	0.0561	0.3430	0.8844
0.3870000	0.0580	0.3859	0.8531
0.4300000	0.0603	0.4287	0.8182
0.4730000	0.0631	0.4714	0.7796
1.9880879	1.9887	0.0118	2.9560

$\varphi = 1.570000$ rad.

μ	Re (a_2)	Im (a_2)	a_3
0.0000000	0.0130	0.0000	1.0001
0.0020000	0.0130	0.0020	1.0001
0.0040000	0.0130	0.0040	1.0001
0.0060000	0.0130	0.0060	1.0001
0.0080000	0.0130	0.0080	1.0001
0.0100000	0.0130	0.0100	1.0000
0.0120000	0.0130	0.0120	1.0000
0.0140000	0.0130	0.0140	1.0000
0.0160000	0.0130	0.0160	0.9999
0.0180000	0.0130	0.0180	0.9998
0.0200000	0.0130	0.0200	0.9997
0.0220000	0.0130	0.0220	0.9997
0.0260000	0.0130	0.0260	0.9995
0.0300000	0.0130	0.0300	0.9992
0.0340000	0.0130	0.0340	0.9990
0.0380000	0.0130	0.0380	0.9987
0.0490000	0.0130	0.0490	0.9977
0.0600000	0.0130	0.0600	0.9965
0.0800000	0.0130	0.0800	0.9938
0.1000000	0.0131	0.1000	0.9902
0.1500000	0.0132	0.1500	0.9777
0.2000000	0.0135	0.2000	0.9602
0.3000000	0.0142	0.2999	0.9102
0.4000000	0.0153	0.3999	0.8402
0.6000000	0.0198	0.5998	0.6403
0.8000000	0.0345	0.7994	0.3607
1.0000000	0.7141	0.7001	0.0243
1.2000000	1.1998	0.0268	0.4408
1.4000000	1.4000	0.0119	0.9605
1.6000000	1.6000	0.0070	1.5603
1.8000000	1.8000	0.0046	2.2403
1.9974989	1.9975	0.0033	2.9902

BIBLIOGRAPHY

1. L. Bieberbach, "Über die Koeffizienten derjenigen Potenzreihen, welche eine schlichte Abbildung des Einheitskreises vermitteln," K. Preuss. Akad. Wiss., Berlin, Sitzungsberichte, 1916, 940–955.
2. C. Carathéodory, "Über den Variabilitätsbereich der Fourier'schen Konstanten von positiven harmonischen Funktionen," Rend. Circ. Mat. Palermo, 32 (1911), 193–217.
3. R. Courant, "The existence of minimal surfaces of given topological structure under prescribed boundary conditions," Acta Mathematica, 72 (1940), 51–98.
4. J. Dieudonné, "Sur les fonctions univalentes," Comptes Rendus (Paris), 192 (1931), 1148–1150.
5. G. Faber, "Neuer Beweis eines Koebe-Bieberbachschen Satzes über konforme Abbildung," K. B. Akad. Wiss. München, Sitzungsberichte der math.-phys. kl., 1916, 39–42.
6. G. M. Golusin, (a) "Sur les théorèmes de rotation dans la théorie des fonctions univalentes," Rec. Math. [Mat. Sbornik] N.S.1, 1 (1936), 293–296.

   (b) "Interior problems of the theory of schlicht functions," Office of Naval Research Report, April, 1947 (translation of the Russian article, Uspekhi Matematicheskikh Nauk, 6 (1939), 26–89, by T. C. Doyle, A. C. Schaeffer, and D. C. Spencer).

   (c) "Method of variations in conformal mapping, I," Rec. Math. [Mat. Sbornik] N.S.1, 19 (1946), 203–236.

   (d) "Method of variations in conformal mapping, II," Rec. Math. [Mat. Sbornik] N.S. 1, 21 (1947), 83–117.

   (e) "Method of variations in conformal mapping, III," Rec. Math. [Mat. Sbornik] N.S. 1, 21 (1947), 119–132.
7. T. H. Gronwall, (a) "Some remarks on conformal representation," Annals of Mathematics, 16 (1914–1915), 72–76.

   (b) "Sur la déformation dans la représentation conforme," Comptes Rendus (Paris), 162 (1916), 249–252.

   (c) "Sur la déformation dans la représentation conforme sous des conditions restrictives," Comptes Rendus (Paris), 162 (1916), 316–318.
8. H. Grunsky, "Koeffizientenbedingungen für schlicht abbildende meromorphe Funktionen," Math. Zeitschrift, 45 (1939), 29–61.
9. K. Joh, "Theorems on 'schlicht' functions, II," Proc. Physico-Mathematical Soc. Japan, 20 (1938), 591–610.
10. P. Koebe, "Über die Uniformisierung beliebiger analytischer Kurven," Nachr. Ges. Wiss. Göttingen, 1907, 191–210.
11. M. Z. Krzywoblocki, "A local maximum property of the fourth coefficient of schlicht functions," Duke Math. Journal, 14 (1947), 109–128.
12. J. E. Littlewood, (a) "On inequalities in the theory of functions," Proc. London Math. Soc. (2), 23 (1925), 481–519.

    (b) "Lectures on the theory of functions," Oxford University Press, 1944.
13. K. Löwner, "Untersuchungen über schlichte konforme Abbildungen des Einheitskreises, I," Math. Annalen, 89 (1923), 103–121.
14. F. Marty, "Sur le module des coefficients de MacLaurin d'une fonction univalente," Comptes Rendus (Paris), 198 (1934), 1569–1571.
15. P. Montel, "Leçons sur les familles normales de fonctions analytiques et leurs applications," Gauthier-Villars (Paris), 1927.

16. E. PESCHL, "Zur Theorie der schlichten Funktionen," Journal für die reine und angewandte Mathematik (Crelle), 176 (1937), 61–94.

17. G. PICK, "Über den Koebeschen Verzerrungssatz," Sächsische Akad. Wiss., Leipzig, Berichte, 68 (1916), 58–64.

18. J. PLEMELJ, "Über den Verzerrungssatz von P. Koebe," Gesellschaft deutscher Naturforscher und Aerzte, Verhandlungen, 85 (1913), II, 1, 163.

19. H. PRAWITZ, "Über Mittelwerte analytischer Funktionen," Arkiv för Mat., Astron. och Fysik, 20 (1927–1928), no. 6, 1–12.

20. R. M. ROBINSON, "Bounded univalent functions," Trans. Amer. Math. Soc., 52 (1942), 426–449.

21. W. ROGOSINSKI, "Über positive harmonische Entwicklungen und typisch-reelle Potenzreihen," Math. Zeitschrift, 35 (1932), 93–121.

22. A. C. SCHAEFFER AND D. C. SPENCER, (a) "The coefficients of schlicht functions," Duke Math. Journal, 10 (1943), 611–635.

(b) "The coefficients of schlicht functions, II," Duke Math. Journal, 12 (1945), 107–125.

(c) "The coefficients of schlicht functions, III," Proc. Nat. Acad. Sciences, USA, 32 (1946), 111–116.

(d) "A general class of problems in conformal mapping," Proc. Nat. Acad. Sciences, USA, 33 (1947), 185–189.

(e) "A variational method in conformal mapping," Duke Math. Journal, 14 (1947), 949–966.

(f) "The coefficients of schlicht functions, IV," Proc. Nat. Acad. Sciences, USA, 35 (1949), 143–150.

23. A. C. SCHAEFFER, M. SCHIFFER, AND D. C. SPENCER, "The coefficient regions of schlicht functions," Duke Math. Journal, 16 (1949), 493–527.

24. M. SCHIFFER, (a) "A method of variation within the family of simple functions," Proc. London Math. Soc. (2), 44 (1938), 432–449.

(b) "On the coefficients of simple functions," Proc. London Math. Soc. (2), 44 (1938), 450–452.

(c) "Variation of the Green function and theory of the p-valued functions," American Journal Math., 65 (1943), 341–360.

(d) "Sur l'équation différentielle de M. Löwner," Comptes Rendus (Paris), 221 (1945), 369–371.

(e) "Hadamard's formula and variation of domain-functions," American Journal Math., 68 (1946), 417–448.

(f) "Faber polynomials in the theory of univalent functions," Bull. Amer. Math. Soc., 54 (1948), 503–517.

25. D. C. SPENCER, "Some problems in conformal mapping," Bull. Amer. Math. Soc., 53 (1947), 417–439.

26. O. SZÁSZ, "Über Funktionen, die den Einheitskreis schlicht abbilden," Jahresbericht der deutschen Mathematiker Vereinigung, 42 (1933), 73–75.

27. O. TEICHMÜLLER, "Ungleichungen zwischen den Koeffizienten schlichter Funktionen," Preuss. Akad. Wiss., Berlin, Sitzungsberichte, 1938, 363–375.

INDEX

Colloquium Publications

1. H. S. White, *Linear Systems of Curves on Algebraic Surfaces;*
F. S. Woods, *Forms of Non-Euclidean Space;*
E. B. Van Vleck, *Selected Topics in the Theory of Divergent Series and of Continued Fractions;*
1905, xii, 187 pp. $3.00

2. E. H. Moore, *Introduction to a Form of General Analysis;*
M. Mason, *Selected Topics in the Theory of Boundary Value Problems of Differential Equations;*
E. J. Wilczynski, *Projective Differential Geometry;*
1910, x, 222 pp. out of print

3_1. G. A. Bliss, *Fundamental Existence Theorems*, 1913; reprinted, 1934, iv, 107 pp. out of print

3_2. E. Kasner, *Differential-Geometric Aspects of Dynamics*, 1913; reprinted, 1934; reprinted, 1948, iv, 117 pp. 2.50

4. L. E. Dickson, *On Invariants and the Theory of Numbers;*
W. F. Osgood, *Topics in the Theory of Functions of Several Complex Variables;*
1914, xviii, 230 pp. out of print

5_1. G. C. Evans, *Functionals and their Applications. Selected Topics, Including Integral Equations*, 1918, xii, 136 pp. out of print

5_2. O. Veblen, *Analysis Situs*, 1922; second ed., 1931; reprinted, 1946, x, 194 pp. 3.35

6. G. C. Evans, *The Logarithmic Potential. Discontinuous Dirichlet and Neumann Problems*, 1927, viii, 150 pp. out of print

7. E. T. Bell, *Algebraic Arithmetic*, 1927, iv, 180 pp. out of print

8. L. P. Eisenhart, *Non-Riemannian Geometry*, 1927; reprinted, 1934; reprinted, 1949, viii, 184 pp. 2.70

9. G. D. Birkhoff, *Dynamical Systems*, 1927; reprinted, 1948, viii, 295 pp. 4.25

10. A. B. Coble, *Algebraic Geometry and Theta Functions*, 1929; reprinted, 1948, viii, 282 pp. 4.00

11. D. Jackson, *The Theory of Approximation*, 1930; reprinted, 1946, viii, 178 pp. 3.35

12. S. Lefschetz, *Topology*, 1930, x, 410 pp. out of print

13. R. L. Moore, *Foundations of Point Set Theory*, 1932, viii, 486 pp; out of print

14. J. F. Ritt, *Differential Equations from the Algebraic Standpoint*, 1932; reprinted, 1948, x, 172 pp. 3.00

15. M. H. Stone, *Linear Transformations in Hilbert Space and their Applications to Analysis*, 1932; reprinted, 1946; reprinted, 1948, vii, 622 pp. out of print

16. G. A. Bliss, *Algebraic Functions*, 1933; reprinted, 1948, x, 220 pp. 3.75

17. J. H. M. Wedderburn, *Lectures on Matrices*, 1934; reprinted, 1944; reprinted, 1946; reprinted, 1949, x, 205 pp. 3.35

18. M. Morse, *The Calculus of Variations in the Large*, 1934; reprinted, 1948, x, 368 pp. 5.35

19. R. E. A. C. Paley and N. Wiener, *Fourier Transforms in the Complex Domain*, 1934; reprinted, 1949, viii, 184 pp. + portrait plate 3.75

20. J. L. Walsh, *Interpolation and Approximation by Rational Functions in the Complex Domain*, 1935, x, 382 pp. 6.00

21. J. M. Thomas, *Differential Systems*, 1937, x, 118 pp. 2.00

22. C. N. Moore, *Summable Series and Convergence Factors*, 1938, vi, 105 pp. 2.00

23. G. Szegö, *Orthogonal Polynomials*, 1939; reprinted, 1949, x, 403 pp. 6.00

24. A. A. Albert, *Structure of Algebras*, 1939, xii, 210 pp. 4.00

25. G. Birkhoff, *Lattice Theory*, 1940; enlarged and completely revised ed., 1949, xiv 283 pp. 6.00

26. N. Levinson, *Gap and Density Theorems*, 1940, viii, 246 pp. 4.00

27. S. Lefschetz, *Algebraic Topology*, 1942; reprinted, 1948, vi, 393 pp. 6.00

28. G. T. Whyburn, *Analytic Topology*, 1942; reprinted, 1948, x, 281 pp. 4.75

29. A. Weil, *Foundations of Algebraic Geometry*, 1946, xx, 288 pp. 5.50

30. T. Radó, *Length and Area*, 1948, vi, 572 pp. 6.75

31. E. Hille, *Functional Analysis and Semi-Groups*, 1948, xii, 528 pp. 7.50

32. R. L. Wilder, *Topology of Manifolds*, 1949, x, 402 pp. 7.00

33. J. F. Ritt, *Differential Algebra*, 1950, viii, 184 pp. 4.40

34. J. L. Walsh, *The Location of Critical Points of Analytic and Harmonic Functions*, 1950, viii, 384 pp. 6.00

35. A. C. Schaeffer and D. C. Spencer, *Coefficient Regions for Schlicht Functions*, with a chapter on *The Region of Values of the Derivative of a Schlicht Funcíion* by Arthur Grad, 1950, xvi, 311 pp. 6.00

AMERICAN MATHEMATICAL SOCIETY
New York 27, N. Y., 531 West 116th Street
Cambridge, England, 1 Trinity Street, Bowes and Bowes